Sustainable Wastewater Management and Treatment

Sustainable Wastewater Management and Treatment

Editors

José Luis Campos
Anuska Mosquera Corral
Ángeles Val del Río
Alba Pedrouso Fuentes

MDPI • Basel • Beijing • Wuhan • Barcelona • Belgrade • Manchester • Tokyo • Cluj • Tianjin

Editors

José Luis Campos
Faculty of Engineering and Sciences
Universidad Adolfo Ibáñez
Viña del Mar
Chile

Anuska Mosquera Corral
Department of Chemical Engineering
Universidade de Santiago de Compostela
Santiago de Compostela
Spain

Ángeles Val del Río
Department of Chemical Engineering
Universidade de Santiago de Compostela
Santiago de Compostela
Spain

Alba Pedrouso Fuentes
Department of Chemical Engineering
Universidade de Santiago de Compostela
Santiago de Compostela
Spain

Editorial Office
MDPI
St. Alban-Anlage 66
4052 Basel, Switzerland

This is a reprint of articles from the Special Issue published online in the open access journal *Sustainability* (ISSN 2071-1050) (available at: www.mdpi.com/journal/sustainability/special_issues/sustainable_wastewater_management_treatment).

For citation purposes, cite each article independently as indicated on the article page online and as indicated below:

LastName, A.A.; LastName, B.B.; LastName, C.C. Article Title. *Journal Name* **Year**, *Volume Number*, Page Range.

ISBN 978-3-0365-4988-0 (Hbk)
ISBN 978-3-0365-4987-3 (PDF)

Contents

About the Editors

José Luis Campos

José Luis Campos has been a teaching and research assistant at University Adolfo Ibañez (Chile) since 2013. He obtained his Ph.D. in Chemical Sciences in the Universidade de Santiago de Compostela (USC) in 2000. During the first years of his research career (1994-2001), he worked in the field of nitrogen removal from industrial wastewater, optimization of aquaculture systems and minimization of sludge production. Since 2001, his research has focused on two additional main topics: aerobic granulation, in order to improve the activated sludge systems, and the anammox process to remove ammonia from wastewater with a low C/N ratio. His research is focused on minimizing the energy consumption of wastewater treatment plants, aiming to reduce their greenhouse emissions, and also searching for alternative water sources to cope with the water scarcity associated with climate change. He has participated and/or supervised several projects and contracts with industries where industrial and pilot plants were applied and new technologies were developed to manage and treat different kinds of wastewater and sludge.

Anuska Mosquera Corral

Full Professor of Chemical Engineering at the University of Santiago de Compostela (Spain) since 2022 and associate professor since 2007. Ph.D. in Chemical Sciences from USC in 1998 focused on nitrogen removal from industrial effluents. Postdoctoral researcher at the TU Delft (The Netherlands) in 2000-2001, where she specialized in aerobic granular reactors to remove organic matter (COD) and nitrogen from industrial wastewater and in the application of autotrophic denitrification (anammox, denitrification with sulphur compounds) to effluents with low COD content. Then, in 2005, she began applying molecular biology techniques (FISH and DGGE) for the identification of bacterial populations in bioreactors. In 2011, she began researching biopolymer (polyhydroxyalkanoates) production from liquid wastes using mixed cultures.

Ángeles Val del Río

Angeles Val del Rio received her degree in Chemical Engineering in June 2006 at the University of Santiago de Compostela (USC). Then, she received an FPU fellowship from the Spanish Government to perform her doctoral thesis in the group of Environmental Biotechnology (BioGroup) at USC which was finished in December 2012. She continued working as a post-doctoral researcher in the BioGroup and in 2015 she was the principal investigator of a national project ("Proyectos I+D+i jóvenes investigadores", MINECO). In 2017, she was granted the Xunta de Galicia fellowship at the National University of Ireland and the University of Minho for two years. Her research expertise is focused on the application of novel biological processes for wastewater treatment, like aerobic granular systems and anammox-based processes.

Alba Pedrouso Fuentes

Alba Pedrouso has been a postdoctoral research fellow at the Universidade of Santiago de Compostela (USC, Spain) and a guest researcher at TU Delft (The Netherlands) since 2021. She obtained her Degree in Chemical Engineering at USC in 2013 and an M.Sc. in Chemical Engineering and Bioprocess in 2014. Since then, her research experience has been focused on developing and optimizing innovative technologies and advanced bioprocesses for their application in wastewater treatment plants making them more economical and environmentally sustainable. In December 2019, she defended her Ph.D. thesis in Chemical and Environmental Engineering entitled "Assessment of the nitritation and anammox processes for mainstream wastewater treatment". Later, Alba worked as Ph.D. Project Manager in a technological Centre, focused on: i) recovering value-added products from agro-food wastes and their membrane concentration and purification, and ii) biologically removing nutrients from wastewater. In July 2021, she returned to USC as a postdoctoral researcher focused on resource recovery from wastewater and autotrophic nitrogen removal.

sustainability

MDPI

Editorial

Sustainable Wastewater Management and Treatment

José Luis Campos [1,*], Anuska Mosquera-Corral [2], Ángeles Val del Rio [2] and Alba Pedrouso [2]

1 Facultad de Ingeniería y Ciencias, Universidad Adolfo Ibáñez, Avda. Padre Hurtado 750, Viña del Mar 2503500, Chile
2 CRETUS Institute, Department of Chemical Engineering, Universidade de Santiago de Compostela, Rua Lope Gómez de Marzoa s/n, E-15782 Santiago de Compostela, Spain; anuska.mosquera@usc.es (A.M.-C.); mangeles.val@usc.es (Á.V.d.R.); alba.pedrouso@usc.es (A.P.)
* Correspondence: jluis.campos@uai.cl

This Special Issue of *Sustainability* aims to report the recent developments in Sustainable Wastewater Management and Treatment, mainly those focused on improving the overall performance of wastewater treatment plants (WWTPs) in terms of both reducing their environmental impact and integrating them into the urban circular economy. The works presented here show new technologies, processes and operational strategies that lead to a paradigm shift in wastewater management, where the minimization of energy consumption and recovery of valuable resources are key aspects.

Most urban wastewater treatment plants were designed to meet local requirements for effluent quality, while environmental aspects such as energy consumption and resource recovery were not considered [1]. However, in recent years, many efforts have been dedicated to achieving a positive energy balance. This might be achieved by promoting organic matter removal through anaerobic routes, increasing biogas production, reducing aeration requirements, and recovering nutrients [2,3]. The main challenges that existing technologies face in fulfilling these goals are the low concentrations of organic matter and nutrients present in urban wastewater. In this respect, various solutions have been proposed, such as segregation at the source to keep resources as concentrated as possible [4] or the concentration of resources within the WWTP by means of membrane systems [5]. Microfiltration, ultrafiltration, or forward osmosis can be applied to separate the organic matter from the WWTP mainstream and divert it to an anaerobic digester in order to improve the WWTP energy balance (Contribution 1). The stricter the filtration process, the better the effluent quality obtained. However, the retention of salts is also promoted, which could inhibit anaerobic digestion of the retained organic matter. In addition to improving the energy balance of WWTPs, the removal of organic matter through anaerobic processes has the additional advantage of lower sludge production than aerobic processes. This is essential, since sludge management is one of the most critical issues in WWTP operation, representing 25% to 65% of the overall operating costs [6]. Thus, anaerobic digestion is the most used process to indirectly decrease costs associated with sludge dewatering, transport, or drying by reducing sludge production. However, anaerobic sludge digestion technologies are generally present only in WWTPs with a treatment capacity larger than 40,000 inhabitant equivalents. Moreover, its efficiency is limited to 30–40% of dry solids reduction. For this reason, other technologies for reducing sludge generated in situ that could be applied in small WWTPs are gaining attention. Among those technologies, ozonation, which enhances the hydrolysis rate of particulate matter (the limiting step of the degradation of the solid) by sludge disintegration, has been successfully applied at full-scale. In fact, if the implementation of an ozonation unit to reduce sludge production is considered during the design of small WWTPs, overall wastewater treatment costs would significantly decrease (Contribution 2).

Another alternative to diminishing WWTPs' energy consumption is using systems based on the combination of microalgae and bacteria to remove both organic matter and

Citation: Campos, J.L.; Mosquera-Corral, A.; Val del Rio, Á.; Pedrouso, A. Sustainable Wastewater Management and Treatment. *Sustainability* **2022**, *14*, 9137. https://doi.org/10.3390/su14159137

Received: 12 July 2022
Accepted: 19 July 2022
Published: 26 July 2022

nitrogen. In these systems, photosynthesis directly provides oxygen, and no external aeration is needed. Contrary to the activated sludge systems, where bacteria remove nitrogen through nitrification/denitrification processes, with the microalgae–bacteria consortia, nitrogen is removed via assimilation for biomass growth without greenhouse gas emissions [7]. The processes based on microalgae are generally carried out in open raceway ponds requiring larger land areas than conventional systems. Therefore, despite being eco-friendly and low-cost technologies for wastewater treatment, their application would be restricted to small communities, where enough land is available. The potential application of microalgae should focus not only on nutrient removal but also on the removal of organic pollutants due to the ability of these microorganisms to remove personal care products, pharmaceuticals, pesticides, and industrial products [8]. For example, Contribution 3 shows that the marine microalgae *Tetraselmis suecica* can remove phenolic compounds, which are hardly biodegradable in conventional biological systems.

WWTPs are designed and operated in such a way that they can comply with the discharge limits under fluctuating influent conditions. In the past, this purpose was achieved by oversizing the WWTPs and overdosing on the oxygen required. Nevertheless, the development of mathematical models and real-time control systems have allowed optimizing the design and operation conditions of WWTPs to maintain a compromise between the effluent quality and the treatment costs. Models are valuable tools to understand how the operational parameters affect the effluent quality, as Contribution 4 shows. They can also be used to upgrade or redesign WWTPs to maximize the valorization of products recovered from the wastewater [9]. For example, aeration accounts for 50% to 60% of the energy consumption during wastewater treatment. For this reason, the research on real-time control systems is mainly focused on aeration to save energy while maintaining a suitable effluent quality. Since the airflow rate can be fitted to oxygen demand, implementing dissolved oxygen (DO) and aeration control systems can save 25 to 40% more energy than manually controlled systems [10]. The recent availability of reliable ammonium sensors has led to their incorporation into aeration control loops, allowing the development of more advanced aeration control strategies that enhances energy savings [11]. However, these sensors are expensive, so they could be replaced by soft sensors developed using multiple linear regression, neural networks, and oxidation-reduction potential and pH sensors (Contribution 5).

In addition to WWTPs' energy efficiency improvements, much effort is being carried out to recover the resources contained in wastewater, such as phosphorus, nitrogen, organic fertilizers, water, and methane, due to the increasing awareness of the exhaustion of non-renewable natural resources. As previously mentioned, the organic matter contained in wastewater is currently mainly revalued as methane through the anaerobic digestion progress. Even though anaerobic digestion is a mature and sustainable technology for organic matter revaluation, recent studies indicate the convenience of exploring other technologies aiming to produce higher-value end-products, such as polyhydroxyalkanoates (PHAs) [12]. PHAs are biodegradable polymers synthesized by numerous bacteria, accumulating as intracellular carbon and energy reservoirs that could be used to replace conventional plastics. Currently, the large-scale production of PHAs is carried out using pure culture microorganisms and high-cost substrates, making their production price non-competitive compared to petrochemical-based plastics. All these issues can be overcome by using mixed microbial cultures (MMC), which do not require sterilized conditions and are able to produce PHAs from the organic matter present in wastewater. Up to now, this option has been tested at a pilot-scale using biomass with PHA-accumulation potential from municipal wastewater treatment and fermented waste sludge as feedstock for PHA accumulation [13]. PHAs production using MMC is not only limited to recovering organic compounds from urban wastewaters, but it can also be carried out using industrial wastewaters. In fact, Contributions 6 and 7 demonstrate the feasibility of enriching a MMC with a PHA storage ability to valorize fish-canning industrial wastewater, which is generally difficult to treat by biological processes due to its high salinity.

Growing demand for high-quality water, together with the impacts of climate change, has increased the number of regions in the world that suffer water scarcity or stress. This scenario has promoted the development of technologies able to use non-conventional water resources, such as seawater and treated urban wastewater, to generate drinking water at a full-scale worldwide [14]. In the case of countries with access to the sea, where both seawater and treated urban wastewater are available, the criteria for choosing between one or the other water source could be based on economic considerations. In this sense, Contribution 8 proposes a methodology to determine which of both non-conventional water resources provides the lowest production costs, depending on the wastewater treatment system used, the capacity of the water generation plant, and the distance to the water source. Water reuse is also promoted in many industrial sectors. Those using cooling tower systems are of special interest due to their high water requirement. Technologies such as reverse osmosis and electrocoagulation, among others, could be used to treat water purged from cooling towers, obtaining a high-quality effluent for reuse (Contribution 9). As a general rule, the technologies that allow achieving a better quality of effluent consume the most energy, so the use of renewable energies would perhaps condition their possible implementation.

WWPTs reduce the concentration of pollutants present in the wastewater to obtain effluents that meet the discharge limits. Nevertheless, WWTPs' effluents still cause important impacts on the ecosystems of receiving waters. To avoid such impacts, in the future, effluents should receive post-treatment through semi-natural wetland systems, as proposed in Contribution 10, or the treatment efficiency of WWTPs should be improved by implementing new units or changing existing technologies. The treatment efficiency improvement should not be only focused on conventional pollutants but also emerging pollutants. Among them, antibiotics stand out due to the growing concern about the increase of the antibiotic resistance of human pathogenic bacteria in aquatic environments caused by the emission of effluents containing antibiotic-resistant bacteria, antibiotic resistance genes, and antibiotics themselves [15].

Urban WWTPs are a key part of the anthropogenic water cycle. Until now, their role has been to decrease the pollutant load of wastewater in order to reduce the impact of the effluents on the ecosystems. However, their design in the future should be improved to promote resource recovery (energy, fresh water, and other valuable materials) and enhance their treatment efficiency, considering emerging pollutants. In this way, the WWTP concept would evolve from a "treatment system" to a "biofactory" to become an essential part of the sustainable circular economy.

List of Contributions:

1. Ortega-Bravo, J.C.; Pavez, J.; Hidalgo, V.; Reyes-Caniupán, I.; Torres-Aravena, Á.; Jeison, D. Biogas Production from Concentrated Municipal Sewage by Forward Osmosis, Micro and Ultrafiltration. https://doi.org/10.3390/su14052629
2. Crutchik, D.; Franchi, O.; Jeison, D.; Vidal, G.; Pinto, A.; Pedrouso, A.; Campos, J.L. Techno-Economic Evaluation of Ozone Application to Reduce Sludge Production in Small Urban WWTPs. https://doi.org/10.3390/su14052480.
3. Meza-Escalante, E.R.; Lepe-Martinié, L.; Díaz-Quiroz, C.; Serrano-Palacios, D.; Álvarez-Valencia, L.H.; Rentería-Mexía, A.; Gortáres-Moroyoqui, P.; Ulloa-Mercado, G. Capacity of Marine Microalga Tetraselmis suecica to Biodegrade Phenols in Aqueous Media. https://doi.org/10.3390/su14116674.
4. Mu'azu, N.D.; Alagha, O.; Anil, I. Systematic Modeling of Municipal Wastewater Activated Sludge Process and Treatment Plant Capacity Analysis Using GPS-X. https://doi.org/10.3390/su12198182.
5. Fox, S.; McDermott, J.; Doherty, E.; Cooney, R.; Clifford, E. Application of Neural Networks and Regression Modelling to Enable Environmental Regulatory Compliance and Energy Optimisation in a Sequencing Batch Reactor. https://doi.org/10.3390/su14074098.

6. Pedrouso, A.; Fra-Vazquez, A.; Val del Rio, A.; Mosquera-Corral, A. Recovery of Polyhydroxyalkanoates from Cooked Mussel Processing Wastewater at High Salinity and Acidic Conditions. https://doi.org/10.3390/su122410386.
7. Carvalho, J.M.; Marreiros, B.C.; Reis, M.A.M. Polyhydroxyalkanoates Production by Mixed Microbial Culture under High Salinity. https://doi.org/10.3390/su14031346.
8. Crutchik, D.; Campos, J.L. Municipal Wastewater Reuse: Is it a Competitive Alternative to Seawater Desalination? https://doi.org/10.3390/su13126815.
9. Soliman, M.; Eljack, F.; Kazi, M.-K.; Almomani, F.; Ahmed, E.; El Jack, Z. Treatment Technologies for Cooling Water Blowdown: A Critical Review.
10. de-los-Ríos-Mérida: J.; Guerrero, F.; Arijo, S.; Muñoz, M.; Álvarez-Manzaneda, I.; García-Márquez, J.; Bautista, B.; Rendón-Martos, M.; Reul, A. Wastewater Discharge through a Stream into a Mediterranean Ramsar Wetland: Evaluation and Proposal of a Nature-Based Treatment System. https://doi.org/10.3390/su13063540.

Funding: This research received no external funding.

Conflicts of Interest: The authors declare no conflict of interest.

References

1. Stalter, D.; Magdeburg, A.; Quednow, K.; Botzat, A.; Oehlmann, J. Do Contaminants Originating from State-of-the-art Treated Wastewater Impact the Ecological Quality of Surface Waters? *PLoS ONE* **2013**, *8*, e60616. [CrossRef] [PubMed]
2. Mo, W.; Zhang, Q. Energy-nutrients-water Nexus: Integrated Resource Recovery in Municipal Wastewater Treatment Plants. *J. Environ. Manag.* **2013**, *127*, 255–267. [CrossRef] [PubMed]
3. Liu, Y.J.; Gu, J.; Liu, Y. Energy Self-sufficient Biological Municipal Wastewater Reclamation: Present Status, Challenges and Solutions Forward. *Bioresour. Technol.* **2018**, *269*, 513–519. [CrossRef] [PubMed]
4. de Graaff, M.S.; Temmink, H.; Zeeman, G.; Buisman, C.J. Energy and Phosphorus Recovery from Black Water. *Water Sci. Technol.* **2011**, *63*, 2759–2765. [CrossRef] [PubMed]
5. Xiong, J.; Yu, S.; Hu, Y.; Yang, Y.; Wang, X.C. Applying a Dynamic Membrane Filtration (DMF) Process for Domestic Wastewater Preconcentration: Organics Recovery and Bioenergy Production Potential Analysis. *Sci. Total Environ.* **2019**, *680*, 35–43. [CrossRef] [PubMed]
6. Pérez-Elvira, S.I.; Nieto Diez, P.; Fdz-Polanco, F. Sludge Minimisation Technologies. *Rev. Environ. Sci. Biotechnol.* **2006**, *5*, 375–398. [CrossRef]
7. Qiao, S.; Hou, C.; Wang, X.; Zhou, J. Minimizing Greenhouse Gas Emission from Wastewater Treatment Process by Integrating Activated Sludge and Microalgae Processes. *Sci. Total Environ.* **2020**, *732*, 139032. [CrossRef] [PubMed]
8. Touliabah, H.E.-S.; El-Sheekh, M.M.; Ismail, M.M.; El-Kassas, H. A Review of Microalgae and Cyanobacteria-Based Biodegradation of Organic Pollutants. *Molecules* **2022**, *27*, 1141. [CrossRef] [PubMed]
9. Solon, K.; Volcke, E.I.P.; Spérandio, M.; van Loosdrecht, M.C.M. Resource Recovery and Wastewater Treatment Modelling. *Environ. Sci. Water Res. Technol.* **2019**, *5*, 631–642. [CrossRef]
10. Maktabifard, M.; Zaborowska, E.; Makinia, J. Achieving Energy Neutrality in Wastewater Treatment Plants through Energy Savings and Enhancing Renewable Energy Production. *Rev. Environ. Sci. Biotechnol.* **2018**, *17*, 655–689. [CrossRef]
11. Medinilla, V.R.; Sprague, T.; Marseilles, J.; Burke, J.; Deshmukh, S.; Delagah, S.; Sharbatmaleki, M. Impact of Ammonia-Based Aeration Control (ABAC) on Energy Consumption. *Appl. Sci.* **2020**, *10*, 5227. [CrossRef]
12. Kleerebezem, R.; Joosse, B.; Rozendal, R.; van Loosdrecht, M.C.M. Anaerobic Digestion without Biogas? *Rev. Environ. Sci. Biotechnol.* **2015**, *14*, 787–801. [CrossRef]
13. Bengtsson, S.; Werker, A.; Visser, C.; Korving, L. *PHARIO: Stepping Stone to a Sustainable Value Chain for PHA Bioplastic Using Municipal Activated Sludge*; Stichting Toegepast Onderzoek Waterbeheer: Amersfoort, The Netherlands, 2017; pp. 1–93.
14. Gerrity, D.; Pecson, B.; Trussell, R.S.; Trussell, R.R. Potable Reuse Treatment Trains Throughout the World. *J. Water Supply Res. T.* **2013**, *62*, 321–338. [CrossRef]
15. Wang, R.; Ji, M.; Zhai, H.; Guo, Y.; Liu, Y. Occurrence of Antibiotics and Antibiotic Resistance Genes in WWTP Effluent-receiving Water Bodies and Reclaimed Wastewater Treatment Plants. *Sci. Total Environ.* **2021**, *796*, 148919. [CrossRef] [PubMed]

sustainability

Article

Biogas Production from Concentrated Municipal Sewage by Forward Osmosis, Micro and Ultrafiltration

Juan Carlos Ortega-Bravo [1,2], Javier Pavez [2], Víctor Hidalgo [2], Isaac Reyes-Caniupán [2], Álvaro Torres-Aravena [3] and David Jeison [3,*]

1 Centro de Gestión y Tecnologías del Agua, Facultad de Ingeniería y Ciencias, Universidad de La Frontera, Av. Francisco Salazar 01145, Temuco 4811230, Chile; juan.ortega@ufrontera.cl
2 Chemical Engineering Department, Universidad de La Frontera, Av. Francisco Salazar 01145, Temuco 4811230, Chile; javierpavezjara@gmail.com (J.P.); v.hidalgo01@ufromail.cl (V.H.); isaac.reyes@ufrontera.cl (I.R.-C.)
3 Escuela de Ingeniería Bioquímica, Pontificia Universidad Católica de Valparaíso, Av. Brasil 2085, Valparaíso 2362803, Chile; alvaro.torres@pucv.cl
* Correspondence: david.jeison@pucv.cl

Abstract: Direct application of anaerobic digestion to sewage treatment is normally only possible under tropical weather conditions. This is the result of its diluted nature and temperatures far from those suitable for anaerobic conversion of organic matter. Then, direct application of anaerobic treatment to sewage would require changing temperature, concentration, or both. Modification of sewage temperature would require much more energy than contained in the organic matter. Then, the feasible alternative seems to be the application of a pre-concentration step that may be accomplished by membrane filtration. This research studied the pre-concentration of municipal sewage as a potential strategy to enable the direct anaerobic conversion of organic matter. Three different membrane processes were tested: microfiltration, ultrafiltration and forward osmosis. The methane potential of the concentrates was determined. Results show that biogas production from the FO-concentrate was higher, most likely because of a higher rejection. However, salt increase due to rejection and reverse flux of ions from the draw solution may affect anaerobic digestion performance.

Keywords: sewage; forward osmosis; microfiltration; ultrafiltration; concentration; biogas

Citation: Ortega-Bravo, J.C.; Pavez, J.; Hidalgo, V.; Reyes-Caniupán, I.; Torres-Aravena, Á.; Jeison, D. Biogas Production from Concentrated Municipal Sewage by Forward Osmosis, Micro and Ultrafiltration. *Sustainability* **2022**, *14*, 2629. https://doi.org/10.3390/su14052629

Academic Editor: Shervin Hashemi

Received: 14 January 2022
Accepted: 20 February 2022
Published: 24 February 2022

1. Introduction

Anaerobic digestion is nowadays an established environmental technology, providing a reliable and affordable treatment alternative for a wide variety of wastewaters. However, direct application of anaerobic digestion to sewage treatment is normally only possible under tropical weather conditions as a result of wastewater temperature and a low organic matter concentration. Then, sewage treatment is generally conducted by aerobic processes. The traditional aerobic treatment is an effective technology for reducing the organic matter concentration of sewage. However, it implies an extensive use of energy. Indeed, aeration represents the most relevant operational cost for aerobic sewage treatment.

Figure 1a presents a theoretical chemical oxygen demand (COD) balance for what could be considered a traditional sewage treatment facility: aerobic treatment by activated sludge, combined with anaerobic digestion of secondary sludge. It is clear that only a small fraction of the chemical energy contained in sewage may be potentially recovered as methane in such a process. Data provided by this figure include theoretical methane yields, so practical energy recovery is normally much lower. A process based on the direct application of anaerobic digestion, followed by an aerobic post-treatment, would greatly increase the potential energy recovery in the form of methane, as is clear when comparing Figure 1a,b.

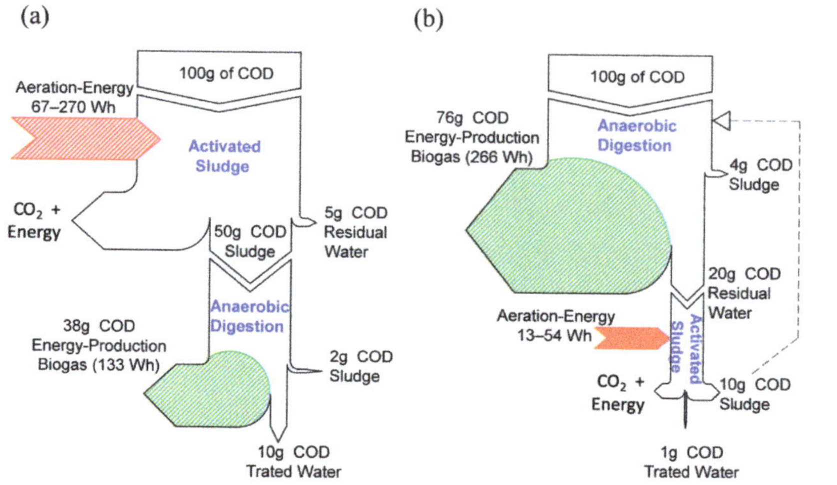

Figure 1. Theoretical COD balance of potential sewage treatment processes: (**a**) aerobic–anaerobic and (**b**) anaerobic–aerobic configuration. Balance considers a theoretical methane production of 0.35 L CH_4 $gCOD^{-1}$, and a methane (lower) heating value of 10 Wh L^{-1}.

The direct application of anaerobic treatment to sewage would require the modification of temperature and/or concentration since both factors are known to determine the performance of anaerobic digestion. The increase in temperature seems an unlikely approach since it would require much more energy than that contained in the organic matter of sewage. Then, the feasible alternative seems to be the application of a pre-concentration step. Water freezing has been reported as a tool for wastewater concentration. By crystallizing water into ice, pollutants can be concentrated in the remaining liquid [1–3]. However, this approach may only be an alternative in geographical zones presenting very low ambient temperatures. Water evaporation is another alternative [4,5]. However, the elevated enthalpy of the evaporation of water would demand high energetic requirements for diluted wastewaters, such as sewage. Moreover, it may induce emissions of volatile substances, such as volatile organic compounds (VOC). Membrane processes are more likely to represent an alternative for sewage concentration. Indeed, the application of membrane technology to wastewater treatment is not new; the application of membrane bioreactors is a clear example [6,7]. The development of membrane processes for environmental applications during the last 3 decades has produced relevant improvements in membrane manufacture and has steadily decreased their costs. Moreover, a series of emerging membrane technologies has arisen as alternatives for water reclamation [8]. As a result, membrane separation techniques are nowadays applied in fields where their use may have been considered unlikely not so long ago, such as the treatment of municipal and industrial wastes [9–11]. The potential advantages of membrane processes, when compared with other separation techniques, are: low energy consumption, environmental friendliness, and high quality permeate.

The concentration of wastewater by membranes, which has normally been referred to as direct membrane filtration (DMF), has been already reported in literature. Indeed, recently, Hube et al. [12] reviewed the use of direct membrane filtration for wastewater treatment, identifying DMF as a tool for resource recovery from wastewaters. In general, moderate to high levels of organic matter retention have been reported. Kimura et al. [13] reported 75% of organic matter recovery using a microfiltration membrane of 0.1 µm of pore size. Gong et al. [14] reported similar levels of recovery, using a pilot-scale hollow-fiber PVDF (polyvinylidene fluoride) membrane with 0.02 µm pore diameter. Similar performance was reported by Lateef et al. [15] when working with a microfiltration (MF) concentration unit. On the other hand, Ravazzini et al. [16] only achieved up to 40% retention of chemical oxygen demand (COD) using ultrafiltration (UF). Differences in performance are expected, because of different sewage characteristics, since COD removal will largely depend on the contribution of solids, colloids and solutes to total COD.

Although MF and UF can retain part of the COD, they are ineffective in retaining nitrogen and phosphorus since a large part of these nutrients are in sewage as soluble compounds [17]. An efficient sewage concentration would need a process enabling retention of all pollutants contained in sewage, including solids, colloids and soluble material. Forward osmosis (FO) is a concentration process based on the natural phenomenon of osmosis. It has gathered increasing attention during the last decade as an alternative for traditional pressure-driven membrane processes, such as reverse osmosis (RO) or nanofiltration (NF). Recently, it has also been proposed as an alternative for sewage treatment or pre-concentration [18]. Osmosis is a natural process that promotes the transport of water between two fluids presenting different osmotic potentials, separated by a selective membrane: the feed and a draw solution. Then, as a result of water migration, the solution presenting higher osmotic potential is diluted, and the solution of lower osmotic potential is concentrated. A secondary process is also required in order to re-concentrate the diluted draw solution so that it can be used in a closed-loop cycle. Although first developments dealing with FO were based on the use of RO membranes, membranes specially developed for FO processes are now readily available in the market [10,19]. The FO-based concentration of sewage has been already reported [20–22]. However, few reports include assays to directly determine the methane potential of the produced concentrates. Moreover, reported research does not usually involve the testing of different membrane technologies with the same sewage. Sewage properties can change depending on the particular characteristics of the community it comes from, making performance comparison between different reports difficult.

This research studied the performance of MF, UF and FO processes for pre-concentration of municipal sewage. Biogenic methane potential tests (BMP) were used in order to determine the potential energy that could be recovered from the concentrated sewage.

2. Materials and Methods

2.1. Sewage

Experiments were conducted with real pre-settled sewage from the city of Temuco, Chile. Sewage was collected at the city's sewage treatment plant from the effluent of the primary settler. After collection, municipal sewage was stored at 4 °C. Pre-concentration processes started within the first 24 h following sewage collection. Pre-settled sewage presented COD concentration, total solids (TS) and volatile solids (VS) of 0.25, 0.56 and 0.34 g L^{-1}, respectively.

2.2. Sewage Concentration Setup

Sewage was concentrated by 3 membrane processes: MF, UF and FO. MF concentration was performed using a tubular ceramic membrane with a pore diameter of 0.2 μm. The membrane tube was 60 cm long and had an internal diameter of 0.5 cm. UF was carried out using a tubular polymeric membrane of 30 nm pore diameter. The membrane tube was 70 cm long and had an internal diameter of 0.8 cm. In both cases, filtration modules contained a sole membrane tube. Table 1 presents the general characteristics of membranes used in this study. Before filtration experiments, MF and UF membranes were subjected to 30 min of oxidative cleaning using NaOCl, providing a free chlorine concentration of 500 mg L^{-1}. During MF and UF operation, transmembrane pressure (TMP) was measured online by means of a pressure transducer. Permeate was collected by means of a peristaltic pump, which provided the required TMP. During MF and UF concentration processes, the flux was automatically controlled in order to keep it nearby the critical permeate flux. A tool previously reported was used for that purpose [23]. Both membranes were operated in cross-flow mode, according to the setup shown in Figure 2a. Initial membrane resistances (measured with clean water) were 9.4×10^{10} and 2.7×10^{11} m^{-1} for MF and UF, respectively.

The FO concentration process was conducted using cellulose triacetate with an embedded polyester screen flat-sheet membrane (CTA) with 0.03 m^2 of membrane area. A new membrane was used during this assay. The membrane was used with the active layer facing the feed side (sewage). The system was operated in cross-flow mode, and no spacers were used. A 0.6 M solution of NaCl was used as a draw solution. Figure 2b presents a schematic representation of the FO filtration setup.

Initial sewage volumes were between 6 and 7 L for the tested membrane processes. The fed (sewage) was kept refrigerated during the concentration process. Once the concentration process was finished, water was flushed through the setup as a way to remove organics that may have remained in the system as fouling layers.

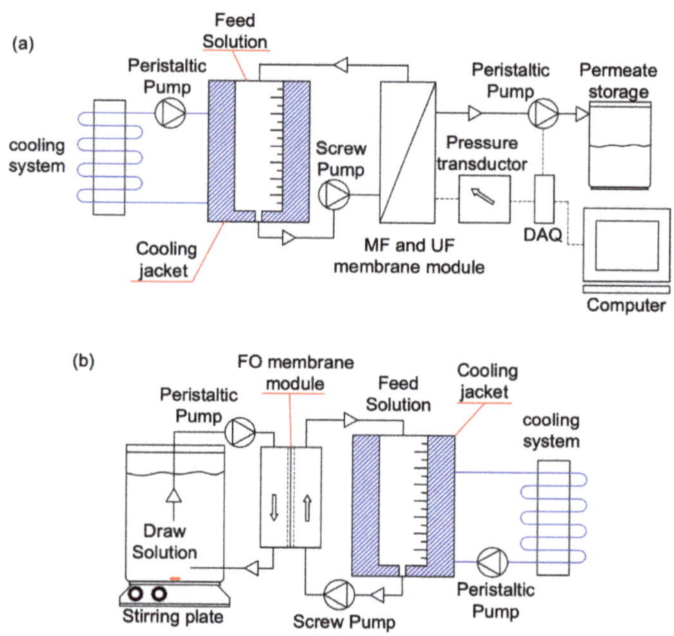

Figure 2. Filtration setups used in this study. (**a**) Micro and ultrafiltration. (**b**) Forward osmosis setup.

Table 1. General characteristics of the membranes used in this study.

	MF	UF	FO
Provider	Atech Innovations, Germany	X-Flow Norit, The Netherlands	Hydration Technology Innovations, USA
Material	Ceramic aluminum oxide (Al_2O_3)	Polymeric	Cellulose triacetate with embedded polyester screen flat-sheet membrane (CTA)
Pore size	0.2 µm	20 nm	-
Configuration	Tubular, inside/out	Tubular, inside/out	Flat sheet
Filtration area	0.0094 m^2	0.0176 m^2	0.03 m^2

2.3. *Biogenic Methane Potential*

BMP assays were conducted to determine the energetic potential of concentrated sewage through MF, UF and FO processes. Assays were conducted in serum bottles of 120 and 80 mL of total and reaction volume, respectively. Granular sludge from a full-scale UASB reactor treating brewery wastewater was used as inoculum. Substrate (concentrated sewage) was supplemented with 200 mg L^{-1} yeast extract, 65 mg L^{-1} NH_4Cl, 18.5 mg L^{-1} KH_2PO_4, 4 mg L^{-1} $CaCl_2 \cdot 2H_2O$ and 5.7 mg L^{-1} $MgSO_4 \cdot 7H_2O$. Sodium bicarbonate was also added to provide buffer capacity (5 g L^{-1}). Reactor bottles were flashed for a duration of 30 s with a mixture of N_2/CO_2 (70/30% respectively) in order to displace the oxygen present in the headspace and then were incubated at 35 °C. Methane production was determined following pressure increase and gas composition in the headspace of the serum bottles. Pressure was determined with a pressure transducer (Cole Parmer, model 200). Biogas composition was measured by gas chromatography with a thermal conductivity detector (Clarus 580, Perkin Elmer, Waltham, MA, USA). BMP is presented, and the volume of produced methane (in standard conditions) per mass of VS is present in the concentrated sewage. Assays were conducted in triplicate, and blank tests without substrate (concentrated sewage) were included to determine endogenous methane production.

A modified Gompertz model [24] was used to analyze and compare data provided by BMP tests. The model is given by the following expression:

$$P = P_m \cdot exp\left(-exp\left(\frac{R_m \cdot e}{P_m}(\lambda - t) + 1\right)\right)$$

where P is the produced methane at time t ($mLCH_4$ gVS^{-1}), P_m is maximum methane production ($mLCH_4$ gVS^{-1}), R_m is the methane production rate ($mLCH_4$ gVS^{-1} d^{-1}) and λ is the lag time (d). Calculation of parameters was achieved through a non-linear regression of experimental data. The coefficient of determination (R^2) was computed to evaluate the goodness of fit of experimental data to the model.

2.4. Specific Methanogen Activity (SMA)

During FO concentration, reverse salt flux was observed from the draw solution to the feed. Specific methanogenic activity (SMA) tests were carried out to determine the influence of NaCl concentration on the activity of anaerobic sludge used for BMP tests. Five different NaCl concentrations were tested: 0, 2.5, 5, 7.5, and 10 g L^{-1}. The SMA assays were carried out in 120 mL serum bottles containing 50 mL of media. Biomass concentration was 1 gVS L^{-1}. A solution of acetic acid was used as a substrate, which was previously neutralized to pH 7 with sodium hydroxide. The composition of the culture media and the procedure to determine the methane production were the same already described for BMP tests. SMA was evaluated, determining the maximum production of methane observed during each assay. Assays were conducted in triplicate.

2.5. Analyses

COD concentration, total solids (TS) and volatile solids (VS) were measured according to Standard Methods 5520D (closed reflux colorimetric method), 2540B and 2540E, respectively [25]. Na^+ concentration was analyzed using an ICP Mass Spectrometer (NexION 350D, Perkin Elmer, Waltham, MA, USA).

3. Results and Discussion

Figure 3 presents permeate flux during MF and UF concentration of sewage. MF provided higher levels of flux than UF, most likely as a result of a lower fouling tendency. TMP oscillated on both cases between 0.1 and 0.4 Bar. In the case of FO, observed flux was in the range of range 2.0–2.5 L m^{-2} h^{-1}, which can be considered low, even though it is in the same order of magnitude as others reported when applying CTA membranes to wastewater treatment [21,26]. Direct concentration by membrane processes such as MF, UF and FO will be most likely limited by cake layer formation and fouling. Several authors have suggested the application of pre-treatments to address this potential problem. For example, several authors have pointed out the convenience of applying DMF to pre-settled sludge. Settling is a simple and reliable way to concentrate a large part of the organic matter contained in sewage and can reduce chances of membrane fouling during DMF, as was observed by a few authors [16,27]. Indeed, during this research, pre-settled sewage was used according to the findings of reported experiences. Other authors have proposed the combination of coagulation with membrane filtration, with positive results [28–30]. Several fouling mitigation alternatives have been tested when using membranes for water treatment applications [31]. For example, fouling during the operation of membrane bioreactors has been extensively studied [32,33]. The particularities of membrane fouling during DMF, and the best strategies for its mitigation, definitively require further research.

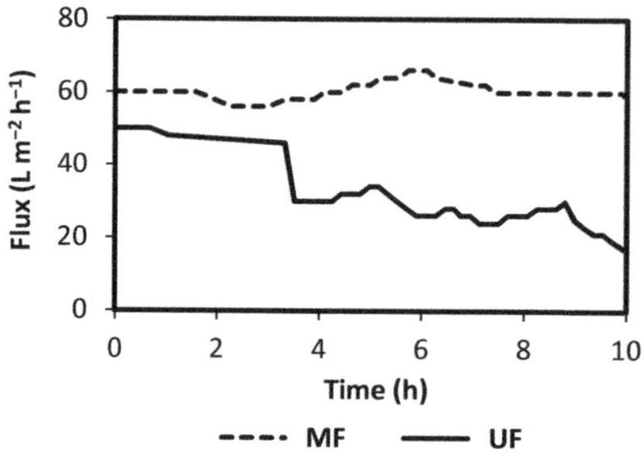

Figure 3. Permeate flux during sewage concentration by MF and UF.

Volume reduction factors during sewage concentration were 8.2, 10.9 and 6.1, for MF, UF and FO, respectively. The time required to achieve such volume reductions was 10 h for MF and UF and 74 h for FO. Assays were originally conceived to produce a volume reduction of eight or higher, which was achieved by MF and UF. However, in the case of FO, the assay was stopped before, considering the time required for concentration as a result of the low flux, to prevent excessive COD decomposition. The high time required for FO-based concentration is the result of the low flux observed for that membrane process. Time of operation could be decreased by simply increasing membrane area or by adjusting operational conditions. For example, increasing cross flow velocity improves hydrodynamic mixing, reducing the external concentration polarization phenomenon [34]. The use of different FO membranes could also be an alternative. For example, membranes such as the Thin Film Composite (TFC) have been reported to provide higher water fluxes [9]. Liang et al. [35] reported the use of vertically oriented porous substrates as supports for minimizing or eliminating the ICP effect of TFC membranes, resulting in water fluxes exceeding 50 L m^{-2} h^{-1}. On the other hand, Wu et al. [36] tested a graphene oxide-modified film nanocomposite (TFN) FO membrane for sewage concentration, reporting an improved filtration performance. Relevant developments in membrane materials are expected to take place during the following years that may provide membranes with enhanced filtration characteristics.

Figure 4 presents COD and VS concentrations of sewage and the concentrates produced by tested membrane processes. Sewage presents low levels of COD, which is the result of the fact that samples were taken after primary settling. The COD content of the concentrated sewage coming from MF and UF presented similar values: 1.46 and 1.52 gCOD L^{-1}, respectively. The COD concentration in the case of the FO process was lower (0.9 gCOD L^{-1}) as a result of a lower volume reduction. Permeate COD concentrations were 0.044 and 0.049 gCOD L^{-1} for MF and UF, respectively. COD concentration in the permeate of FO was in the range of the detection limit of the tested method, most likely because the high rejection of FO membranes can provide for dissolved organic carbon [37,38]. Performance of MF and UF membranes in terms of COD rejection were similar, despite the differences in pore size. Most likely, this is the result of the effect that fouling can have on the rejection of porous membranes since fouling layers can act as a secondary active membrane [39].

Recovered COD, i.e., the proportion of COD present in the sewage that could be found in the concentrate, was 71, 59 and 62% for MF, UF and FO, respectively. These values are in the range of or are slightly below those reported by other researchers when studying DMF with different membrane processes [13,14,21]. Values of recovered COD were influenced by the permanence of a relevant amount of COD in the setup, as fouling layers and contained in the liquid remaining in tubes. This phenomenon is the result of the use of a clean starting system and a batch operation. Continuous operation is expected to minimize COD retention in the system as fouling layers with respect to the fed COD. Moreover, in the case of FO, the long operation time most likely favored biodegradation.

Table 2 presents the Na^+ concentration of sewage and concentrated sewage. As expected, no big changes in Na^+ are observed when the sewage is pre-concentrated with MF and UF processes as a result of the null sodium rejection that these membranes can accomplish. However, Na^+ concentration in concentrated sewage by the FO process increased nine times. Such an increase is 50% higher than that expected, considering full Na^+ rejection. The difference must be then associated with reverse salt flux from draw solution, which could contaminate concentrate sewage, potentially affecting subsequent anaerobic digestion process.

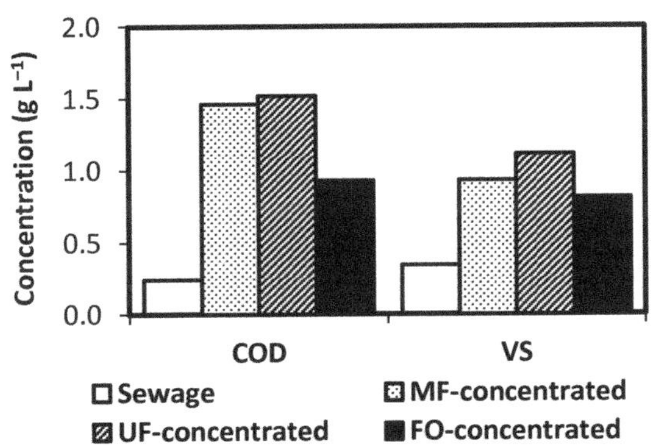

Figure 4. Sewage and concentrated-sewage characterization.

Table 2. Sodium concentration on sewage and concentrated sewage.

Sewage	Na^+ (g L^{-1})
Original Sewage	0.22
UF concentrated	0.20
MF concentrated	0.19
FO concentrated	1.94

Methane production during BMP assays using concentrates from MF, UF and FO are presented in Figure 5a. BMP was very similar for MF and UF, probably as a result of a similar rejection, as already commented. BMP values were lower than the ones reported by Gong et al. [14] and Hafuka et al. [40], who applied DMF using UF and MF membranes, respectively. This could be partially the result of differences in sewage properties. Interestingly, the BMP was higher for FO concentrate; however, the methane production rate was notoriously slower. Such observations are confirmed by the kinetic parameters calculated using the modified Gompertz model, presented in Table 3. P_m for FO was 30% higher than those for MF and UF, which were almost coincident. On the other hand, R_m for FO was only 11% of that evaluated for MF. Even though concentrates produced by MF and UF presented different values for R_m, BMP kinetics were enough in both cases to provide maximum production in about 10–15 days. It is possible that the

COD degradation that occurred during the FO concentration process may have affected the nature of a part of the sewage COD, affecting observed P_m calculated for FO. However, a comparison of observed BMP values may also suggest that solids retained by FO and MF/UF were, at least partially, of a different nature, as a result of the FO rejection of solutes that MF and UF were not able to contain. On the other hand, the lower methane production rate observed for FO (represented by a lower R_m) is most likely the result of the presence of Na^+ or other ions, as a result of the high rejection of FO membrane and the reverse salt flux. Figure 5b presents the effect of Na^+ concentration on the SMA of the anaerobic sludge used for BMP determination. A concentration of 2.5 $gNa^+ L^{-1}$ already produces a decrease in SMP of about 20% during acetate conversion. Sodium concentration during BMP tests for FO concentrate was close to 2 $g L^{-1}$, so this factor may have played a role in the methane kinetics observed. This result agrees with the report from Gao et al. [41], who also identified the potential inhibitory effect of NaCl during the anaerobic digestion of FO-concentrated sewage.

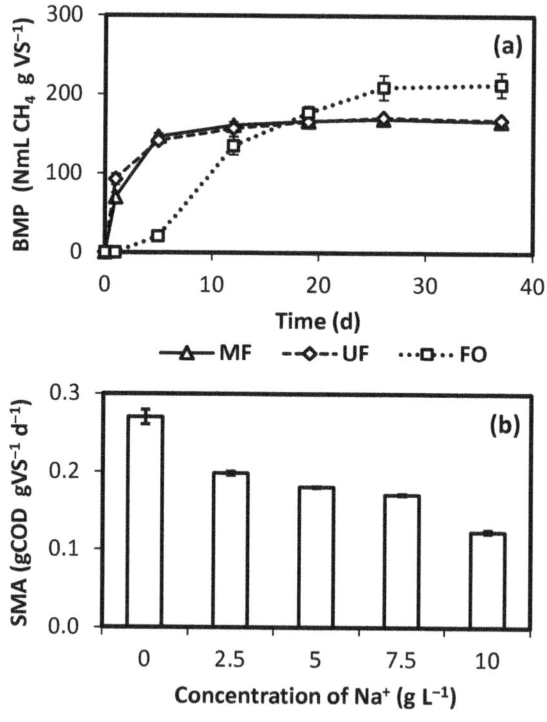

Figure 5. Results of the (**a**) biogenic methane potential (BMP) analyses for membrane concentrated sewage (MF, UF and FO). (**b**) Sludge methane activity (SMA) using acetate as substrate, at different concentrations of NaCl. Error bars indicate standard deviation between replicas.

Table 3. Kinetic parameters obtained from first order exponential model of BMP test for MF, UF and FO process.

	P_m mLCH$_4$ gSV^{-1}	R_m mLCH$_4$ gSV^{-1} d^{-1}	λ d	R^2
MF	161.9	105.2	0.33	0.988
UF	160.5	146.4	0.36	0.977
FO	210.6	16.4	3.94	0.997

Based on the BMP results, the potential energy contained in the concentrates can be estimated, considering the heating value of generated methane. For these calculations, a lower heating value was considered, 5050 kJ kg^{-1} [42]. The energy potential was 1.6, 1.6 and 2.1 kWh per kg of VS contained in the concentrates obtained by MF, UF and FO, respectively. Based on these values, energetic potential can be calculated per m^3 of the original sewage before concentration. Results are 0.19, 0.17 and 0.28 kWh per m^3 of untreated sewage. These values have been computed considering the levels of organic matter recovery observed during this research, which could certainly be improved. Since membrane concentration during this research was conducted with pre-settled sewage, previous values represent only part of the energetic potential of the wastewater. Then, it is inferred that a combination of settling and membrane concentration may be a suitable alternative for the extraction of the energetic potential of sewage. Even though FO is a separation process that provides a much higher rejection of contaminants, if energy recovery is the main objective driving the concentration process, MF and UF may provide similar levels of potential energy recuperation. However, a post-treatment will be required to remove the remaining contaminants in the permeate from those membrane operations. It is interesting to comment that a combination of FO and anaerobic digestion for sewage treatment has also been addressed in the form of FO-based anaerobic membrane bioreactors (FOAnMBR) [43,44]. However, depending on the way a reactor is operated, performance may be limited by the low organic matter concertation of the feed [18]. Then, the combination of membrane-based preconcentration and anaerobic digestion would still present advantages that should sustain future research oriented to the development of this alternative.

4. Conclusions

- Membrane concentration offers the possibility to convert organic matter contained in sewage into biogas, a source of renewable energy.
- Even though MF and UF can provide the required concentration process, enabling biogas production, rejection is normally limited to solids and colloids, so post-treatment may be required.
- Membrane fouling is a key aspect of membrane performance, and it needs to be further studied to enable the long-term operation of a direct membrane filtration operation.
- FO can provide a better rejection than MF and UF, which may result in higher methane yields. It is inferred then that FO's sewage concentration may be technically feasible. However, some drawbacks need to be overcome, such as low water flux and potentially inhibitory salt concentration, resulting from high solute rejection and reverse salt flux from the draw solution.

Author Contributions: Conceptualization, J.C.O.-B., I.R.-C. and D.J.; methodology, V.H. and J.P.; formal analysis, V.H. and J.P.; investigation, J.C.O.-B., V.H., I.R.-C. and J.P.; data curation, J.C.O.-B.; writing—original draft preparation, J.C.O.-B., Á.T.-A. and D.J.; writing—review and editing, D.J., Á.T.-A.; supervision, D.J. All authors have read and agreed to the published version of the manuscript.

Funding: This research was funded by FONDECYT-ANID, grant number 1150982 and 3160398, CRHIAM CENTER grant number 15130015 (ANID/FONDAP).

Institutional Review Board Statement: Not applicable.

Informed Consent Statement: Not applicable.

Acknowledgments: The authors want to thank the financial support provided by FONDECYT projects number 1150982, 3160398 (ANID, Chile) and by CRHIAM center (ANID/FONDAP/15130015).

Conflicts of Interest: The authors declare no conflict of interest.

References

1. Lorain, O.; Thiebaud, P.; Badorc, E.; Aurelle, Y. Potential of freezing in wastewater treatment: Soluble pollutant applications. *Water Res.* **2000**, *35*, 541–547. [CrossRef]
2. Gay, G.; Lorain, O.; Azouni, A.; Aurelle, Y. Wastewater treatment by radial freezing with stirring effects. *Water Res.* **2003**, *37*, 2520–2524. [CrossRef]
3. Gao, W.; Shao, Y. Freeze concentration for removal of pharmaceutically active compounds in water. *Desalination* **2009**, *249*, 398–402. [CrossRef]
4. Algehed, J.; Berntsson, T. Evaporation of black liquor and wastewater using medium-pressure steam: Simulation and economic evaluation of novel designs. *Appl. Therm. Eng.* **2003**, *23*, 481–495. [CrossRef]
5. Younggreen, W.; Piazza, J. Wastewater evaporation in oil refineries. *INFORM* **2009**, *20*, 473–475.
6. Shimizu, Y.; Okuno, Y.-I.; Uryu, K.; Ohtsubo, S.; Watanabe, A. Filtration characteristics of hollow fiber microfiltration membranes used in membrane bioreactor for domestic wastewater treatment. *Water Res.* **1996**, *30*, 2385–2392. [CrossRef]
7. Visvanathan, C.; Ben Aim, R.; Parameshwaran, K. Membrane Separation Bioreactors for Wastewater Treatment. *Crit. Rev. Environ. Sci. Technol.* **2000**, *30*, 1–48. [CrossRef]
8. Xie, M.; Shon, H.K.; Gray, S.R.; Elimelech, M. Membrane-based processes for wastewater nutrient recovery: Technology, challenges, and future direction. *Water Res.* **2016**, *89*, 210–221. [CrossRef]
9. Zhao, S.; Zou, L.; Tang, C.Y.; Mulcahy, D. Recent developments in forward osmosis: Opportunities and challenges. *J. Membr. Sci.* **2012**, *396*, 1–21. [CrossRef]
10. Lutchmiah, K.; Verliefde, A.; Roest, K.; Rietveld, L.; Cornelissen, E. Forward osmosis for application in wastewater treatment: A review. *Water Res.* **2014**, *58*, 179–197. [CrossRef]
11. Linares, R.V.; Li, Z.; Sarp, S.; Bucs, S.; Amy, G.; Vrouwenvelder, J. Forward osmosis niches in seawater desalination and wastewater reuse. *Water Res.* **2014**, *66*, 122–139. [CrossRef] [PubMed]
12. Hube, S.; Eskafi, M.; Hrafnkelsdóttir, K.F.; Bjarnadóttir, B.; Bjarnadóttir, M.Á.; Axelsdóttir, S.; Wu, B. Direct membrane filtration for wastewater treatment and resource recovery: A review. *Sci. Total Environ.* **2020**, *710*, 136375. [CrossRef] [PubMed]
13. Kimura, K.; Yamakawa, M.; Hafuka, A. Direct membrane filtration (DMF) for recovery of organic matter in municipal wastewater using small amounts of chemicals and energy. *Chemosphere* **2021**, *277*, 130244. [CrossRef] [PubMed]
14. Gong, H.; Jin, Z.; Xu, H.; Yuan, Q.; Zuo, J.; Wu, J.; Wang, K. Enhanced membrane-based pre-concentration improves wastewater organic matter recovery: Pilot-scale performance and membrane fouling. *J. Clean. Prod.* **2018**, *206*, 307–314. [CrossRef]
15. Lateef, S.K.; Soh, B.Z.; Kimura, K. Direct membrane filtration of municipal wastewater with chemically enhanced backwash for recovery of organic matter. *Bioresour. Technol.* **2013**, *150*, 149–155. [CrossRef]
16. Ravazzini, A.; van Nieuwenhuijzen, A.; van der Graaf, J. Direct ultrafiltration of municipal wastewater: Comparison between filtration of raw sewage and primary clarifier effluent. *Desalination* **2005**, *178*, 51–62. [CrossRef]
17. Mezohegyi, G.; Bilad, M.R.; Vankelecom, I.F. Direct sewage up-concentration by submerged aerated and vibrated membranes. *Bioresour. Technol.* **2012**, *118*, 1–7. [CrossRef]
18. Wu, X.; Lau, C.; Pramanik, B.; Zhang, J.; Xie, Z. State-of-the-Art and Opportunities for Forward Osmosis in Sewage Concentration and Wastewater Treatment. *Membranes* **2021**, *11*, 305. [CrossRef]
19. McCutcheon, J.R.; McGinnis, R.L.; Elimelech, M. Desalination by ammonia–carbon dioxide forward osmosis: Influence of draw and feed solution concentrations on process performance. *J. Membr. Sci.* **2006**, *278*, 114–123. [CrossRef]
20. Lutchmiah, K.; Cornelissen, E.; Harmsen, D.J.H.; Post, J.; Lampi, K.; Ramaekers, H.; Rietveld, L.; Roest, K. Water recovery from sewage using forward osmosis. *Water Sci. Technol.* **2011**, *64*, 1443–1449. [CrossRef]
21. Zhang, X.; Ning, Z.; Wang, D.; da Costa, J.C.D. Processing municipal wastewaters by forward osmosis using CTA membrane. *J. Membr. Sci.* **2014**, *468*, 269–275. [CrossRef]
22. Gao, Y.; Fang, Z.; Liang, P.; Huang, X. Direct concentration of municipal sewage by forward osmosis and membrane fouling behavior. *Bioresour. Technol.* **2018**, *247*, 730–735. [CrossRef] [PubMed]
23. Jeison, D.; Van Lier, J. On-line cake-layer management by trans-membrane pressure steady state assessment in Anaerobic Membrane Bioreactors for wastewater treatment. *Biochem. Eng. J.* **2006**, *29*, 204–209. [CrossRef]
24. Donoso-Bravo, A.; Pérez-Elvira, S.I.; Fdz-Polanco, F. Application of simplified models for anaerobic biodegradability tests. Evaluation of pre-treatment processes. *Chem. Eng. J.* **2010**, *160*, 607–614. [CrossRef]
25. American Public Health Association (APHA); American Water Works Association (AWWA); Water Environment Federation (WEF). *Standard Methods for the Examination of Water and Wastewater*; American Public Health Association: Washington, DC, USA, 2017; ISBN 9780875532875.
26. Soler-Cabezas, J.; Mendoza-Roca, J.-A.; Vincent-Vela, M.-C.; Luján-Facundo, M.; Pastor-Alcañiz, L. Simultaneous concentration of nutrients from anaerobically digested sludge centrate and pre-treatment of industrial effluents by forward osmosis. *Sep. Purif. Technol.* **2018**, *193*, 289–296. [CrossRef]
27. Urbanowska, A.; Kabsch-Korbutowicz, M. The Use of Flat Ceramic Membranes for Purification of the Liquid Fraction of the Digestate from Municipal Waste Biogas Plants. *Energies* **2021**, *14*, 3947. [CrossRef]
28. Diaz, S.D.; Peña, L.V.; Cabrera, E.G.; Soto, M.M.; Cabezas, L.M.V.; Sánchez, L.R.B. Effect of previous coagulation in direct ultrafiltration of primary settled municipal wastewater. *Desalination* **2012**, *304*, 41–48. [CrossRef]

29. Huang, B.-C.; Guan, Y.-F.; Chen, W.; Yu, H.-Q. Membrane fouling characteristics and mitigation in a coagulation-assisted microfiltration process for municipal wastewater pretreatment. *Water Res.* **2017**, *123*, 216–223. [CrossRef]
30. Jin, Z.; Meng, F.; Gong, H.; Wang, C.; Wang, K. Improved low-carbon-consuming fouling control in long-term membrane-based sewage pre-concentration: The role of enhanced coagulation process and air backflushing in sustainable sewage treatment. *J. Membr. Sci.* **2017**, *529*, 252–262. [CrossRef]
31. El Batouti, M.; Alharby, N.F.; Elewa, M.M. Review of New Approaches for Fouling Mitigation in Membrane Separation Processes in Water Treatment Applications. *Separations* **2021**, *9*, 1. [CrossRef]
32. Hamedi, H.; Ehteshami, M.; Mirbagheri, S.A.; Rasouli, S.A.; Zendehboudi, S. Current Status and Future Prospects of Membrane Bioreactors (MBRs) and Fouling Phenomena: A Systematic Review. *Can. J. Chem. Eng.* **2018**, *97*, 32–58. [CrossRef]
33. Du, X.; Shi, Y.; Jegatheesan, V.; Haq, I.U. A Review on the Mechanism, Impacts and Control Methods of Membrane Fouling in MBR System. *Membranes* **2020**, *10*, 24. [CrossRef] [PubMed]
34. Gebreyohannes, A.Y.; Curcio, E.; Poerio, T.; Mazzei, R.; Di Profio, G.; Drioli, E.; Giorno, L. Treatment of Olive Mill Wastewater by Forward Osmosis. *Sep. Purif. Technol.* **2015**, *147*, 292–302. [CrossRef]
35. Liang, H.-Q.; Hung, W.-S.; Yu, H.-H.; Hu, C.-C.; Lee, K.-R.; Lai, J.-Y.; Xu, Z.-K. Forward osmosis membranes with unprecedented water flux. *J. Membr. Sci.* **2017**, *529*, 47–54. [CrossRef]
36. Wu, X.; Tanner, J.; Ng, D.; Acharya, D.; Xie, Z. Sewage concentration via a graphene oxide modified thin-film nanocomposite forward osmosis membrane: Enhanced performance and mitigated fouling. *Chem. Eng. J.* **2020**, *420*, 127718. [CrossRef]
37. Nguyen, N.C.; Chen, S.-S.; Yang, H.-Y.; Hau, N.T. Application of forward osmosis on dewatering of high nutrient sludge. *Bioresour. Technol.* **2013**, *132*, 224–229. [CrossRef] [PubMed]
38. Almoalimi, K.; Liu, Y.-Q. Fouling and cleaning of thin film composite forward osmosis membrane treating municipal wastewater for resource recovery. *Chemosphere* **2021**, *288*, 132507. [CrossRef]
39. Jeison, D.; Dias, I.; Van Lier, J.B. Anaerobic membrane bioreactors: Are membranes really necessary? *Electron. J. Biotechnol.* **2008**, *11*, 1–2. [CrossRef]
40. Hafuka, A.; Takahashi, T.; Kimura, K. Anaerobic digestibility of up-concentrated organic matter obtained from direct membrane filtration of municipal wastewater. *Biochem. Eng. J.* **2020**, *161*, 107692. [CrossRef]
41. Gao, Y.; Fang, Z.; Liang, P.; Zhang, X.; Qiu, Y.; Kimura, K.; Huang, X. Anaerobic digestion performance of concentrated municipal sewage by forward osmosis membrane: Focus on the impact of salt and ammonia nitrogen. *Bioresour. Technol.* **2019**, *276*, 204–210. [CrossRef]
42. Cengel, Y.A.; Boles, M.A. *Thermodynamics: An Engineering Approach*, 8th ed.; McGraw-Hill: New York, NY, USA, 2015.
43. Gu, Y.; Chen, L.; Ng, J.-W.; Lee, C.; Chang, V.W.-C.; Tang, C.Y. Development of anaerobic osmotic membrane bioreactor for low-strength wastewater treatment at mesophilic condition. *J. Membr. Sci.* **2015**, *490*, 197–208. [CrossRef]
44. Chen, L.; Gu, Y.; Cao, C.; Zhang, J.; Ng, J.-W.; Tang, C. Performance of a submerged anaerobic membrane bioreactor with forward osmosis membrane for low-strength wastewater treatment. *Water Res.* **2014**, *50*, 114–123. [CrossRef] [PubMed]

sustainability

MDPI

Article

Techno-Economic Evaluation of Ozone Application to Reduce Sludge Production in Small Urban WWTPs

Dafne Crutchik [1,*], Oscar Franchi [1], David Jeison [2], Gladys Vidal [3], Alicia Pinto [4], Alba Pedrouso [5] and José Luis Campos [1]

[1] Faculty of Engineering and Sciences, Universidad Adolfo Ibáñez, Av. Diagonal Las Torres 2640, Santiago 7941169, Chile; oscar.franchi@edu.uai.cl (O.F.); jluis.campos@uai.cl (J.L.C.)
[2] Escuela de Ingeniería Bioquímica, Pontificia Universidad Católica de Valparaíso, Av. Brasil 2085, Valparaíso 2362803, Chile; david.jeison@pucv.cl
[3] Engineering and Biotechnology Environmental Group, Environmental Science Faculty & EULA—Chile Center, Universidad de Concepción, Concepción 4070411, Chile; glvidal@udec.cl
[4] ESVAL S.A., Cochrane 751, Valparaíso 2361810, Chile; apintoc@esval.cl
[5] CRETUS Institute, Department of Chemical Engineering, Universidade de Santiago de Compostela, E-15782 Santiago de Compostela, Spain; alba.pedrouso@usc.es
* Correspondence: dafne.crutchik@uai.cl; Tel.: +56-(2)-2331-1934

Abstract: In Chile, small wastewater treatment plants (WWTPs) (treatment capacity of less than 4,800 m^3/d) are normally not designed with consideration for the potential valorization of generated sludge. For this reason, they are generally operated at high solids residence times (SRT) (15 d) to promote the decay of biomass, promoting less sludge production and reducing the costs associated with biomass management. Operation at high SRT implies the need for a larger activated sludge system, increasing capital costs. The implementation of a sludge-disintegration unit by ozonation in future WWTPs could enable operation at an SRT of 3 d, with low sludge generation. In this work, we evaluate how the implementation of a sludge-ozonation system in small WWTPs (200–4000 m^3/d) would affect treatment costs. Four scenarios were studied: (1) a current WWTP operated at an SRT of 15 d, without a sludge ozonation system; (2) a WWTP operated at an SRT of 15 d, with a sludge-ozonation system that would achieve zero sludge production; (3) a WWTP operated at an SRT of 3 d, with a sludge-ozonation system that would provide the same sludge production as scenario 1; (4) a WWTP operated at an SRT of 15 d, with a sludge-ozonation system that would achieve zero sludge production. Economic analysis shows that the treatment costs for scenarios 1 and 2 are similar, while a reduction in cost of up to 47% is obtained for scenarios 3 and 4.

Keywords: disintegration process; ozonation; sludge reduction; sludge retention time

Citation: Crutchik, D.; Franchi, O.; Jeison, D.; Vidal, G.; Pinto, A.; Pedrouso, A.; Campos, J.L. Techno-Economic Evaluation of Ozone Application to Reduce Sludge Production in Small Urban WWTPs. *Sustainability* **2022**, *14*, 2480. https://doi.org/10.3390/su14052480

Academic Editor: Agostina Chiavola

Received: 30 December 2021
Accepted: 26 January 2022
Published: 22 February 2022

1. Introduction

The management of sewage sludge is an important issue for wastewater treatment systems. In fact, although the volume of sludge produced in urban WWTPs is around 1% of their influent flow, sludge management represents around 50–60% of the total operating costs of WWTPs [1,2]. To address this in a more sustainable manner, sewage-sludge management has evolved from an approach involving only treatment and disposal to one considering conversion into value-added products (i.e., bioenergy or biobased materials). The latter alternative has the potential to reduce the quantity of sludge that ultimately needs to be disposed of and can reduce overall operating costs [3,4]. However, this approach usually only represents a viable alternative for valorizing sewage sludge in WWTPs that serve a population equivalent (PE) greater than 24,000 (4800 m^3/d) [5].

In Chile, as in many Latin American countries, about 80% of urban cities have a PE less than 24,000 [6], and consequently, the number of small WWTP is significant [7]. Moreover, progressive population growth and consequent urbanization is expected to increase in the number of urban WWTPs and therefore the quantity of sewage sludge that will require proper management [7]. At present, 97% of the Chilean urban population is served by approximately 300 WWTPs, and around 62% of these plants have design capacities of less than 4800 m^3/d [6,7]. Around 60% of these WWTPs use activated sludge technology to remove organic matter from wastewater, generating about 345,000 dry tons per year that must be managed [7]. Operation of activated sludge systems at high SRT values (15 d) is the most used strategy for reducing sludge production in Chilean WWTPs. As SRT increases, the maintenance energy in bacterial metabolism increases, which leads to a reduction in sludge production [8–11]. A reduction of about 60% of excess sludge production can be achieved when SRT is increased from 2 to 18 days [12]. However, the operation of activated sludge systems with longer SRTs results in high operating costs associated with aeration required by aerobic biomass [9,12]. Additionally, this sludge-reduction strategy promotes the conversion of ammonium nitrogen into nitrate, as a result of a nitrification process, increasing the aeration requirements. Required energy consumption for conventional activated sludge WWTPs is between 0.14 and 0.16 kWh/m^3 and can increase to 0.65–2.28 kWh/m^3 when nitrification processes take place during biological wastewater treatment [9,13].

An alternative method to reduce sludge production in WWTPs is the application of in situ sludge-reduction technologies, such as ozonation, ultrasonic methods, alkaline treatment and thermal processes [2,14,15]. Until now, only ozonation and ultrasonic methods have been applied in full-scale urban WWTPs [16]. Ozonation promotes the disintegration and solubilization of biodegradable and non-biodegradable compounds in the sludge [17,18]. This process can be applied to a fraction of the sludge recycled from the secondary decanter to the biological reactor. In this case, ozonated sludge is returned to the biological reactor, where the biodegradable fraction of the hydrolyzed sludge is assimilated by active microorganisms. This procedure can generate a relevant net reduction in the amount of sludge that must be managed [10,12]. The ozone dosage is typically in the range of 0.01–0.74 g O_3/g total suspended solids (TSS), resulting in a sludge reduction of between 10% and 100% [16,17,19]. It is important to note that literature has reported that the ozone-dose range should be between 0.03 and 0.05 g O_3/g TSS to obtain an appropriate balance between sludge-reduction efficiency and operating costs [20,21].

The implementation of a sludge-ozonation system in existing small urban WWTPs (operated at an SRT of 15 d) would be economically feasible when sludge management (dewatering and disposal) costs are higher than the costs associated with ozonation-unit operation. These costs are the sum of the investment associated with required equipment and the increase in operating costs related to energy consumption for ozone production, as well as the additional aeration required to cope with the increase in chemical oxygen demand (COD) resulting from the ozonated sludge [10,15]. In the case of future small urban WWTPs, in addition to the aspects already mentioned, it should be considered that the implementation of a sludge-ozonation system would allow for the design of WWTPs to be operated at short SRTs (3 d) with a reduced sludge production. Additionally, operation at such low SRTs would allow for a reduction in investment costs associated with new WWTPs, mainly due to the lower volume required for both a biological reactor and secondary decanter. In this context, both sludge-minimization strategies have advantages and disadvantages when compared to one another, and there is no obviously superior sludge-reduction method for urban WWTPs with treatment capacities of less than 24,000 PE. For these reasons, this work is focused on studying the economic viability of the installation of sludge-ozonation systems in existing and future small WWTPs.

This work is intended to represent a contribution to existing literature by providing information that can be useful in the evaluation and implementation of potential of sludge-ozonation units as a way to reduce sludge production in small WWTPs. Previous studies have concentrated on the techno-economic evaluation of sludge-reduction technologies in the water line and/or sludge line of large WWTPs [2,15–17], while economic assessments of sludge-reduction strategies for small urban WWTPs have not been studied.

2. Materials and Methods

2.1. Description of Studied WWTP Configurations

Four scenarios for the treatment of urban wastewater in small cities are studied in this work:

- Scenario 1: This scenario is based on a conventional WWTPs that include preliminary treatment (i.e., screening and grit-removal units), followed by an activated sludge unit operated at an SRT of 15 days in order to remove organic matter from wastewater, as well as a tertiary treatment based on chlorination. Generated sludge is dewatered by means of a decanting centrifuge.
- Scenario 2: This scenario is similar to the first one, but a sludge-ozonation unit is implemented to obtain zero sludge production during wastewater treatment. With this alternative, a fraction of the mixed-liquor sludge is continuously transferred to the sludge-ozonation unit for disintegration and is then returned to the biological treatment system for biodegradation.
- Scenario 3: A WWTP is designed with an activated sludge system operated at an SRT of 3 days, combined with a sludge-ozonation unit to produce the same quantity of sludge as in the first scenario.
- Scenario 4: This scenario is similar to the third one, but in this case, the ozonation unit is designed to achieve zero sludge production, so decanting-centrifuge requirements are not considered.

For all studied scenarios, the dewatered sludge is disposed of in a landfill, without any valorization process. In this work, WWTPs with capacity to treat sewage from a population equivalent of between 1000 and 20,000 PE (between 200 and 4000 m^3/d) were studied.

2.2. Mass and Energy Balances

For the studied scenarios, mass and energy balances were performed considering typical operating conditions of WWTPs (Table 1). The total influent COD concentration was 500 mg/L (soluble biodegradable COD (S_S): 150 mg/L; soluble non-biodegradable COD (S_I): 50 mg/L; particulate biodegradable COD (X_S): 200 mg/L; particulate non-biodegradable COD (X_I): 100 mg/L). Mass and energy balances enabled determination of oxygen consumption, energy consumption, ozone consumption and sludge production, as well as sizing of the required unit for wastewater and sludge treatments (preliminary treatment, biological treatment, sludge ozonation, dewatering and chlorination). Operational conditions of wastewater and sludge treatment units were considered constant, i.e., assuming no daily or seasonal variations occur during the operation of WWTPs.

Table 1. Summary of values used to perform both mass and energy balances.

Unit Operation	Values
Mass balances	
Activated sludge process	Hydraulic retention time (HRT): 0.25 d Biomass concentration in aeration tank: 4 kg VSS/m^3 Biomass yield ($Y_{x/s}$): 0.43 kg VSS/kg $COD_{consumed}$ [22] Decay coefficient (k_d): 0.24 d^{-1} [22] Volatile suspended solids concentration in the effluent ($VSS_{effluent}$): 0.02 kg VSS/m^3 Non-biodegradable fraction of heterotrophic biomass (X_P/X_H): 0.15 [22] 1.42 kg COD/kg VSS for X_P and X_H [22] 1.55 kg COD/kg VSS for X_I COD fraction [23] Oxygen requirement for ammonium oxidation: 4.57 kg O_2/kg N [22]
Ozonation unit	X_H, X_I and X_P were solubilized into S_S
Sludge dewatering	25% dry matter TSS/VSS ratio: 0.75 kg/kg
Energy balances	
Wastewater influent pumping	0.0385 kWh/$m^3_{influent}$ [22]
Screens	0.0004 kWh/$m^3_{influent}$ [22]
Grit removal	0.008 kWh/$m^3_{influent}$ [22]
Aeration	1 kWh/kg O_2 [22]
Chlorination	0.00055 kWh/$m^3_{influent}$ [22]
Sludge pumping	0.01 kWh/$m^3_{influent}$ [22]
Centrifuge	0.3 kWh/kg TSS [24]
Ozone generation	15 kWh/kg O_3 [25]

TSS and VSS are the concentration of total suspended solids and volatile suspended solids, respectively; and X_H and X_P are the concentration of heterotrophic biomass concentration and endogenous residues from decay, respectively.

For scenario 1, the daily amount of sludge produced from the activated sludge unit was calculated according to the methodology described by Crutchik et al. (2020) [26]. In the case of scenarios 2, 3 and 4, where the activated sludge system was combined with a sludge-ozonation system (Figure 1), sludge generation (kg TSS/d) was calculated using Equations (1) to (5).

$$Produced\ sludge = \frac{(VSS_{XIr} + VSS_{XPr} + VSS_{XHr})}{0.75} \times Q_0 \times HRT \times \left(\frac{1}{SRT} - \alpha \times \beta\right) \quad (1)$$

$$VSS_{XIr} = \frac{X_{I0} \times SRT}{HRT \times 1.55} \quad (2)$$

$$VSS_{XPr} = k_d \times 0.15 \times X_{Hr} \times SRT \quad (3)$$

$$VSS_{XPr} = \frac{\frac{(S_{s0}+X_{s0})\times Y_{x/s}}{1+k_d\times SRT}}{\frac{HRT}{SRT} + Y_{x/s} \times \frac{Q_{ozonated\ sludge}}{Q_0} \times \frac{\alpha\times(1-0.15)}{1+k_d\times SRT}} \quad (4)$$

$$\beta = \frac{Q_{ozonated\ sludge}}{V_R} \quad (5)$$

where VSS_{XIr}, VSS_{XPr} and VSS_{XHr} are the concentration of VSS inside the reactor associated to X_I, X_P and X_H, respectively (kg/m^3); Q_0 is the inlet flow rate (m^3/d); $Q_{ozonated\ sludge}$ is the inlet flow rate of the sludge-ozonation system (m^3/d); V_R is the volume of the aeration tank (m^3); α is the fraction of sludge that was solubilized during ozonation; and β is the amount of daily ozonated sludge with regard to the total amount of sludge in the activated sludge unit. In this work, a specific ozonation dosage (SOD) was set at 0.03 kg O_3/kg TSS [27], resulting in a sludge-solubilization degree of 25% (α) [27]. According to Equation (1), once the SOD is fixed and therefore the solubilization sludge degree is established, the amount of generated sludge can be controlled by manipulation of the fraction of daily ozonated sludge (β).

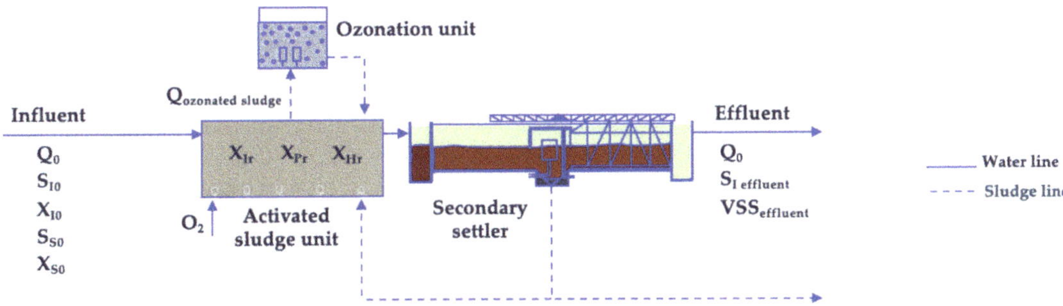

Figure 1. Schematic layout of the activated sludge system combined with an sludge-ozonation unit.

2.3. Methodology for Economic Assessment

Economic analysis of wastewater treatment for the studied scenarios was carried out considering total capital costs, as well as operating and maintenance (O&M) costs of the WWTP. Total capital costs, as well as operating and maintenance costs for treatment units are detailed below:

(a) Total capital costs include those related to the purchase of main equipment (preliminary treatment, activated sludge system, sludge-ozonation, sludge-dewatering and chlorination units) and for required for piping, instrumentation/electricity, engineering costs and civil work. The ozonation unit involved the incorporation of an ozone generator includes an ozonation tank and two pumps. Equipment capital costs were calculated based on the data and cost functions reported in studies found in the literature [28–30]. Costs related to the required equipment for piping, instrumentation/electricity, engineering costs and civil work were estimated as 15%, 25%, 10%, 34%, and 12% of total equipment costs, respectively.

(b) Operating and maintenance costs include energy consumption (due to mixing and pumping, oxygen requirements for biological treatment system, sludge dewatering and ozonation generation), reagents, labor and maintenance. Energy consumption related to the ozonation unit includes the cost of oxygen supply, the energy requirement for the production of ozone and pumping of ozonated sludge. Energy consumption was corrected based on the WWTP size, taking into account results obtained by Trapote et al. (2014) [31]. The price of electricity used was USD 0.095 /kWh [32], and costs associated with sludge disposal were calculated as USD 100 /Ton [33]. The amount of reagents needed for the sludge-dewatering (polyelectrolytes: USD 2/kg) and chlorination processes (sodium hypochlorite: USD 0.52 /kg) was calculated, taking into account a dose of 5 g/kg TSS [22] and 5.1 mg/L, respectively. Maintenance costs were calculated as fixed percentages of the capital cost (1%). The labor cost of operators was assumed to be USD 5.45 /person hour.

The minimum cost of wastewater treatment (USD/m^3) was estimated as the value that results in a net present value (NPV) of zero (Equation (6)):

$$NPV = \sum_{t=1}^{T} \frac{C_t \times (1+i)^t}{(1+r)^t} - total\ capital\ costs \tag{6}$$

where C_t is the sum of the operating and maintenance costs, i is the inflation rate (3%), r is the interest rate (5%) and T is the payback time (20 years).

Uncertainty and variability can be present in the input variables used in economic analysis. Therefore, a sensitivity analysis was carried out. For this purpose, two economic parameters (price of energy and sludge-management costs) were considered. A range of ± 15% was considered for each parameter.

3. Results and Discussion

3.1. Effects of Sludge Ozonation on Sludge Production and Energy Consumption

Sludge-mass balance was applied to a WWTP with a sludge-ozonation system in order to determine how the fraction of daily ozonated sludge (β) affects sludge production (Figure 2A). The operation of the WWTP without ozone application was used as baseline to calculate the reduction in sludge production. As can be observed in Figure 2A, when a certain SOD (kg O_3/kg TSS) is applied in the ozonation tank, the reduction in sludge production depends linearly on the fraction of daily ozonated sludge. Therefore, the degree of sludge reduction does not depend on the applied SOD, as is generally reported in the literature [10]. Instead, it depends on the product of the applied SOD and the fraction of daily ozonated sludge, i.e., the daily amount of applied ozone per mass of solids (kg O_3/kg TSS·d) [34]. The SRT applied in the activated sludge system is another factor that affects sludge reduction (Figure 2A). An increase in SRT promotes biomass decay, and therefore, less sludge is generated compared to systems operated at low SRT. This implies that an increase in SRT in the activated sludge system will produce a reduction in the amount of ozone required to obtain a given reduction in sludge production. In fact, when a given SOD is applied to the sludge, the observed biomass yield coefficient (g $SS_{generated}$/g $COD_{removed}$) is lower than that of an activated sludge system operated at a higher SRT [35]. Generally, the literature provides values of both SOD and percentage of sludge reduction. However, there is insufficient attention given to the daily fraction of ozonated sludge and the SRT. This could explain the discrepancies in sludge reduction reported for similar values of SOD [10,21]. In some works, the flow rate of returned ozonated sludge is provided, but these data alone do not allow for calculation of the daily fraction of ozonated sludge [17,36]. As a consequence, the obtained results cannot be extrapolated to other operating conditions.

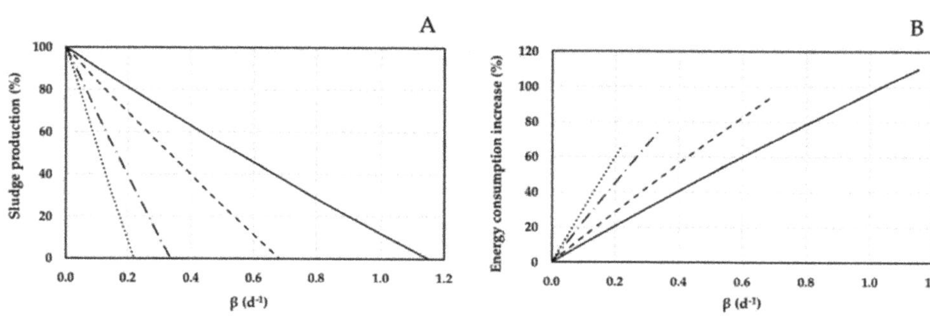

Figure 2. Effect of daily fraction of ozonated sludge (β) on (**A**) sludge production and (**B**) increase in energy consumption (SRT: —— 3 d; ----- 5 d; —·— 10 d; ······ 15 d). An ozone dose of 0.03 kg O_3/kg TSS, resulting in a sludge solubilization of 25% was used for calculations [26].

The specific daily ozone dose (kg O_3/kg TSS·d) applied to the sludge depends on both the SOD and the daily fraction of ozonated sludge. Therefore, a given specific daily ozone dose could be applied by supplying a high-SOD to a low daily fraction of ozonated sludge, or vice versa. This second strategy is preferable, since the application of a high SOD (>0.05 g O_3/g TSS) promotes a preferable reaction with dissolved organic matter and/or radical scavengers released from the sludge instead of solids [37–39]. This entails an inefficient use of the ozone to achieve a reduction in excess sludge production [19], which is not economically feasible [40]. In this work, the application of a low SOD (0.03 g O_3/g TSS) was proposed in order to favor disintegration of solids so that the amount of generated excess sludge could be controlled by changing the daily fraction of reactor sludge subjected to the solubilization process. This operating strategy has already been applied by Yasui and Shibata [41] and Yasui et al. [42], who found a linear reduction in the production of excess sludge in a pilot plant, as well as a reduction in the daily fraction of reactor sludge subjected to an SOD of 0.05 g O_3/g TSS. These authors observed that for different SRTs, zero excess sludge production could be achieved by changing the daily fraction of reactor sludge subjected to ozonation. This would confirm the results obtained in the present work, i.e., that there is a linear relationship between β and reduction in excess sludge production and that, furthermore, for a given SRT, the value of β can be adjusted such that zero sludge production is achieved.

The quality of sludge in terms of COD-type composition (fractions of X_H, X_p and X_I) also affects the degree of solubilization obtained for a given daily specific ozone dose and therefore the percentage of sludge reduction. It is expected that sludges containing a high fraction of X_H would be hydrolyzed easier than those with a high predominance of X_p and X_I fractions. Therefore, the operation of WWTPs at low SRTs would generate sludge requiring a lower ozone dose to be solubilized. However, the few studies that have investigated this topic show contradictory results [43,44]. For this reason, in this work, the solubilization degree was considered constant, independent of the sludge age.

Another parameter that could explain the discrepancy in results found in the literature when ozone is applied to reduce sludge production at full scale WWTPs is the ozone mass-transfer efficiency to the bulk liquid [45], which depends on several operating variables, such as gas flow rate, TSS concentration and ozone concentration [46–48]. These parameters are the key to improving the efficiency of the ozonation process and therefore to decreasing operating and capital costs associated with sludge reduction. For this reason, the obtained results should be expressed as a function of the amount of transferred ozone rather than the amount of applied ozone [34].

In the model used to carry out mass balance, it was assumed that all organic matter solubilized by ozone was biodegradable. This assumption was based on previous results reported by Boehler and Siegrist (2006) [35], who observed that 90% of solubilized organic matter in ozonated sludge was biodegradable. This fact is also supported by results provided by Gardoni et al. (2011) [36] and Torregrossa et al. (2012) [49]. They found that the effluent quality of WWTPs where ozone was used to promote sludge lysis hardly changed in terms of organic-matter concentration. This implies that the application of ozone increases oxygen consumption of the WWTP for two reasons: (a) the promotion of heterotrophic biomass lysis, which releases particulate biodegradable organic matter (*Xs*) contained within bacteria to the bulk liquid; and (b) the lysis of the non-biodegradable particulate fractions of organic matter (*Xi* and *Xp*), which allows them to be converted into a biodegradable carbon source for bacteria. As can be observed in Figure 2B, the application of ozone to promote sludge lysis has a greater impact on the energy consumption of WWTPs when they are operated at a low SRT. Basically, this is because operation at low SRT values promotes the conversion of biodegradable organic matter into biomass, which implies high sludge production and a low aeration requirement (low energy consumption) compared to WWTPs operated at high SRTs. Therefore, more ozone and "extra aeration" are needed to solubilize and oxidize the sludge generated when the WWTPs are operated under low-SRT conditions.

3.2. Economic Impact of Sludge Ozonation on Treatment Costs

Capital and O&M costs were calculated for each studied scenario and for different WWTP capacities (Figure 3). The implementation of a sludge-ozonation unit in an existing WWTP that operates at an SRT of 15 days implies an increase of 0.5–1.2% in capital costs (Scenarios 1 and 2, Figure 3A) and a decrease in O&M costs of 0.1–5.2% (Figure 3B), resulting in a reduction in sludge production to zero. The overall economic evaluation, considering both capital and O&M costs, shows that the treatment costs of scenarios 1 and 2 are almost the same. That is, the cost saving related to sludge management and the extra capital and O&M costs of sludge ozonation are similar under the current sludge-disposal and energy costs (Figure 4). Of course, the convenience of reducing sludge production by means of ozonation strongly depends on the sludge-disposal taxes of each country. For example, in countries with high sludge-disposal taxes, reduced sludge production by ozonation is an economically viable strategy [17,35]. Even in cases where treatment costs are similar, the achievement of zero sludge production by ozonation could be an interesting operating strategy, since sludge-disposal regulations could change over time, and sludge disposal in landfills or agricultural applications of sludge could be restricted.

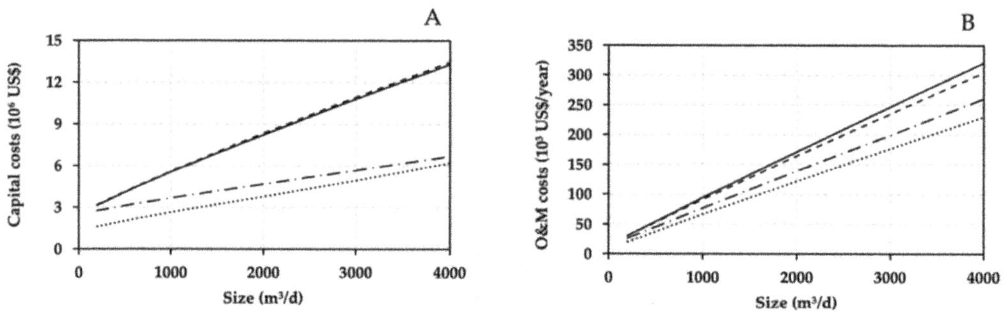

Figure 3. Capital (**A**) and O&M (**B**) costs estimated for each scenario and for WWTPs of different sizes (—— Scenario 1; ----- Scenario 2; —·— Scenario 3; ······· Scenario 4).

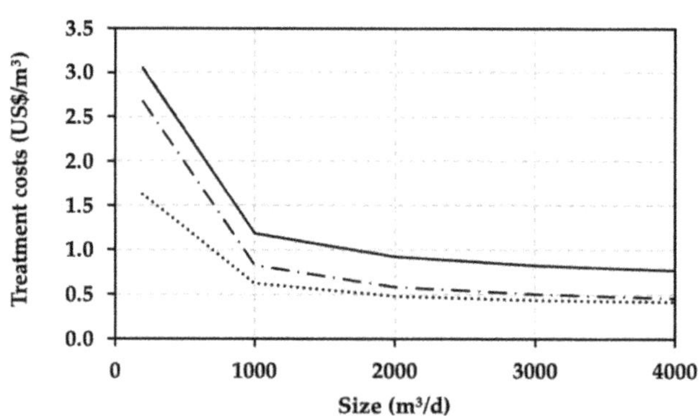

Figure 4. Treatment costs estimated for each scenario and for WWTPs of different sizes (—— Scenario 1; ----- Scenario 2; —·— Scenario 3; ······· Scenario 4).

Moreover, scenario 3 consisted of a WWTP designed to operate at an SRT of 3 days, including an ozonation unit to reduce sludge production to such values as those obtained when WWTPs are operated at an SRT of 15 (Scenario 1). Therefore, sludge production for scenarios 1 and 3 are similar. Scenario 3 would see a reduction in both capital (12.3–19.1%) and operating costs (11.9–49.7%) compared to scenario 1 (Figure 3A,B), leading to a significant decrease in treatment costs (Figure 4). The reduction in capital costs is mainly due to the reduced volumes of the activated sludge reactor and the secondary decanter. On the other hand, the reduction in operating costs is mainly attributed to aeration requirements, since the nitrification process does not take place when WWTPs are operated at an SRT of 3 d. This process increases the energy consumed by activated sludge systems by about 70% compared to activated sludge systems in which only organic matter is removed [9].

Treatment costs could be minimized if the WWTP is designed to operate at an SRT of 3 days with zero sludge production (Scenario 4, Figure 4). This scenario is especially advantageous in the case of WWTPs with the lowest studied treatment capacity, since a dewatering centrifuge would be no longer needed, generating relevant savings in capital costs (Figure 3A). As treatment capacity increases, these cost saving offset by the need for ozonation units with a higher capacity. As a result, the treatment costs for scenarios 3 and 4 tend to match.

Scenarios 3 and 4, where WWTPs are operated at an SRT of 3 days, provide the lowest treatment costs. This results from their low capital costs in comparison to WWTPs operated at an SRT of 15 days and their lower operating costs derived from the absence of nitrification and therefore the oxygen requirement associated with that process. In the case of scenario 4, the treatment costs are around 47% lower than those of scenario 1 for all the WWTP capacities studied in this work.

Sensitivity analysis shows that the operating costs of scenarios 1 and 3 present similar variations (around 3.4–4.1%) for changes of 15% in the prices of both sludge disposal and energy (Figure 5A,C). Logically, the operating costs of scenarios 2 and 4 do not depend on sludge-disposal price, while changes of 15% in energy prices cause variations of 7.2 and 7.0%, respectively (Figure 5B,D).

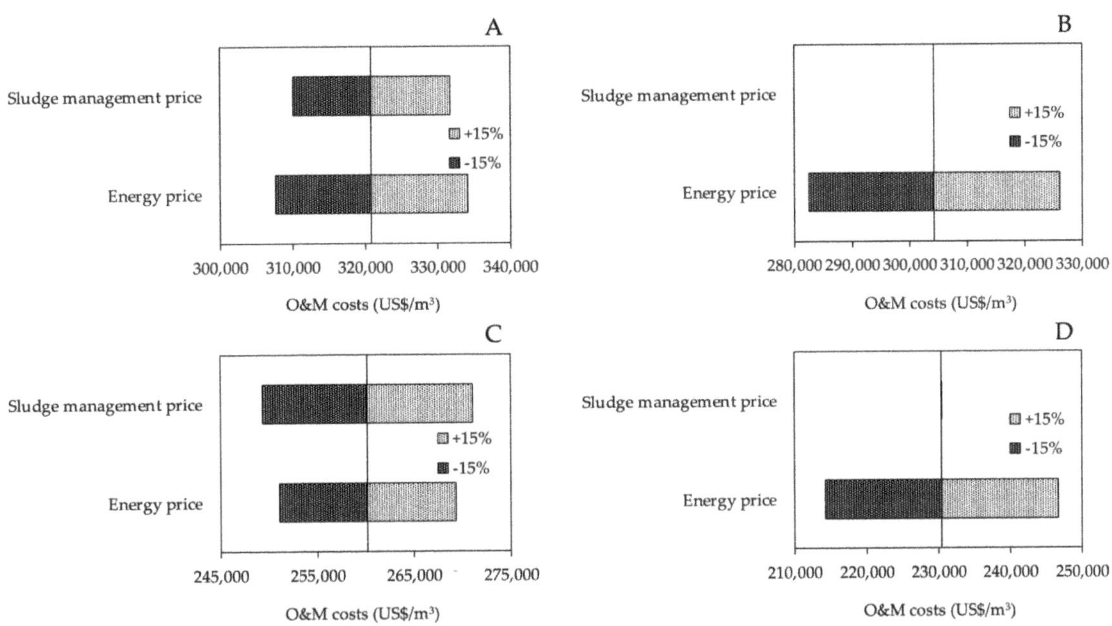

Figure 5. Sensitivity analysis for (**A**) Scenario 1; (**B**) Scenario 2; (**C**) Scenario 3 and (**D**) Scenario 4.

The economic feasibility of reducing sludge production through ozonation depends mainly on the cost of sludge management, as well as capital and operating costs associated with the installation of an ozonation system. To reduce the cost of the last two factors, it is necessary that the ratio of g $TSS_{hydrolyzed}$/g $O_{3applied}$ be as high as possible in order to minimize the capacity of the ozone generator and resultant ozone consumption. In this sense, in applications carried out at an industrial level, specific doses of ozone less than 0.05 g O_3/g TSS are applied [17,36,50] since they optimize the ratio of g $TSS_{hydrolyzed}$/g $O_{3applied}$. However, factors such as ozone concentration used during the sludge-hydrolysis process [46] or the efficiency of O_3 transfer to the liquid phase [10,18] are still poorly studied, and their optimization could improve the economics of the process. On the other hand, the ozone required to hydrolyze a given amount of secondary sludge depends on its composition, which could be determined based on its COD fractions (X_H, X_P and X_I) [43]. These fractions depend on both the characteristics of the wastewater and the SRT imposed in the activated sludge system. Therefore, a deeper understanding of how the hydrolysis of each of the COD fractions occurs is needed in order to determine the amount of O_3 required in each case. Hydrolysis of the sludge by ozonation could cause a worsening of the effluent quality. The literature agrees that the implementation of sludge-ozonation systems in WWTPs does not significantly influence the concentration of COD and SS In their effluent [10], although contradictory results have been reported regarding the concentration of nitrogen compounds [17,51]. According to Isazadeh et al. [52], the efficiency of N removal in WWTPs depends more on operating conditions than on the presence of an ozonation system. Therefore, when nitrogenous compounds must be removed from wastewater to satisfy discharge regulations, it is necessary to define the operating conditions of the WWTP that would allow a compromise to be reached between minimizing excess sludge production and obtaining a suitable effluent quality.

Currently, the operation of WWTPs is focused on recovering resources from wastewater in order to minimize the environmental impact associated with wastewater treatment. In this sense, the use of sludge generated in agriculture, instead of its disintegration through ozonation, could have a beneficial environmental impact due to the reduced use of chemical fertilizers. In addition, it must be considered that for a given SRT, the implementation of an ozone-disintegration unit will imply greater energy consumption by a WWTP, associated with both the generation of ozone and an increased aeration requirement. Therefore, not only the techno-economic aspects associated with sludge ozonation should be taken into account but also its environmental impacts, as it has already been done in the case of sludge hydrolysis prior to anaerobic digestion [53].

4. Conclusions

In WWTPs where a sludge-ozonation system is implemented, the reduction in sludge generation and the increase energy consumption depend not only on the specific ozone dose supplied (kg O_3/kg TSS) but also on the daily fraction of sludge ozonated, as well as the SRT applied in the activated sludge reactor. For a fixed daily specific ozone dose (kg O_3/kg TSS·d), the reduction in sludge generation and the increase in energy consumption is less with an increased SRT.

In the case of Chile, the implementation of sludge-ozonation systems in existing small WWTPs operated at an SRT of 15 d would not provide a reduction in treatment costs, since savings related to sludge management are offset by the necessary investment costs, as well as the increase in energy consumption. However, if a sludge-ozonation system is considered in the design of a new WWTP, operation at an SRT of 3 d could be considered, which would provide lower sludge production. This would result in an important decrease in capital costs, which would have a considerable impact on treatment costs.

In any case, the implementation of a sludge-ozonation system would not have a negative impact on treatment costs. Additionally, the operation of WWTPs would not be affected by future changes to sludge-management regulations.

Author Contributions: Conceptualization, J.L.C. and D.C.; methodology, D.C., G.V. and J.L.C.; writing-original draft, data curation and formal analysis, D.C. and J.L.C.; writing—review and editing, O.F., A.P. (Alba Pedrouso), D.J. and A.P. (Alicia Pinto). All authors contributed to the writing and proofreading of the paper. All authors have read and agreed to the published version of the manuscript.

Funding: This research was funded by the Chilean Government through the Projects ANID/FONDECYT/1200850 and CRHIAM Centre grant number ANID/FONDAP/15130015, by the Spanish Government through TREASURE [CTQ2017-83225-C2-1-R] and by the European Commission LIFE ZERO WASTE WATER [LIFE19ENV/ES/000631] projects. The authors from Universidade de Santiago de Compostela belong to CRETUS Strategic Partnership [ED431E 2018/01] and the Galician Competitive Research Group [GRC ED431C-2021/37]. All the Spanish programs are co-funded by FEDER (EU).

Institutional Review Board Statement: Not applicable.

Informed Consent Statement: Not applicable.

Data Availability Statement: Not applicable.

Conflicts of Interest: The authors declare no conflict of interest.

References

1. Foladori, P.; Andreottola, G.; Ziglio, G. Sludge composition and production in full-plants. In *Sludge Reduction Technologies in Wastewater Treatment Plants*; IWA Publishing: London, UK, 2010; pp. 7–19.
2. Campos, J.L.; Figueroa, M.; Vázquez, J.R.; Mosquera-Corral, A.; Roca, E.; Méndez, R. Evaluation of in-situ sludge reduction technologies for wastewater treatment plants. In *Sludge: Types, Treatment Processes and Disposal*; Nova Science Publishers Inc.: Hauppauge, NY, USA, 2009; pp. 161–186.
3. Alloul, A.; Ganigué, R.; Spiller, M.; Meerburg, F.; Cagnetta, C.; Rabaey, K.; Vlaeminck, S.E. Capture–ferment-upgrade: A three-step approach for the valorization of sewage organics as commodities. *Environ. Sci. Technol.* **2018**, *52*, 6729–6742. [CrossRef] [PubMed]
4. Amen, T.W.M.; Eljamal, O.; Khalil, A.M.E.; Matsunaga, N. Biochemical methane potential enhancement of domestic sludge digestion by adding pristine iron nanoparticles and iron nanoparticles coated zeolite compositions. *J. Environ. Chem. Eng.* **2017**, *5*, 5002–5013. [CrossRef]
5. Gretzschel, O.; Schmitt, T.G.; Hansen, J.; Siekmann, K.; Jakob, J. Sludge digestion instead of aerobic stabilisation—A cost benefit analysis based on experiences in Germany. *Water Sci. Technol.* **2014**, *69*, 430–437. [CrossRef] [PubMed]
6. INE. Ciudades, Pueblos, Aldeas y Caseríos. 2019. Available online: https://geoarchivos.ine.cl/File/pub/Cd_Pb_Al_Cs_2019.pdf (accessed on 23 November 2021).
7. SISS. Superintendencia de Servicios Sanitarios. *Informe de Gestión del Sector Sanitario 2019.* Available online: https://www.siss.gob.cl/586/articles-17955_recurso_1.pdf (accessed on 14 October 2021).
8. Di Iaconi, C.; De Sanctis, M.; Altieri, V.G. Full-scale sludge reduction in the water line of municipal wastewater treatment plant. *J. Environ. Manag.* **2020**, *269*, 110714. [CrossRef] [PubMed]
9. Sid, S.; Volant, A.; Lesage, G.; Heran, M. Cost minimization in a full-scale conventional wastewater treatment plant: Associated costs of biological energy consumption versus sludge production. *Water Sci. Technol.* **2017**, *76*, 2473–2481. [CrossRef]
10. Semblante, G.U.; Hai, F.I.; Ngo, H.H.; Guo, W.; You, S.; Price, W.E.; Nghiem, L.D. Sludge cycling between aerobic, anoxic and anaerobic regimes to reduce sludge production during wastewater treatment: Performance, mechanisms and implications. *Bioresour. Technol.* **2014**, *155*, 395–409. [CrossRef]
11. Van Loosdrecht, M.C.M.; Henze, M. Maintenance, endogeneous respiration, lysis, decay and predation. *Water Sci. Technol.* **1999**, *39*, 107–117. [CrossRef]
12. Guo, J.; Fang, F.; Yan, P.; Chen, Y. Sludge reduction based on microbial metabolism for sustainable wastewater treatment. *Bioresour. Technol.* **2020**, *297*, 122506. [CrossRef]
13. Siatou, A.; Manali, A.; Gikas, P. Energy consumption and internal distribution in activated sludge wastewater treatment plants of Greece. *Water* **2020**, *12*, 1204. [CrossRef]
14. Campbell, K.; Wang, J.; Tucker, R.; Struemph, C. Implementation of long solids retention time activated sludge process for rural residential community. *Water Environ. Res.* **2020**, *93*, 174–185. [CrossRef]
15. Collivignarelli, M.C.; Abbà, A.; Miino, M.C.; Torreta, V. What advanced treatments can be used to minimize the production of sewage sludge in WWTPs? *Appl. Sci.* **2019**, *9*, 2650. [CrossRef]
16. Wang, Q.; Wei, W.; Gong, Y.; Yu, Q.; Li, Q.; Sun, J.; Yuan, Z. Technologies for reducing sludge production in wastewater treatment plants: State of the art. *Sci. Total Environ.* **2017**, *587–588*, 510–521. [CrossRef] [PubMed]
17. Chiavola, A.; Salvati, C.; Bongirolami, S.; Di Marcantonio, C.; Boni, M.R. Techno-economic evaluation of ozone-oxidation for sludge reduction at the full-scale. Comparison between the application to the return activated sludge (RAS) and the sludge digestion unit. *J. Water Process. Eng.* **2021**, *42*, 102114. [CrossRef]
18. Hashimoto, K.; Kubota, N.; Okuda, T.; Nakai, S.; Nishijima, W.; Motoshige, H. Reduction of ozone dosage by using ozone in ultrafine bubbles to reduce sludge volume. *Chemosphere* **2021**, *274*, 129922. [CrossRef] [PubMed]
19. Foladori, P.; Andreottola, G.; Ziglio, G. Ozonation. In *Sludge Reduction Technologies in Wastewater Treatment Plants*; IWA Publishing: London, UK, 2010; pp. 249–294.
20. Demir, O.; Filibeli, A. Fate of return activated sludge after ozonation: An optimization study for sludge disintegration. *Environ. Technol.* **2012**, *33*, 1869–1878. [CrossRef] [PubMed]
21. Chu, L.; Yan, S.; Xing, X.; Sun, X.; Jurcik, B. Progress and perspectives of sludge ozonation as a powerful pretreatment method for minimization of excess sludge production. *Water Res.* **2009**, *43*, 1811–1822. [CrossRef]
22. Tchobanoglous, G.; Stensel, H.D.; Tsuchihashi, R.; Burton, F.L.; Abu-Orf, M.; Bowden, G.; Pfrang, W.; Metcalf & Eddy Inc. *Wastewater Engineering: Treatment & Resource Recovery*, 5th ed.; McGraw Hill: New York, NY, USA, 2014.
23. Takács, I. Experiments in Activated Sludge Modelling. Ph.D. Thesis, Ghent University, Gent, Belgium, 2008.
24. Greenfield, P.F.; Batstone, D.J. Anaerobic digestion: Impact of future greenhouse gases mitigation policies on methane generation and usage. *Water Sci. Technol.* **2005**, *52*, 39–47. [CrossRef]

25. Chiavola, A.; D'Amato, E.; Gori, R.; Lubello, C.; Sirini, P. Techno-economic evaluation of the application of ozone-oxidation in a full-scale aerobic digestion plant. *Chemosphere* **2013**, *91*, 656–662. [CrossRef]
26. Crutchik, D.; Franchi, O.; Caminos, L.; Jeison, D.; Belmonte, M.; Pedrouso, A.; Val del Rio, A.; Mosquera-Corral, A.; Campos, J.L. Polyhydroxyalkanoates (PHAs) production: A feasible economic option for the treatment of sewage sludge in municipal wastewater treatment plants? *Water* **2020**, *12*, 1118. [CrossRef]
27. Manterola, G.; Uriarte, I.; Sancho, L. The effect of operational parameters of the process of sludge ozonation on the solubilization of organic and nitrogenous compounds. *Water Res.* **2008**, *42*, 3191–3197. [CrossRef]
28. Acampa, G.; Giustra, M.G.; Parisi, C. Water Treatment Emergency: Cost Evaluation Tools. *Sustainability* **2019**, *11*, 2609. [CrossRef]
29. Van Starveren, J. *Cost Estimating Relationships on Wastewater Treatment Plant Projects: An Analysis to Increase Accuracy on Cost Estimating for Wastewater Treatment Plant Projects*; University of Twente: Enschede, The Netherlands, 2019; Available online: http://essay.utwente.nl/80779/1/Staveren-Job-van-%20geanonimiseerd.pdf (accessed on 23 November 2021).
30. Sharma, J.R. Development of a Preliminary Cost Estimation Method for Water Treatment Plants. Master's Thesis, The University of Texas at Arlington, Arlington, TX, USA, 2010.
31. Trapote, A.; Albaladejo, A.; Simón, P. Energy consumption in an urban wastewater treatment plant: The case of Murcia Region (Spain). *Civ. Eng. Environ. Syst.* **2014**, *31*, 304–310. [CrossRef]
32. ENEL. Tarifas Vigentes. Available online: https://www.enel.cl/es/clientes/informacion-util/tarifas-y-reglamentos/tarifas.html (accessed on 23 September 2021).
33. Aguas Nuevas, S.A.; Santiago, Chile. Report of operating management. Private communication, 2021.
34. Richardson, E.E.; Hanson, A.; Hernández, J. Ozonation of continuous flow activated sludge for reduction of waste solids. *Ozone Sci. Eng.* **2009**, *31*, 247–256. [CrossRef]
35. Boehler, M.; Siegrist, H. Potential of activated sludge disintegration. *Water Sci. Technol.* **2006**, *53*, 207–216. [CrossRef] [PubMed]
36. Gardoni, D.; Ficara, E.; Fornarelli, R.; Parolini, M.; Canziani, R. Long-term effects of the ozonation of the sludge recycling stream on excess sludge reduction and biomass activity at full-scale. *Water Sci. Technol.* **2011**, *63*, 2032–2038. [CrossRef]
37. Yeom, I.T.; Lee, K.R.; Lee, Y.H.; Ahn, K.H.; Lee, S.H. Effects of ozone treatment on the biodegradability of sludge from municipal wastewater treatment plants. *Water Sci. Technol.* **2002**, *46*, 421–425. [CrossRef]
38. Zhang, G.; Yang, J.; Liu, H.; Zhang, J. Sludge ozonation: Disintegration, supernatant changes and mechanisms. *Bioresour. Technol.* **2009**, *100*, 1505–1509. [CrossRef]
39. Yan, S.T.; Chu, L.B.; Xing, X.H.; Yua, A.F.; Sunc, X.L.; Jurcik, B. Analysis of the mechanism of sludge ozonation by a combination of biological and chemical approaches. *Water. Res.* **2009**, *43*, 195–203. [CrossRef]
40. Richardson, E.E.; Hanson, A. Regression modeling of ozonation process in wastewater treatment plants for reduction of waste activated sludge. *Ozone Sci. Eng.* **2014**, *36*, 451–464. [CrossRef]
41. Yasui, H.; Shibata, M. An innovative approach to reducing excess sludge production in the activated sludge process. *Water Sci. Technol.* **1994**, *30*, 11–20. [CrossRef]
42. Yasui, H.; Nakamura, K.; Sakuma, S.; Iwasaki, M.; Sakai, Y. A full-scale operation of a novel activated sludge process without excess sludge production. *Water Sci. Technol.* **1996**, *34*, 395–404. [CrossRef]
43. Fall, C.; Silva-Hernández, B.C.; Hooijmans, C.M.; Lopez-Vazquez, C.M.; Esparza-Soto, M.; Lucero-Chávez, M.; van Loosdrecht, M.C.M. Sludge reduction by ozone: Insights and modeling of the dose response effects. *J. Environ. Manag.* **2018**, *206*, 103–112. [CrossRef] [PubMed]
44. Labelle, M.A.; Ramdani, A.; Deleris, S.; Gadbois, A.; Dold, P.; Comeau, Y. Ozonation of endogenous residue and active biomass from a synthetic activated sludge. *Water Sci. Technol.* **2011**, *63*, 297–302. [CrossRef]
45. Hashimoto, K.; Nakai, S.; Motoshige, H.; Nishijima, W. Sludge reduction in a full-scale wastewater treatment plant using ultrafine- and micro-bubble. *Ozone Sci. Eng.* **2021**, *43*, 127–135. [CrossRef]
46. Sun, X.; Liu, B.; Zhang, L.; Aketagawa, K.; Xue, B.; Ren, Y.; Bai, J.; Zhan, Y.; Chen, S.; Dong, B. Partial ozonation of returned sludge via high-concentration ozone to reduce excess sludge production: A pilot study. *Sci. Total Environ.* **2022**, *807*, 150773. [CrossRef]
47. Gardoni, D.; Ficara, E.; Vergine, P.; Canziani, R. A full-scale plug-flow reactor for biological sludge ozonation. *Water Sci. Technol.* **2015**, *71*, 560–565. [CrossRef]
48. Zhou, H.; Smith, D.W. Ozone mass transfer in water and wastewater treatment: Experimental observations using a 2D laser particle dynamics analyzer. *Water Res.* **2000**, *34*, 909–921. [CrossRef]
49. Torregrossa, M.; Di Bella, G.; Di Trapani, D. Comparison between ozonation and the OSA process: Analysis of excess sludge reduction and biomass activity in two different pilot plants. *Water Sci. Technol.* **2012**, *66*, 185–192. [CrossRef]
50. Meng, X.; Liu, D.; Yang, K.; Song, X.; Zhang, G.; Yu, J.; Zhang, J.; Tang, Y.; Li, K. A full scale anaerobic-anoxic-aerobic process coupled with low-dose ozonation for performance improvement. *Bioresour. Technol.* **2013**, *146*, 240–246. [CrossRef]
51. Sui, P.; Nishimura, F.; Tsuno, H. Nitrogen behavior during sludge ozonation: A long-term observation by pilot experiments. *Water Sci. Technol.* **2014**, *70*, 289–296. [CrossRef]
52. Isazadeh, S.; Urbina Rivas, L.E.; Ozdural Ozcer, P.; Frigon, D. Reduction of waste biosolids by RAS-ozonation: Model validation and sensitivity analysis for biosolids reduction and nitrification. *Environ. Model. Softw.* **2015**, *65*, 41–49. [CrossRef]
53. Carballa, M.; Duran, C.; Hospido, A. Should We Pretreat Solid Waste Prior to Anaerobic Digestion? An Assessment of Its Environmental Cost. *Environ. Sci. Technol.* **2011**, *45*, 10306–10314. [CrossRef] [PubMed]

sustainability

MDPI

Article

Capacity of Marine Microalga *Tetraselmis suecica* to Biodegrade Phenols in Aqueous Media

Edna R. Meza-Escalante [1], Larissa Lepe-Martinié [1], Carlos Díaz-Quiroz [2], Denisse Serrano-Palacios [1], Luis H. Álvarez-Valencia [3], Ana Rentería-Mexía [2], Pablo Gortáres-Moroyoqui [2] and Gabriela Ulloa-Mercado [2,*]

1 Department of Water and Environment Sciences, Sonora Institute of Technology, 5 de Febrero 818 South, Ciudad Obregón 85000, Mexico; edna.meza@itson.edu.mx (E.R.M.-E.); larissa_gisela@hotmail.com (L.L.-M.); denisse.serrano@itson.edu.mx (D.S.-P.)

2 Department of Biotechnology and Food Sciences, Sonora Institute of Technology, 5 de Febrero 818 South, Ciudad Obregón 85000, Mexico; carlos.diazq@itson.edu.mx (C.D.-Q.); ana.renteria@itson.edu.mx (A.R.-M.); pablo.gortares@itson.edu.mx (P.G.-M.)

3 Department of Agronomic and Veterinary Sciences, Sonora Institute of Technology, 5 de Febrero 818 South, Ciudad Obregón 85000, Mexico; luis.alvarez@itson.edu.mx

* Correspondence: ruth.ulloa@itson.edu.mx

Abstract: Phenolic compounds are toxic and dangerous to the environment and human health. Although the removal of phenols and their derivatives is very difficult, it has been achieved by applying some biological processes. The capacity of microalga to remove phenolic compounds has been demonstrated; however, few reports of the removal of these compounds in a mixture have been published. The removal of phenol, *p*-cresol and *o*-cresol was performed by batch kinetics at 50 and 100 mg L^{-1}, and the simultaneous degradation of phenol, *p*-cresol and *o*-cresol was carried out in a mixture at 40 mg L^{-1} using the marine microalga *Tetraselmis suecica*. The kinetic study was carried out for 192 h. For concentrations of 50 mg L^{-1} and 100 mg L^{-1}, phenolic compound consumption efficiencies greater than 100% and 85%, respectively, were obtained, and up to 73.6% removal in the mixture. The results obtained indicate that the marine microalga carries out a process of the oxidation of organic matter and phenolic compounds, mineralizing up to 31.4% to CO2 in the mixture. Biological treatments using the marine microalga *T. suecica* can be considered feasible to treat effluents with concentrations similar to those of the present study.

Keywords: biodegradation; marine microalgae; phenolic compounds; kinetics

Citation: Meza-Escalante, E.R.; Lepe-Martinié, L.; Díaz-Quiroz, C.; Serrano-Palacios, D.; Álvarez-Valencia, L.H.; Rentería-Mexía, A.; Gortáres-Moroyoqui, P.; Ulloa-Mercado, G. Capacity of Marine Microalga *Tetraselmis suecica* to Biodegrade Phenols in Aqueous Media. *Sustainability* **2022**, *14*, 6674. https://doi.org/10.3390/su14116674

Academic Editors: José Luis Campos, Anuska Mosquera Corral, Ángeles Val del Río and Alba Pedrouso Fuentes

Received: 20 April 2022
Accepted: 24 May 2022
Published: 30 May 2022

1. Introduction

Wastewater is a natural sub-product from any productive process. Phenol is one of the main contaminants in the effluents of chemical industries and petrochemicals, coke oven plants, coal mining and pharmaceuticals, with concentrations of 2 mg L^{-1} to up to 6000 mg L^{-1} [1].

Phenol and its derivatives are corrosive and toxic compounds that are highly dangerous, even at low levels, posing a risk to human health and the environment. Therefore, the management of phenol-polluted wastewater represents major environmental challenges, due to its chemical complexity [2]. Hence, strategies for its treatment and elimination have been studied. In this regard, physical and chemical methods are the most frequently applied processes to remove phenols from the environment and industrial wastewater. Among them, electro-oxidation, activated carbon adsorption and photocatalytic degradation are the most common [3].

Microalgae have gained attention for their wide applications in different areas, including the removal of polluting compounds in water [4], due to their ability to remove toxic compounds [5] such as phenol [6], as well as various phenolic compounds [7]. In particular, treatment with microalgae species such as Chlorella, Scenedesmus, Dunaliella and Porphyridium [8], which are freshwater or marine microalgae, has been studied for water treatment [9]. Few papers have mentioned the marine genus Tetraselmis [10]. In general, microalgae exhibit a low tolerance to high phenol concentration, and consequently, low degradation rates are achieved in this condition [11].

The main advantages of using microalgae as biological treatments to remove phenols from wastewater (freshwater or marine) are their characteristics in terms of growth and composition. Microalgae are diverse, fast-growing and adaptable to various environments, with cultivation options in either natural or artificial conditions; they use sunlight as an energy source and, in addition, some species have a mixotrophic metabolism [12]. Moreover, the cultures are low-cost and easy to carry out in comparison with electro-oxidation or activated carbon adsorption.

Degradation studies of phenolic compounds have been reported using Chlorella pyrenoidosa, at concentrations up to 800 mg L^{-1}, achieving a removal of up to 97% [2]. In addition, *Chlorella vulgaris* has been used in the degradation of a mixture of phenol and *p*-cresol in the treatment of coal gasification wastewater. The degradation was improved after the addition of $NaHCO_3$ [11]. The synergistic effect of glucose (as co-substrate) and phenol was studied using this microalga, increasing the tolerance of microalga to phenol up to 10 g L^{-1} [13]. An interesting study revealed the metabolic pathways of *Tetraselmis suecica* when it is cultured under nitrogen depletion and starvation. This study highlighted that transcripts involved in signal transduction pathways, stress and antioxidant responses, and solute transport were strongly up-regulated when *T. suecica* was cultured under nitrogen starvation, with incidence in the biochemical composition and photosynthesis activity [14]. This can be a mechanism used by *T. suecica* to remove contaminants from water, although in some cases, the contaminant removal is associated with many other contributing factors. Among them, Forootanfar et al. [15] described how *T. suecica* was able to remove 67% of the *p*-chlorophenol at an initial concentration of 20 mg L^{-1} from the medium within a 10-day period. However, the efficacy of the process was dependent on the *p*-chlorophenol concentration at 60 mg L^{-1}, which was inhibitory. The adaptation phase was *p*-chlorophenol-dependent; in this regard, at high concentration, the lag phase growth was the longest, suggesting that *p*-CP was not used as a primary growth substrate by the algae because *p*-CP elimination was increased by enhancing the cell numbers during the 10-day period. In this way, Forootanfar et al. [15] established that besides cell adsorption, absorption is another biological pathway for removing organic contaminants by bioaccumulation, biodegradation and biotransformation. In addition, these authors used the purified laccase enzyme for the degradation of substituted phenols. Moreover, Petroutsos et al. [16] discovered another mechanism in *T. marina* to eliminate 2,4-dichlorophenol, the conjugation of a phenolic compound with glucose (glucosidation) and the malonylation of phenols such as quercetin, which reduced the toxicity of this hazardous compound.

On the other hand, Pop et al. [17] studied the bisphenol effect at 50, 100 and 200 ppm in the aquatic plant *Lemna minor*. The results show chlorosis for the 200 and 100 ppm groups, and no budding formation at all concentrations evaluated. At 100 ppm, non-fermenting Gram-negative bacteria (*Klebsiella aerogenes*) and Escherichia coli were found in the root of the plants; it is maybe associated with the plant chlorosis. However, an interesting protective effect to oxidative stress was found in the 100 and 200 ppm groups. The above shows Lemna minor has the potential to withstand and metabolize Bisphenol A up to 100 ppm.

However, to date, there are just a few reports on the removal of phenolic compounds in a mixture of phenol, *p*-cresol and *o*-cresol, which is the way in which they are normally found in wastewater. Therefore, the aim of this study was to evaluate the ability of the marine microalgae *Tetraselmis suecica* to remove phenol, *p*-cresol and *o*-cresol separately and simultaneously in a mixture.

2. Materials and Methods

2.1. *Tetraselmis suecica* Culture

The *T. suecica* microalgae strain was acquired from the Northwest Biological Research Center (CIBNOR) of Baja California Sur, Mexico. The culture medium for biomass production was seawater with a salinity of 35‰, previously sterilized at 15 lb in^{-2} for 15 min, adding a sterile nutrient solution, an "algal" medium composed of macro elements and trace elements [18].

The microalga was inoculated at an initial cell density of 1×10^6 cells mL^{-1} in glass serological culture bottles with 800 mL of medium. The microalga was placed in an incubation room with a light intensity of 180 μmol m^{-2} s^{-1}, in cycles of 12 h of light–darkness at a temperature of 20 °C and an airflow of 0.55 vvm (air volume/volume of medium per minute), enriched with a pulse of CO_2 to maintain the pH between 7 and 8. Biomass harvesting was carried out once the exponential phase of growth was achieved.

2.2. Degradation Kinetics

Biological degradation kinetic studies show the consumption of a nutrient and/or contaminant over time, in order to establish the efficiency of removal and rate of consumption. The kinetics were performed in 160 mL serological bottles, using a working volume of 60 mL, consisting of 55 mL of algal medium, to which 1 g L^{-1} $NaHCO_3$ was added as a carbon source. The bottles were inoculated, in duplicate, with 1 g of Total Suspended Solids (TSS) L^{-1} of biomass. Subsequently, the phenolic compounds were added separately (phenol, *p*-cresol and *o*-cresol), each at concentrations of 50 and 100 mg L^{-1}. The experimental units remained in support with agitation at 200 rpm (Thermo Scientific Model No. 2345 Vernon Hills, IL, USA) and with an incidence of light (180 μmol m^{-2} s^{-1}). The samples were taken at 0, 6, 12, 24, 48, 72, 96, 120, 144, 168 and 192 h after starting the degradation kinetics. Finally, degradation kinetics was carried out with the mixture of the three phenolic compounds at a concentration of 40 mg L^{-1} each, under the same conditions of inoculation, agitation and light incidence as the previous tests.

Control kinetics were performed. First, abiotic kinetics in serological bottles with the algal medium enriched with $NaHCO_3$ were carried out under the same conditions as the biotic kinetics, testing concentrations of 50 and 100 mg L^{-1} of each phenolic compound in order to rule out chemical oxidation. Measurement of the phenol concentration was at 0, 15, 24, 48, 72, 96, 120 and 144 h. Total organic carbon COD was measured at the beginning at 0, 74 and 144 h. In addition, a control test was carried out with inert biomass and 50 mg L^{-1} of each phenolic compound to rule out the adsorption of contaminants in the cell wall of the microalgae. For this test, the biomass was exposed to a temperature of 180 °C for 24 h to cause the destruction of the microalgae. The other conditions of these kinetics are the same as those described for biotic kinetics. Measurement of the phenol concentration was at 0, 6, 24, 72, 120 and 192 h. COD was measured at 0, 24, 120 and 192 h.

2.3. Determination of Phenolic Compounds

For the quantification of the phenolic compounds, firstly, the wavelength of maximum absorbance for phenol and *p*-cresol was determined by spectrophotometric scanning, defined at 271 nm and 291 nm, respectively. A calibration curve was prepared in a concentration range of 0 to 200 mg L^{-1} of each compound, separately. The samples were filtered through a 0.45 μm nylon membrane, before being read in the spectrophotometer. On the other hand, the wavelength of maximum absorbance for *o*-cresol was determined as proposed by APHA [19]. A total of 125 μL of 0.5 N NH_4OH was added to 5 mL of the sample and the pH was adjusted to 7.9 ± 0.1 with a phosphate buffer. Next, 50 μL of 4-amino-antipyridine was added, followed by 50 μL of potassium ferricyanide $K_3[Fe(CN)_6]$. After 15 min the absorbance was registered at 500 nm in a spectrophotometer. A calibration curve in the range of 0 to 5 mg L^{-1} of *o*-cresol was used for quantification.

The determination of the phenolic compounds in the mixture was carried out by means of high-performance liquid chromatography (HPLC, Aligent Varian ProStar model) with the modified methodology of Meza-Escalante et al. [20]. A C-18 column was used at 40 °C with a mobile phase of acetonitrile water (30:70), a flow rate of 1.2 mL min^{-1} and a run time of 15 min at a pressure of 80 atm. The measurement was performed at a wavelength of 291 nm with an injection volume of 20 µL.

The quantification was carried out by comparing the areas of the calibration curves (in the range of 15 to 150 mg L^{-1}) of the individual phenolic compounds.

From the concentrations of the phenolic compounds at different times, the removal efficiency was calculated. In addition, the specific consumption rate was determined based on the following equation [21]:

$$qs = \frac{ds}{dt} \cdot \frac{1}{X} = \frac{Ln\ So - Ln\ S}{\Delta t} \cdot \frac{1}{X}$$

2.4. Determination of COD and Inorganic and Organic Carbon

Complete oxidation of the phenolic compounds in a mixture was analyzed through total carbon (TC), inorganic carbon (IC) measurements, and Chemical Oxygen Demand (COD). COD was determined with a spectrophotometer (Spectroquant Pharo 300, Swedesboro, NJ, USA), following the procedure described in standard methods [19]. For quantification, a calibration curve of potassium biphthalate from 0 to 500 mg L^{-1} was performed. TC and IC were measured in a TOC-L Shimadzu total organic carbon analyzer (Shimadzu Co., Kyoto, Japan). Calibration curves were made with potassium biphthalate for TC and the IC curve was made with sodium carbonate and sodium bicarbonate, both from 0 to 500 mg L^{-1}. The samples were taken on the third and eighth day of incubation.

3. Results

In the present study, the capacity of the microalga *T. suecica* for the elimination of phenolic compounds (phenol, *o*-cresol, and *p*-cresol), individually and in a mixture, was evaluated. These compounds are mainly present in industrial wastewater of pharmaceutical, tannery and petrochemical origin.

3.1. Kinetic Study of Phenol, p-Cresol and o-Cresol at 50 and 100 mg L^{-1}

Firstly, the control kinetics carried out (biotic and with inert biomass) did not show removal of the compounds (coefficient of variation <10%), which rules out chemical oxidation processes and the presence of the adsorption of contaminants in the cell wall of the microalgae. The above suggests that any removal of the phenolic compounds present in the tests was a biological process carried out by microalgae.

The biotic kinetics of the separate phenolic compounds at 50 and 100 mg L^{-1} are shown in Figure 1. In Figure 1A, the total removal of *o*-cresol was achieved at 75 h, phenol at 96 h and *p*-cresol at 120 h, when the phenolic compounds were added at 50 mg L^{-1}. However, it can be seen that *o*-cresol was immediately removed, achieving a concentration of 5 mg L^{-1} at 24 h of bioassay, which is 90% removal, while phenol and *p*-cresol barely reached about 50% removal. A significant difference was observed between *o*-cresol and the other treatments ($p \leq 0.05$). On the other hand, in Figure 1B, a similar behavior was observed in the consumption of the separate phenolic compounds at 100 mg L^{-1}, highlighting the removal of *o*-cresol at 24 h, which was similar to phenol and *p*-cresol removal at up to 48 h, with a non-significant difference observed between them. The maximal removal of 94.5% was reached at 192 h (Table 1). The high removal of *o*-cresol was according to the highest consumer-specific rate (q_s) in both concentrations (Table 1).

These results suggest that the increase in the concentration of phenolic compounds inhibits the growth and metabolic activity of *T. suecica*; the same behavior was noticed in *Chlorella vulgaris* cultured in the presence of phenol [13] when its concentration was increased. In addition, Joseph and Joseph [8] showed that, although chlorophyte microalgae, such as *T. suecica*, are the most resistant to phenolic compounds, in some species, inhibition may occur at concentrations from 100 mg L^{-1}. Moreover, at a higher initial concentration of phenol, it takes more time to be degraded completely by *Chlorella pyrenoidosa* [2], which accords with the behavior shown by *T. suecica* in this study.

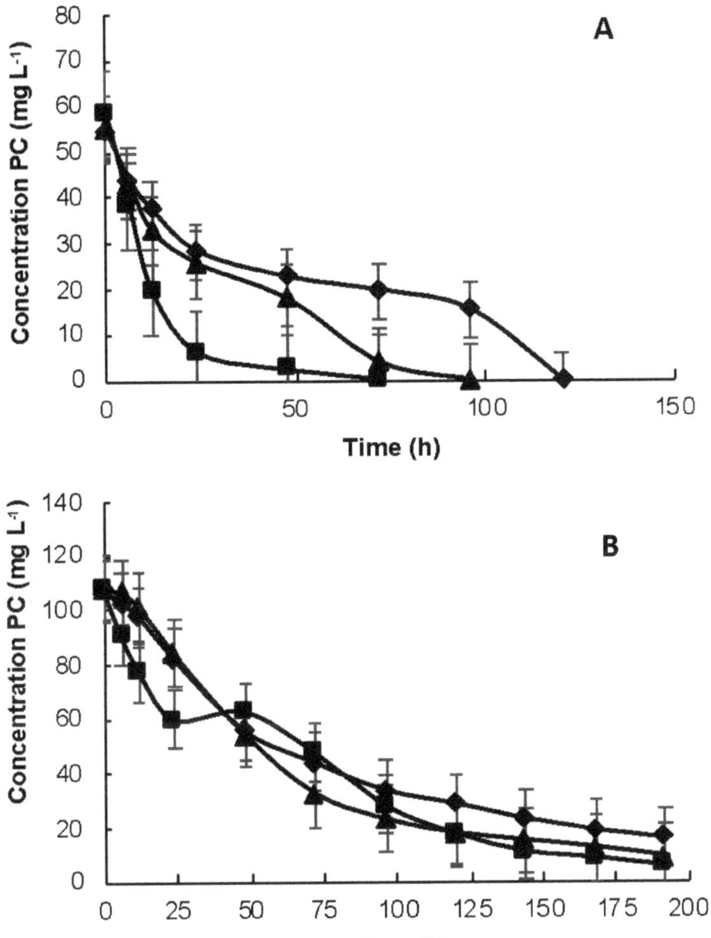

Figure 1. Consumption profile of phenolic compounds (PC) at 50 mg L^{-1} (**A**) and 100 mg L^{-1} (**B**). The symbols represent the concentration of each compound: phenol (triangle), *p*-cresol (rhombus) and *o*-cresol (square).

Table 1. Specific consumption rates (q_s) and removal efficiencies (RE) obtained by marine microalgae at two different concentrations of phenolic compounds (PC). The result is given as a percentage.

Compound	**50 mg L^{-1}**		**100 mg L^{-1}**	
	q_s	**RE**	q_s	**RE**
Phenol	1.86 ± 0.22	100 ± 0	1.12 ± 0.01	91.69 ± 0.77
p-cresol	0.93 ± 0.05	100 ± 0	1.11 ± 0.05	85.20 ± 0.66
o-cresol	3.27 ± 0.03	100 ± 0	1.97± 0.31	94.50 ± 0.35

The q_s values are given in mg PC (g SST · h)$^{-1}$ ± the standard deviation. Data were calculated using at least three points. The removal efficiency at 100 ppm of the PC was determined for 192 h of incubation.

In this regard, the ability of an improved strain of *Chlorella* sp. was demonstrated, which was capable of removing 500 to 700 mg L^{-1} of phenol completely within seven days at an initial biomass of 0.6 g/L^{-1} and under continuous illumination (118 µmol m^{-2} s^{-1}) [21]. These findings highlight the importance of the adaptability of strains to increase the removal capacity of phenolic compounds. On other hand, Wang et al. [22] showed that phenol, at concentrations of less than 100 mg L^{-1}, could be completely degraded in 4 days (96 h) using the microalga, *Isochrysis galbana*.

Regarding the elimination efficiencies (Table 1), it is observed that in the kinetics at 50 mg L^{-1}, 100% consumption efficiency was achieved for all the phenolic compounds. On the other hand, at a concentration of 100 mg L^{-1}, the consumption efficiencies achieved 85.2% for *p*-cresol and 94.5% for *o*-cresol.

Moreover, the specific rate of consumption (q_s) was calculated to both bioassays (Table 1). The *qs* at 50 mg L^{-1} of each compound are indicated, where it can be seen that *o*-cresol shows the highest specific rate of consumption (q_s), with 3.27 ± 0.03 mg *o*-cresol g TSS $^{-1}$ h^{-1}, followed by phenol and, finally, *p*-cresol, which obtain consumption rates of approximately 1.7 and 3.5 times lower. By increasing the concentration to 100 mg L^{-1} of phenolic compounds, a higher rate of consumption was again obtained for *o*-cresol, with a q_s = 1.97 ± 0.31 mg *o*-cresol g TSS^{-1} h^{-1}, followed by phenol and *p*-cresol, which obtain specific rates of consumption of approximately 1.75 and 1.77 times lower with respect to *o*-cresol.

There are no reports on the specific rates of consumption of phenolic compounds with microalgae, but there is information on the q_s of *p*-cresol in a denitrifying biological process, which are higher than those reported in the present study, with values of at least two and six times higher at 50 ppm [20,23] and three times for 100 ppm [24]. However, the effect of increasing the concentration of the phenolic compound caused, in all studies, a decrease in the rate of consumption.

The order of degradability observed in this work was *o*-cresol > phenol > *p*-cresol. This was contrary to that obtained by Papazi and Kotzabasis [7], who show a degradation of *o*-cresol < *p*-cresol, due to the position of the methyl group with respect to the hydroxyl group. In this regard, the inhibition was demonstrated by the decreases in *qs* when phenolic compounds were increased, except in the case of *p*-cresol (from 0.93 ± 0.05 to 1.11 ± 0.05 mg *p*-cresol g TSS^{-1} h^{-1}). In the case of phenol and *o*-cresol, *qs* decreased 1.7 times when phenolic concentration was increased from 50 to 100 mg L^{-1}.

The degradation of phenolic compounds has been studied using marine algae, microalgae and bacteria. In addition, degradation was studied considering the coexistence or mixture of phenol and isomers of cresol, such as *o*-cresol, *m*-cresol and *p*-cresol. In this regard, the presence of *p*-cresol inhibited the degradation of phenol or *o*-cresol, or in other cases, *m*-cresol degradation was inhibited by the presence of phenol [25]. On the other hand, a marine alga, *Ochromonas dánica*, removed 100% of phenols at concentrations of 60 and 375 g L^{-1} [26]. In some cases, bacteria have been used to degrade phenols and cresols in a mixture. In addition, *Chlorella vulgaris* was able to degrade phenol and cresol during co-metabolism; phenol at low concentrations (100 mg L^{-1}) significantly promoted the degradation of *p*-cresol. Moreover, phenol degradation was improved when $NaHCO_3$ was added to the culture media [11] or when a co-substrate such as glucose was used [13].

3.2. Biotic Removal Kinetics for a Mixture of Phenolic Compounds

The capacity of microalgae to remove phenolic compounds has been demonstrated. However, few reports of the removal of these compounds in a mixture have been published. In this regard, the capacity of *T. suecica* to simultaneously remove phenol, *p*-cresol and *o*-cresol was evaluated.

The simultaneous removal of phenol, *p*-cresol and *o*-cresol was evaluated in a mixture, as shown in Figure 2. The total time of the kinetics was 168 h. Up to 36% of *o*-cresol was removed at 6 h; from that time, there was no more removal, with a constant concentration of 25.5 mg L^{-1} remaining. *p*-cresol was slowly and continuously removed over time, achieving a removal of 73.6% at 168 h. Phenol concentration was constant. No removal of phenol was observed until 120 h; from that time, only 12.5% was removed at 168 h. This behavior suggests that the presence of cresol isomers inhibits the microalga's metabolism to remove phenol when it is in contact with the three phenolic compounds in a mixture. On the other hand, the most significantly consumed COD was at 12 h of kinetics (Figure 3). After this time, an inhibition of organic matter consumption was observed, achieving 31.4% of removal efficiency of COD. This result can be compared with TOC and IC quantification, in which TOC decreased, while IC increased within the time (data not shown). This finding could be due to phenol and cresols having been degraded and mineralized to CO_2. Comparing these results with those in Figure 2, it is possible to assume that *p*-cresol is the main phenolic compound degraded to CO_2, due to its removal of 73.6%.

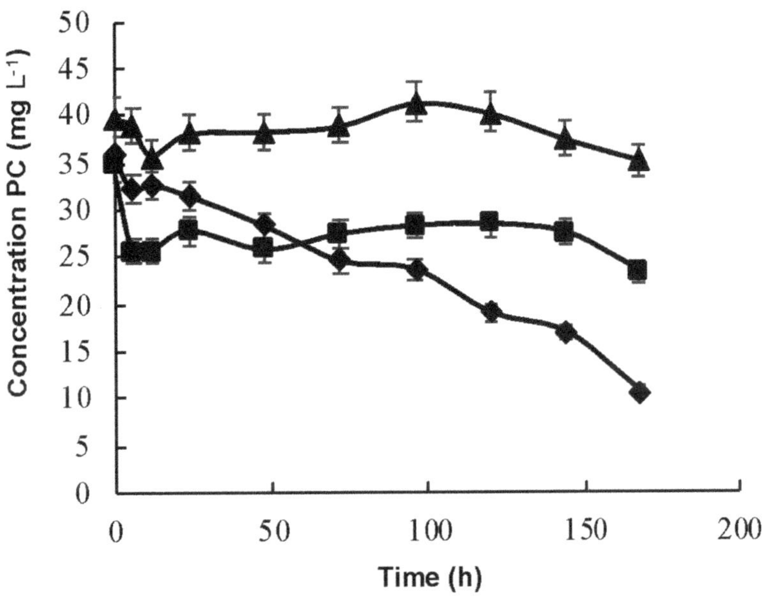

Figure 2. Consumption profile of phenolic compounds (PC) in a mixture at 40 mg L^{-1}. The symbols represent the concentration of each compound: phenol (triangle), *p*-cresol (rhombus) and *o*-cresol (square).

Studies on the biodegradation of cresol isomers have rarely been compared with those about phenol [10]. In this regard, Surkatti and El-Naas [4] studied the degradation of phenol and cresol isomers by bacteria, finding that in the presence of *p*-cresol or *m*-cresol, the *o*-cresol degradation was inhibited. Moreover, when an autochthonous microbial mixture was produced to degrade phenolic compounds, in the presence of phenol, *m*-cresol degradation was inhibited [22]. In the present study, phenol presence affected the *o*-cresol degradation, because when *o*-cresol was studied separately, a removal of 100% was achieved in both evaluated concentrations (50 and 100 mg L^{-1}). According to this, phenol at low concentrations significantly promoted the degradation of *p*-cresol by *Chlorella vulgaris* during cometabolism [13], but in our case, *p*-cresol was the only phenolic compound in the kinetics. When the degradation of phenol, *p*-cresol and *o*-cresol was simultaneously evaluated, the degradation order was phenol < *o*-cresol < *p*-cresol. This behavior was similar to results obtained by Papazi and Kotzabasis [7], who demonstrated that the degradability of cresols by microalgae is affected by the position of the methyl group with respect to the hydroxyl.

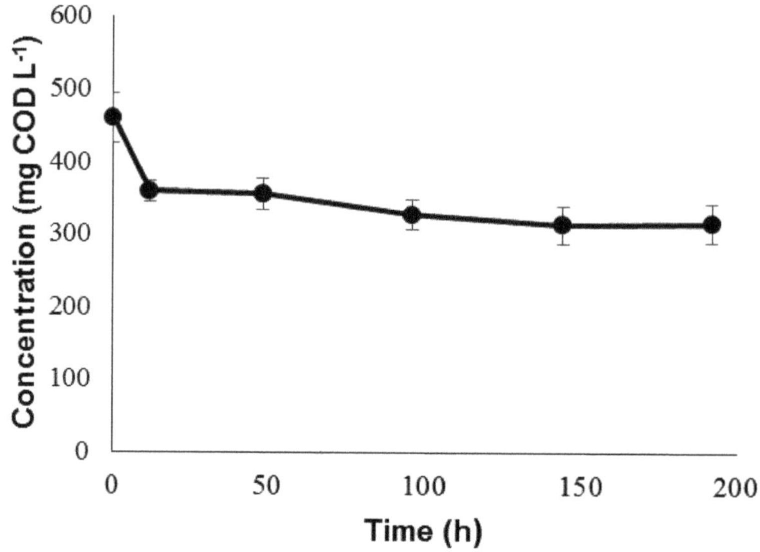

Figure 3. Organic matter consumption profile (COD) in a mixture of phenol, *p*-cresol and *o*-cresol at a concentration of 40 mg L^{-1} each.

Furthermore, adsorption and absorption (it can also involve bioaccumulation, biodegradation and biotransformation) are biological pathways for removing organic contaminants [16]. In this study, only the adsorption mechanism was determined, which was at its minimum in the control assay with an inert biomass of *T. suecica*. Therefore, we assume that absorption was the mechanism used by *T. suecica* to eliminate the phenolic compounds. However, it is possible that other mechanisms are involved in this process, such as the conjugation of phenolic compounds with sugars and conjugated carbohydrates with phenols such as quercetin, which reduce the toxicity of this hazardous compound. This proposed mechanism was discovered in *T. suecica* for the elimination of *p*-chlorophenol by Petroutsos et al. [16].

In Table 2, the capacity of microalgae, marine and freshwater, to remove or eliminate phenolic compounds is summarized, including the results of the present study. It is interesting to remark upon the diversity of phenolic derivates that microalgae can eliminate from different sources and matrices in a variety of concentrations, mainly presenting the mechanism of absorption in most of them, which includes biotransformation, bioaccumulation and biodegradation.

Table 2. Degradation of phenolic compounds by marine and freshwater microalgae.

Microalga	Phenolic Compound	Conditions	Initial Concentration (mg L^{-1})	Removal (%)	Reference
Tetraselmis suecica	phenol, *p*-cresol, *o*-cresol (separately) mixture of phenol, *p*-cresol and *o*-cresol	192 h	50 100 40 (each one in a mixture)	100 85 Up to 73.6	This work
Chlorella pyrenoidosa	phenol [1] *p*-cresol [2]	coal gasification effluent, pH 8	800 [1] 400 [2]	97.4	[2]
Chlorella pyrenoidosa	phenol	refinery wastewater	200	100	[6]

Table 2. *Cont.*

Microalga	Phenolic Compound	Conditions	Initial Concentration (mg L^{-1})	Removal (%)	Reference
Chlorella vulgaris	phenol [1] *p*-cresol [2]	cometabolic $NaHCO_3$	100 [1] 300 [2]	68.2 [1] 64 [2]	[11]
Chlorella vulgaris	phenol	mixotrophic with glucose addition (co-sustrate), 6 days.	Up to 400	Up to 30	[13]
Tetraselmis suecica	*p*-chlorophenol	10-day period in aqueous medium [1]. Immobilized in alginate beads [2]	20	67 [1] 94 [2]	[15]
Tetraselmis marina	2,4-dichlorophenol (2,4-DCP)	6 days glycosidation and malonylation	Not defined	Up to 1 mM	[16]
Chlorella sp	phenol	0.6 g/L^{-1} initial biomass, 7 days	500	100	[22]
Isochrysis galbana	phenol	96 h	<100	100	[23]
Ochromonas dánica	phenol	heterotrophic growth with 2 mM glucose, 2 days	94	100	[26]

Superscript in the same line indicates the phenolic compound with its respective concentration used.

4. Conclusions

The marine microalgae *Tetraselmis suecica* turned out to be an alternative for the removal of the phenolic compounds studied (phenol, *p*-creosol and *o*-cresol), reaching removal efficiencies of 85% at concentrations of up to 100 ppm, which can be an alternative for the elimination of these compounds in the treatment of wastewater. Moreover, the present study constitutes a clear demonstration of the simultaneous elimination by microalgae of those pollutants generated mainly by the petrochemical industry.

Author Contributions: Conceptualization: G.U.-M. and E.R.M.-E.; Methodology: E.R.M.-E. and L.L.-M.; Validation, A.R.-M., L.H.Á.-V. and D.S.-P.; Formal Analysis: G.U.-M., E.R.M.-E. and P.G.-M., Investigation, E.R.M.-E. and L.L.-M.; Resources: G.U.-M. and E.R.M.-E.; Data Curation, G.U.-M. and C.D.-Q.; Writing—Original Draft Preparation, E.R.M.-E. and L.L.-M.; Writing—Review and Editing, G.U.-M.; Project Administration: G.U.-M. and E.R.M.-E. All authors have read and agreed to the published version of the manuscript.

Funding: This research was funded by Sonora Institute of Technology, grant number PROFAPI_2020_0068.

Institutional Review Board Statement: Not applicable.

Informed Consent Statement: Not applicable.

Data Availability Statement: The datasets used and/or analyzed during the current study are available from the corresponding author.

Conflicts of Interest: The authors declare no conflict of interest.

References

1. Barik, M.; Das, C.P.; Verma, A.K.; Sahoo, S.; Sahoo, N.K. Metabolic profiling of phenol biodegradation by an indigenous *Rhodococcus pyridinivorans* strain PDB9T N-1 isolated from paper pulp wastewater. *Int. Biodeterior. Biodegrad.* **2021**, *158*, 105168. [CrossRef]
2. Bakthavatsalam, A. A comparative study on growth and degradation behavior of *C. pyrenoidosa* on synthetic phenol and phenolic wastewater of a coal gasification plant. *J. Environ. Chem. Eng.* **2019**, *7*, 103079.
3. Feng, J.; Qiao, K.; Pei, L.; Lv, J.; Xie, S. Using activated carbon prepared from *Typha orientalis* Presl to remove phenol from aqueous solutions. *Ecol. Eng.* **2015**, *84*, 209–217. [CrossRef]

4. Surkatti, R.; El-Naas, M. Competitive interference during the biodegradation of cresols. *Int. J. Environ. Sci. Technol.* **2018**, *15*, 301–308. [CrossRef]
5. Abdel-Raouf, N.; Al-Homaidan, A.; Ibraheem, I. Microalgae and wastewater treatment. *Saudi J. Biol. Sci.* **2012**, *19*, 257–275. [CrossRef]
6. Das, B.; Mandal, T.K.; Patra, S. A comprehensive study on *Chlorella pyrenoidosa* for phenol degradation and its potential applicability as biodiesel feedstock and animal feed. *Appl. Biochem. Biotechnol.* **2015**, *176*, 1382–1401. [CrossRef]
7. Papazi, A.; Kotzabasis, K. Inductive and resonance effects of substituents adjust the microalgal biodegradation of toxical phenolic compounds. *J. Biotechnol.* **2008**, *135*, 366–373. [CrossRef]
8. Joseph, V.; Joseph, A. Acclimation of algal species following exposure to phenol. *Bull. Environ. Contam. Toxicol.* **1999**, *62*, 87–92. [CrossRef]
9. Aravantinou, A.F.; Theodorakopoulos, M.A.; Manariotis, I.D. Selection of microalgae for wastewater treatment and potential lipids production. *Bioresour. Technol.* **2013**, *147*, 130–134. [CrossRef]
10. Duan, W.; Meng, F.; Cui, H.; Lin, Y.; Wang, G.; Wu, J. Ecotoxicity of phenol and cresols to aquatic organisms: A review. *Ecotoxicol. Environ. Saf.* **2018**, *157*, 441–456. [CrossRef]
11. Xiao, M.; Ma, H.; Sun, M.; Yin, X.; Feng, Q.; Song, H.; Gai, H. Characterization of cometabolic degradation of *p*-cresol with phenol as growth substrate by *Chlorella vulgaris*. *Bioresour. Technol.* **2019**, *281*, 296–302. [CrossRef]
12. Satyanarayana, K.; Mariano, A.; Vargas, J. A review on microalgae, a versatile source for sustainable energy and materials. *Int. J. Energy Res.* **2011**, *35*, 291–311. [CrossRef]
13. Kong, W.; Yang, S.; Guo, B.; Wang, H.; Huo, H.; Zhang, A.; Niu, S. Growth behavior, glucose consumption and phenol removal efficiency of *Chlorella vulgaris* under the synergistic effects of glucose and phenol. *Ecotoxicol. Environ. Saf.* **2019**, *186*, 109762. [CrossRef]
14. Lauritano, C.; De Luca, D.; Amoroso, M.; Benfatto, S.; Maestri, S.; Racioppi, C.; Esposito, F.; Ianora, A. New molecular insights on the response of the green alga *Tetraselmis suecica* to nitrogen starvation. *Sci. Rep.* **2019**, *9*, 1–12. [CrossRef] [PubMed]
15. Forootanfar, H.; Shakibaie, M.; Bagherzadeh, Z.; Aghaie-Khozani, M.; Nafissi-Varcheh, N.; Monsef-Esfahani, H.R.; Faramarzi, M.A. The removal of ρ-chlorophenol in aqueous cultures with free and alginate-immobilized cells of the microalga *Tetraselmis suecica*. *J. Appl. Phycol.* **2013**, *25*, 51–57. [CrossRef]
16. Petroutsos, D.; Katapodis, P.; Samiotaki, M.; Panayotou, G.; Kekos, D. Detoxification of 2, 4-dichlorophenol by the marine microalga *Tetraselmis marina*. *Phytochemistry* **2008**, *69*, 707–714. [CrossRef]
17. Pop, C.-E.; Draga, S.; Măciucă, R.; Niță, R.; Crăciun, N.; Wolff, R. Bisphenol A Effects in Aqueous Environment on *Lemna minor*. *Processes* **2021**, *9*, 1512. [CrossRef]
18. Fabregas, J.; Herrero, C.; Cabezas, B.; Abalde, J. Mass culture and biochemical variability of the marine microalga *Tetraselmis suecica* Kylin (Butch) with high nutrient concentrations. *Aquaculture* **1985**, *49*, 231–244. [CrossRef]
19. Federation, W.E.; Association, A. *Standard Methods for the Examination of Water and Wastewater*; American Public Health Association (APHA): Washington, DC, USA, 2005; p. 21.
20. Meza-Escalante, E.R.; Texier, A.C.; Cuervo-López, F.; Gomez, J.; Cervantes, F.J. Inhibition of sulfide on the simultaneous removal of nitrate and *p*-cresol by a denitrifying sludge. *J. Chem. Technol. Biotechnol. Int. Res. Process Environ. Clean Technol.* **2008**, *83*, 372–377. [CrossRef]
21. Miyanaga, K.; Unno, H. 2.05-Reaction Kinetics and Stoichiometry. In *Comprehensive Biotechnology*, 2nd ed.; Moo-Young, M., Ed.; Academic Press: Burlington, ON, Canada, 2011; pp. 33–46.
22. Wang, Y.; Meng, F.; Li, H.; Zhao, S.; Liu, Q.; Lin, Y.; Wang, G.; Wu, J. Biodegradation of phenol by *Isochrysis galbana* screened from eight species of marine microalgae: Growth kinetic models, enzyme analysis and biodegradation pathway. *J. Appl. Phycol.* **2019**, *31*, 445–455. [CrossRef]
23. González-Blanco, G.; Cuervo-López, F.; Cervantes, F.J.; Beristain-Cardoso, R.; Gómez, J. Nitrite as oxidizing power for *p*-cresol removal using a denitrifying sludge: Kinetic study. *J. Chem. Technol. Biotechnol.* **2013**, *88*, 2176–2180. [CrossRef]
24. Wang, L.; Xue, C.; Wang, L.; Zhao, Q.; Wei, W.; Sun, Y. Strain improvement of *Chlorella sp.* for phenol biodegradation by adaptive laboratory evolution. *Bioresour. Technol.* **2016**, *205*, 264–268. [CrossRef] [PubMed]
25. Saravanan, P.; Pakshirajan, K.; Saha, P. Biodegradation of phenol and m-cresol in a batch and fed batch operated internal loop airlift bioreactor by indigenous mixed microbial culture predominantly *Pseudomonas* sp. *Bioresour. Technol.* **2008**, *99*, 8553–8558. [CrossRef] [PubMed]
26. Semple, K.T.; Cain, R.B. Biodegradation of phenols by the alga *Ochromonas danica*. *Appl. Environ. Microbiol.* **1996**, *62*, 1265–1273. [CrossRef]

sustainability

MDPI

Article

Systematic Modeling of Municipal Wastewater Activated Sludge Process and Treatment Plant Capacity Analysis Using GPS-X

Nuhu Dalhat Mu'azu *, Omar Alagha * and Ismail Anil

Department of Environmental Engineering, College of Engineering, Imam Abdulrahman Bin Faisal University, P.O. Box 1982, Dammam 31451, Saudi Arabia; ianil@iau.edu.sa

* Correspondence: nmdalhat@iau.edu.sa (N.D.M.); oaga@iau.edu.sa (O.A.); Tel.: +966-507532689 (N.D.M.); +966-506616532 (O.A.)

Received: 13 August 2020; Accepted: 26 September 2020; Published: 4 October 2020

Abstract: Mathematical modeling has become an indispensable tool for sustainable wastewater management, especially for the simulation of complex biochemical processes involved in the activated sludge process (ASP), which requires a substantial amount of data related to wastewater and sludge characteristics as well as process kinetics and stoichiometry. In this study, a systematic approach for calibration of the activated sludge model one (ASM1) model for a real municipal wastewater ASP was undertaken in GPS-X. The developed model was successfully validated while meeting the assumption of the model's constant stoichiometry and kinetic coefficients for any plant influent compositions. The influences of vital ASP parameters on the treatment plant performance and capacity analysis for meeting local discharge limits were also investigated. Lower influent chemical oxygen demand in mgO_2/L (COD) could inhibit effective nitrification and denitrification, while beyond 250 mgO_2/L, there is a tendency for effluent quality to breach the regulatory limit. The plant performance can be satisfactory for handling even higher influent volumes up to 60,000 m^3/d and organic loading when Total Suspended Solids/Volatile Suspended Solids (VSS/TSS) and particulate COD (X_{COD})/VSS are maintained above 0.7 and 1, respectively. The wasted activated sludge (WAS) has more impact on the effluent quality compared to recycle activated sludge (RAS) with significant performance improvement when the WAS was increased from 3000 to 9000 m^3/d. Hydraulic retention time (HRT) > 6 h and solids retention time (SRT) < 7 days resulted in better plant performance with the SRT having greater impact compared with HRT. The plant performance could be sustained for a quite appreciable range of COD/5-day Biochemical Oxygen Demand (BOD_5 in mgO_2/L) ratio, Mixed Liquor Suspended Solid (MLSS) of up to 6000 mg/L, and when BOD_5/total nitrogen (TN) and COD/TN are comparatively at higher values. This work demonstrated a systematic approach for estimation of the wastewater treatment plant (WWTP) ASP parameters and the high modeling capabilities of ASM1 in GPS-X when respirometry tests data are lacking.

Keywords: activated sludge model 1 (ASM1); GPS-X Mantis model 1 model; model calibration and validation; municipal wastewater management; stoichiometric and kinetic parameters; treatment performance evaluation

1. Introduction

Activated sludge system (ASS) is one of the critical treatment processes for various wastewaters, with over 90% of the municipal wastewater treatment plant (WWTP) using it as the core part of their treatment scheme [1,2]. The activated sludge process (ASP) uses suspended microbial consortium in the wastewater to remove biodegradable organic matters and nutrients, mainly under aerobic conditions [3,4]. Mathematical modeling has become an indispensable tool, especially for the simulation of complex biochemical processes involved in ASP, which requires a substantial amount of data related to wastewater and sludge characteristics, as well as process kinetics and stoichiometry. Among various dynamic and steady-state mathematical expressions and models describing the ASP, activated sludge models (ASM) such as suggested by the International Water Association (IWA) have most frequently been employed for the design, operation, and optimization of biological wastewater treatment plants [5]. Amongst these models, the ASM number one (ASM1) has become an internationally accepted ASM, describing the biological removal processes of organic matter and nitrogen, including nitrification and denitrification mechanisms [5,6]. The model comprises eight essential processes: (i) aerobic growth of heterotrophic biomass, (ii) anoxic growth of heterotrophic biomass, (iii) aerobic growth of autotrophic biomass, (iv) heterotrophic biomass decay, (v) autotrophic biomass decay, (vi) soluble organic nitrogen ammonification, (vii) hydrolysis of entrapped particulate organic matter, and (viii) hydrolysis of entrapped organic nitrogen [5,7]. ASM1 has been considered as the primary reference model because it prompted the universal recognition of ASS modeling. With high potentials of providing a good depiction of the sludge production process, the ASM1 primarily describes the removal of organic and nitrogenous compounds with concurrent NO_3^- and O_2 consumption as acceptors of the electron [5]. ASM1 model basically adopts COD as a parameter for representing organic matter concentration in wastewater. The ASM models developed by IWA have been incorporated into the various design, simulation, and optimization software such as GPS-X (Hydromantis Environmental Software Solutions, Inc. Hamilton, ON, Canada), BioWin (EnviroSim associates LTD. Hamilton, ON, Canada), ASIM Activated Sludge SIMulation Program (Swiss Federal Institute of Aquatic Science and Technology Eawag, Dübendorf, Switzerland), SIMBA (IFAK-Institut für Automation und Kommunikation e.V. Magdeburg, Germany), WEST (DHI A/S Hørsholm, Denmark), STOAT (WRc plc. Wiltshire, UK) and Sumo (Dynamita, Sigale, France) [8,9]. The learning and prediction performances of the software mentioned herein are significantly dependent on the successful calibration and validation of the model [8].

The parameters that have the greatest impact on the model can be identified by performing the sensitivity analysis with the parameter estimation method during the dynamic or steady-state calibration steps [10,11]. Following the calibration stage, the model must be validated by comparing the actual treatment plant data and predicted data by the calibrated model. The GPS-X software, a comprehensive plant-wide model, developed by Hydromantis Environmental Software Solutions Inc. [12], has recently been receiving great attention due to its wide variety of pre-compiled treatment technologies, ease of use, and easily accessible training materials [13–15]. In steady-state model calibration, parameters such as f_p, b_H, and Y_H that influence long-term behavior are an example of relevant parameters to be considered [16]. Nevertheless, calibration parameters responsible for the short-term (i.e., dynamic behavior), such as μ_{maxH}, μ_{maxA}, the correction that included $\eta_{anoxic,h}$, K_S, K_{NH4}, K_{OH}, and K_{OA}, are equally important to be estimated during the steady-state calibration [10,12,16]. However, one of the major setbacks for ASM calibration is identifying and selecting the most appropriate parameters -amongst several parameters- that are crucial for achieving apt and good calibration for a given ASP [10]. Thus, the appropriate methodology needs to be adopted for success in this regard [10], which is usually achieved via performing analysis to ascertain the ability of the combination of set parameters that would adequately describe the ASP [16]. Finding the right and most appropriate ASM parameters for a given system is challenging due to the complex nature of ASMs, several integrated model's parameters as well as their interdependencies [5,6,16]. This renders the determination of all necessary model parameters and successful ASP calibration time-consuming and expensive processes [10]. Consequently, several methodologies have been proposed for achieving successful ASM calibrations [10,11,15,17–21]. In this regard, sensitivity analysis has been applied in the determination of the most influential ASMs model parameters as it helps in achieving calibration, reduces the difficulty, and to optimize the process [11,22]. Other authors [17,19,23] presented detailed steps for the most relevant parameters for real WWTP data under ASM1. Their analysis resulted in identifying a set of kinetic and stoichiometric coefficients that included Y_H, Y_A, μ_{maxH}, B_H, μ_{maxA}, K_S, K_{OA}, and $\eta_{anoxic,H}$ as the most relevant parameters for their model. Systematic methodologies were also presented for the selection of the most relevant parameters for the ASM2d model [17,24]. Using BioWin software, Liwarska-Bizukojc and Biernacki [10] identified 17 most influential stoichiometric and kinetic and parameters for ASM-based models from 46 stoichiometric and 71 kinetic parameters via sensitivity analysis.

Generally, identifying and selecting the most relevant parameters during modeling WWTP using ASP models' platforms is a cumbersome and challenging task to implement during the calibration process. Approaches presented by other authors using methods such as sensitivity analysis are prone to be challenging as they rely on other technicalities (such as requirements for determination of statistical parameters), which are outside the capabilities and jurisdiction of GPS-X software. This is manifested in its bare implementation, especially for complex ASM GPS-X ASP models, as demonstrated in the reviewed literature. Thus, in this study, a simple and systematic approach for successful steady-state calibration implemented, entirely, under GPS-X based ASP model for a real WWTP modeling is presented. The approach does not rely on any prior information about real influent composition such as organic and nitrogen fractions and other required mass balance data related to state and composite variables and their fractionation; hence it does not require respirometry tests. Characterization of influent via laboratory respirometry protocols is used for determining several stoichiometric and biokinetics for biological process calibration under study [25]. Respiratory protocols are related to oxygen consumption measurements for biomass aerobic activities involving autotrophic and heterotrophic bacteria exogenous-growth (classed into NOB and AOB ammonium oxidizing bacteria) as well as their endogenous-respirations [13,25,26]. These measurement protocols are highly technical, expensive, and often challenging. Moreover, the approach presented herein is supported by the visual as well as real-time output presentation capabilities of GPS-X models. This work targeted employing the high potential modeling capabilities of ASM1 executed in GPS-X software for identifying the most influential and relevant kinetic and stoichiometric parameters and applying them for modeling and understating the influence of operational parameters on real WWTP.

2. Materials and Methods

2.1. Description of Dhahran North Sewage Wastewater Treatment Plant (DWWTP)

Dhahran North Sewage Wastewater Treatment Plant (DWWTP) is located at the Dhahran district of Eastern Province, Saudi Arabia (26°18′31.18″ N 50°9′40.87″ E) (Figure 1). DWWTP is an activated sludge wastewater plant and receives only domestic wastewater with an average flow rate of 52,012 (±4440) m^3/d. The wastewater flow enters the DWWTP via gravity sewers from three areas: Doha residential area, Saudi Aramco Dhahran facility, and King Fahd University of Petroleum and Minerals (KFUPM) campus area. The influent first contacts with mechanically cleaned step screens followed by vortex grit removal chambers. A flow equalization tank is used to balance the peak flows. After that, the screened and de-gritted wastewater flow is distributed to the aeration tanks, where the biological stabilization of organic matter is accomplished. The aerated effluent exiting the aeration tanks is conveyed to secondary settling tanks for the removal of solids. The belt filter press units have been used for sludge dewatering of excessive sludge.

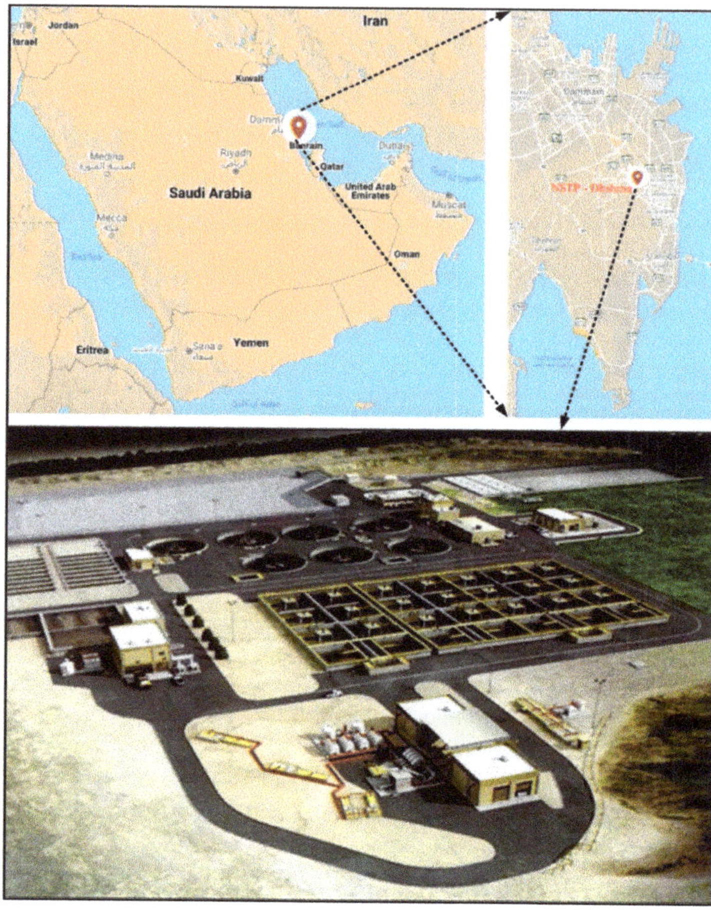

Figure 1. Geographic location and an aerial view of the Dhahran North Sewage Wastewater Treatment Plant (DWWTP).

2.2. *DWWTP Wastewater Quality Characteristics*

The influent and effluent data used in this research are collected from the influent into the ASP from the primary clarifier and effluent from the secondary clarifier of the DWWTP, respectively. The collected samples were transferred to the Environmental Chemistry Laboratory of Environmental Engineering Department, Imam Abdulrahman Bin Faisal University, and analyzed immediately for pH, conductivity, turbidity, dissolved Chemical Oxygen Demand in mgO_2/L (COD), Biochemical Oxygen Demand in a 5-day incubation period in mgO_2/L (BOD_5), NO_2^- NO_3^-, NH_3–N, and Total Nitrogen (TN) in compliance with the procedures in "The Standard Methods for the Examination of Water and Wastewater" [27]. Table 1 summarizes the descriptive statistics such as mean, median, standard deviation (SD), minimum (Min), maximum (Max), kurtosis, and skewness of the analyzed parameters for influent and effluent samples of the DWWTP. During the studied period, the pH values of influent samples were found to be 2.40% (±2.76) higher than those of effluent samples, which could be ascribed to the denitrification process resulting in pH increase [28,29]. The turbidity removal efficiency of the DWWTP was calculated as 96.8% (±2.23), while conductivity change between influent and effluent samples was negligible. The BOD_5 and COD treatment performances of the ASP in DWWTP were 89.2% (±7.79) and 75.4% (±20.74, respectively. The NH_3^+–N was the dominating nitrogen form in the influent samples and treated with 98.6% (±1.74) efficiency, leading to 67.6% (±6.90) of TN removal efficiency while meeting regulatory limits as provided in Table 2.

2.3. Methodology

2.3.1. DWWTP ASP Modeling in GPS-X

GPS-X software version 8 (Academic license) developed by Hydromantis Environmental Software Solutions, Inc. is used in this present work [12]. It is a widely used comprehensive standalone model built with integrated biological wastewater treatment processes for ASP and anaerobic digestion system (ADS), as well as many other involving physical and chemical reactions. The Mantis model integrated into GPS-X software re-adapted ASM1, incorporating some amendments about additional growth processes related to heterotrophic and autotrophic organisms. Additionally, the Mantis model factored aerobic denitrification as part of its components [30]. In this work, the ASP model employed was designed in a carbon and nitrogen custom components library in GPS-X software under the MANTIS and simple1d clarifier model [12]. The model comprises of over 60 composite and state variables along with several libraries of expressions describing the processes with more than 30 stoichiometric and 24 kinetic input and output parameters incorporated [12]. This implies that the modeling approach adopted ASM1 for carbon degradation, nitrification, and denitrification thereby, targeting only removal of COD, BOD_5, and nitrogen components (i.e., NO_2^- + NO_3^-, NH_3–N, and TN).

Table 1. Descriptive statistics summary of the parameters measured for influent and effluent of DWWTP.

Parameter	Unit	Mean	Median	SD	Min	Max	Kurtosis	Skewness
				Influent Values				
pH	–	7.44	7.35	0.247	7.13	8.06	1.11	1.06
COD	mgCOD/L	180	179	73.2	68.0	359	1.30	0.568
BOD_5	mgO_2/L	79.9	72.0	26.6	48.0	144	2.00	1.49
NO_2^-	mgN/L	0.048	0.040	0.025	0.019	0.119	3.68	1.64
NO_3^-	mgN/L	0.729	0.498	0.624	0.141	2.01	−0.183	0.954
NH_3–N	mgN/L	18.7	19.3	6.41	7.03	35.0	1.94	0.664
TN	mgN/L	22.8	24.1	5.87	12.2	27.8	1.76	−1.37
				Effluent Values				
pH	–	7.61	7.57	0.256	7.12	8.23	1.73	0.712
COD	mgCOD/L	44.2	40.7	21.7	15.4	94.2	0.199	0.779
BOD_5	mgO_2/L	8.63	8.00	5.35	2.00	18.0	−0.919	0.500
NO_2^-	mgN/L	0.040	0.033	0.024	0.016	0.096	2.49	1.72
NO_3^-	mgN/L	2.71	2.75	1.50	0.960	6.07	−0.061	0.440
NH_3–N	mgN/L	0.261	0.100	0.370	0.091	1.58	12.3	3.38
TN	mgN/L	7.38	7.82	1.09	5.50	8.32	0.622	−1.25

Table 2. Average monthly influent and effluent quality parameters for DWWTP used for the present study.

Month	1		2		3		4		Local Effluent Discharge Limits [1]
Parameter	Influent	Effluent	Influent	Effluent	Influent	Effluent	Influent	Effluent	
pH	7.75 ± 0.18	7.86 ± 0.32	7.22 ± 0.03	7.56 ± 0.03	7.32 ± 0.08	7.54 ± 0.06	7.48 ± 0.15	7.50 ± 0.23	6.0–8.4
NO_3^-	0.34 ± 0.25	2.42 ± 1.06	1.14 ± 0.07	3.96 ± 1.31	0.88 ± 0.433	3.28 ± 0.36	0.188 ± 0.00	4.20 ± 0.00	10
NH_3–N	18.35 ± 2.71	0.04 ± 0.07	13.66 ± 4.59	0.05 ± 0.05	19.45 ± 2.60	0.57 ± 0.60	23.38 ± 8.42	0.24 ± 0.14	5
COD	215 ± 20.45	21.43 ± 3.53	133.53 ± 71.89	43.32 ± 14.78	217.75 ± 88.55	21.50	152 ± 32.55	129.05 ± 132.03	50
BOD_5	64 ± 10.20	8.50 ± 3.84	90.50 ± 26.92	15.50 ± 2.60	92.50 ± 33.00	4 ± 2.45	72.5 ± 12.48	6.50 ± 2.60	40
DO	5.10 ± 1.68	2.48 ± 1.51	4.60 ± 2.81	4.38 ± 1.26	3.09 ± 1.64	4.42 ± 0.85	4.54 ± 1.65	5.22 ± 0.24	>2
NO_2^-	0.05 ± 0.01	0.03 ± 0.01	0.05 ± 0.02	0.033 ± 0.01	0.054 ± 0.04	0.06 ± 0.30	0.04 ± 0.03	0.03 ± 0.02	–
TN	20.75 ± 0.356	7.49 ± 0.752	21.12 ± 0.951	7.38 ± 1.56	26.59 ± 2.98	6.65 ± 1.54	24.95 ± 2.45	6.64 ± 0.56	10

2.3.2. Model Assumptions

With model constraints as per ASM1 [5], the assumptions are adopted for the development of the model in this study are as follows

- ASP operates at a content temperature
- a constant concentration of dissolved oxygen (DO) is maintained, and there is sufficient mixing within the reactor
- pH is steady and near-neutral value
- The model's coefficients are assumed to be constants for any influent characteristics
- there are enough inorganic nutrients to ensure sufficient growth
- there is simultaneous hydrolysis of organic and nitrogenous compounds.

2.3.3. GPS-X Modeling Approach

Using the data in Table 2, the modeling in this study was undertaken via simulation in the GPS-X environment using the following steps

1. Collection minimum real DWWTP data required for the GPS-X modeling
2. Portraying the existing DWWTP in terms of influent and physical data of the central unit operations
3. Selection of model objects via the construction of the DWWTP layout representation in GPS-X environment
4. Characterization DWWTP influent wastewater quality parameters (inserting the required values of the easy to measure, i.e., COD, NO_2^- + NO_3^-, NH_3–N, and TN)
5. Adjusting and setting influent fractionation of the COD and nitrogen components (not easy to measure influent components) using the GPS-X influent advisor to an acceptable state and composite variables mass balance
6. Running the model and calibration via adjusting kinetic, stoichiometric, and other relevant parameters to fit the model to obtain the best matching between model output and the actual plant effluent quality data
7. Validate the calibrated model using a different set of DWWTP wastewater quality data
8. Running simulations under different scenarios to analyze the effect of relevant operational parameters of the plant capacity and performance in terms of final effluent quality.

2.3.4. Systematic GPS-X Model Calibration and Validation

The influent data for month 1 (as presented in Table 2) was employed as the calibration simulation input data. Meanwhile, for the model validation, the period covering three months (months 2–4) was used as the input data. The schematic steps for the systematic model calibration and validation adopted in this study are illustrated in Figure 2. The standard statistical, as elaborated earlier, was undertaken (as provided in Table 1), and the calculated confidence intervals were found to be within a 90–95% significance level. The simulations were performed under steady-state conditions for both model calibration and validation processes. The best-identified parameters proving collective matching of measured plant effluent quality data, i.e., total COD, BOD_5, $NO2^-$ + NO_3^-, NH_3–N and TN for the calibration, were then applied on the for validation simulations undertaken for the different periods (month 2, 3, and 4). The steps for the DWWTP plant steady-state systematic calibration process are explained as follows.

1. Representation of DWWTP for the biological treatment units and stages flow diagram in the GPS-X environment is depicted in Figure 3.
2. Selection of the GPS-X model library (here Carbon and nitrogen costume library).
3. Using the GPS-X influent advisor, characterize the influent flow via inputting the plant influent quality data: COD, NO_2^- + NO_3^-, NH_3–N, and TN.

4. If GPS-X influent advises mass balance calculations requirements (i.e., organic, nitrogen and MANTIS fractions) are satisfactory for the state, and composite variables move to step 6.
5. If there exists an imbalance in mass balance calculations in step 4 above, manually adjust organic, nitrogen, and MANTIS fractions until satisfactory state and composite variables balance are achieved.
6. Run calibration using month 1 data based on GPS-X defaults kinetics and stoichiometric parameters values. If the model prediction fits all the respective effluent quality parameters data within acceptable limits, calibration is done.
7. If the GPS-X default model fails, initial screening and identifying of the most sensitive parameters is undertaken by running ASP and clarifier model parameters optimizations for four whole months plant data via manually adjusting the relevant defaults parameters values one-by-one while visually observing GPS-X output response in collective predicting the effluent quality parameters in terms of COD, BOD_5, $NO_2^- + NO_3^-$, NH_3–N, and TN.
8. Re-run the calibration of month 1 data via changing the values of the identified and screened parameters in step 7 to optimize their prediction abilities further.
9. Select the parameter values that yielded the best modeling prediction in terms of standard effluent quality parameters. These are the final calibrated parameters.
10. Validate the performance of the above-developed model against a different set of real DWWTP data (i.e., month 2, 3, and 4).

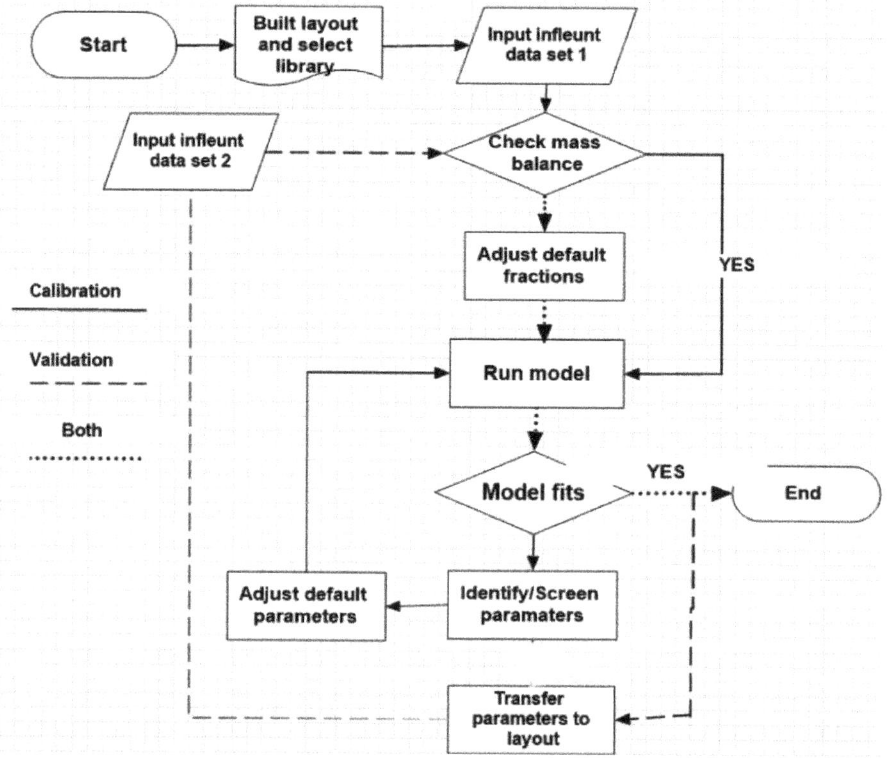

Figure 2. Flowchart for systematic GPS-X model calibration and validation process.

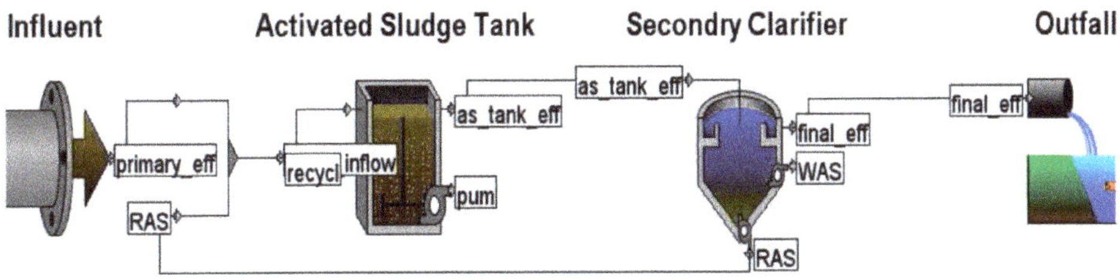

Figure 3. A schematic flow diagram for DWWTP activated sludge process (ASP) representation in GPS-X [12].

2.3.5. WWTP Performance and Capacity Analyses

The capacity of DWWTP under different steady-state operational scenarios of influencing parameters was investigated via simulations using the validated GPS-X model. The influence of primary influent wastewater characteristics, fractioning components of COD and nutrients as well as the physical, operational variables were assessed against collectively, meeting regulatory effluent discharge limits of COD, BOD_5, NO_2^- + NO_3^-, NH_3−N, and TN.

3. Results and Discussion

3.1. DWWTP ASP Model Calibration

The model calibration simulation targeted estimation of the best-fitted parameters for a specific given set of real data acquired from the DWWTP understudied. Data for month 1 was used for initial calibration using GPS-X default parameter values [12]. This was achieved by characterizing the influent composition satisfying the mass balance expressions that required changing the GPS-X default values of the influence fractioning for $BOD_5/BOD_{ultimate}$, S_i, S_s, X_i, X_{BH}, X_{BA}, and S_{nh} from 0.66, 0.05, 0.2, 0.13, 0, and 0 to 0.75, 0.0556, 0.32, 0.12, 0.176 and 0.142, respectively (Table 3). The initial calibration result under this scenario as depicted in Figure 4a was excellent for carbon contents (i.e., COD and BOD); however, it failed to capture the actual values of all the nitrogenous components by grossly overestimating NH_3−N (13.093 against 0.0408 mgN/L) and TN (14.23 against 7.49 mN/L) and underestimating NO_2^- + NO_3^- (0.16 against 2.45 mgN/L). Thus, this necessitated an additional strategy adopted by first identifying the most relevant parameters, and then adjusting them to balance the model's predictability, collectively, for both the carbonous and nitrogenous quality parameters using procedure elaborated under Section 2.3.4 steps 7–9. The calibration results for this second approach presented in Figure 4b shows that there is good agreement between the predicted and actual for all the parameters, collectively. This was achieved after identifying and optimizing the thoroughly mixed tank composite variables stoichiometry, model stoichiometry, kinetics parameter as well as the clarifier parameters as provided in Tables 3 and 4. The quality of the GPS-X calibration results was ascertained after confirming that simulated output was within the actual data values statistical confidence interval. In contrast, the difference between measured and simulated values of the quality parameters was insignificant [10,16,31]. Additionally, most of the values of the settings for the calibrated models are well within the range of values reported in the literature [10,32–37]. The GPS-X default calibration failure was high, attributed to the lower or higher values of the parameters as suggested by GPS-X [12], especially those related to nitrogenous compounds (such as i_{XB}, i_{XP}, K_{NH4}, and k_A) in Table 4. These parameters were found to be very sensitive to slight changes, yet they had to be systematical, increased by 2.47, 3.24, 3.58, and 4 folds to achieve the acceptable calibration (Table 4). This was the case despite initially satisfying major parameters accountable for the long-term behavior (such as, i.e., Y_H, b_H), for NOB ($b_{aerob,N}$) and AOB ($b_{aerob,A}$). This corroborates earlier observation reported by other authors [10] that achieved lower predictions of simulated COD and BOD_5 which they postulated to be as results of overestimation of BioWin model biomass affinity to their studied wastewater (represented by a half-saturation constant for heterotrophic biomass for their model).

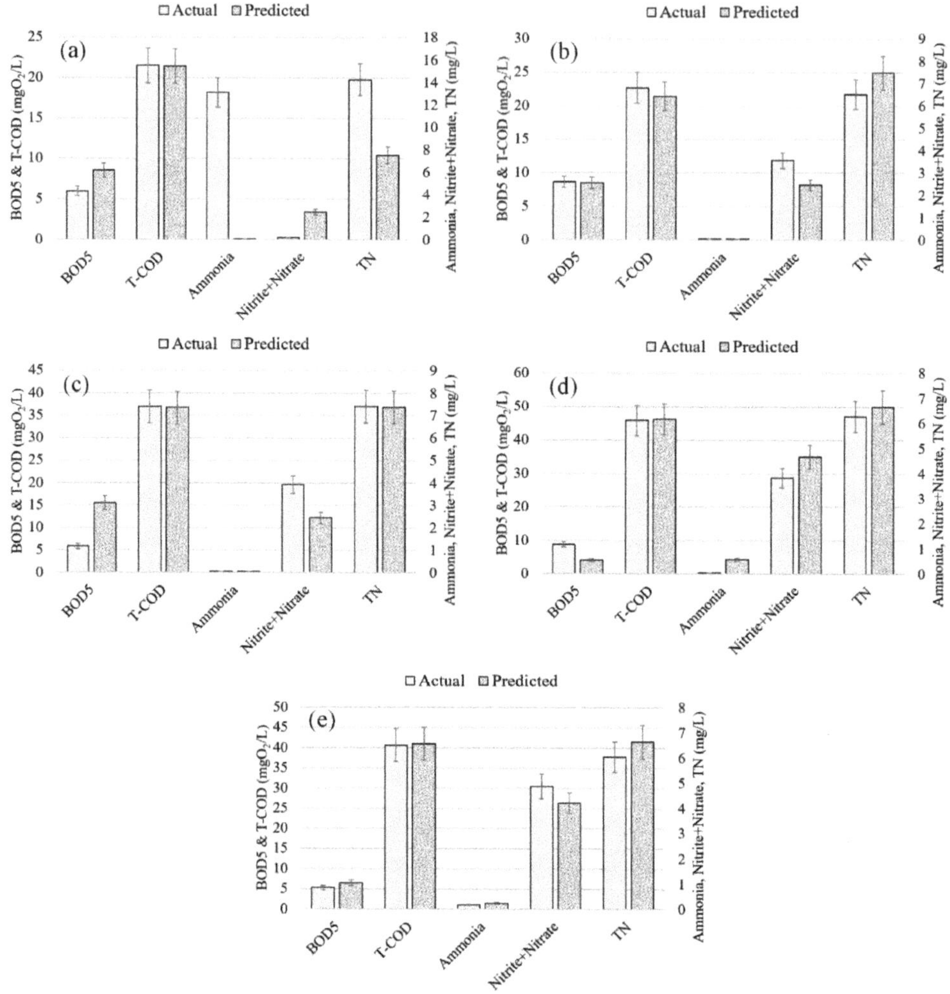

Figure 4. (**a**) GPS-X default (**b**) calibration-month-1 data (**c**) validation-month-2 data (**d**) validation -month-3 data and (**e**) validation-month-4 data.

Table 3. GPS-X Default or adjusted influent stoichiometry parameters based on GPS-X influent advisor.

Influent Stoichiometry Composition			GPS-X Default	Calibration		Validation	
Classification	Parameter	Unit		Month 1	Month 2	Month 3	Month 4
Influent Fractions	i_{cv}	gCOD/gVSS	1.8	1.8	1.8	1.8	1.8
	f_{bod}	–	0.66	0.75	0.75	0.75	0.75
	i_{vt}	gVSS/gTSS	0.75	0.75	0.75	0.8	0.8
Mantis Nutrient Fractions	i_{XB}	gN/gCOD	0.068	0.068	0.0177	0.068	0.068
	i_{XP}	gN/gCOD	0.068	0.068	0.0544	0.068	0.068
Organic Fractions	S_i	–	0.05	0.0556	0.0556	0.08	0.08
	S_S	–	0.2	0.32	0.32	0.32	0.32
	X_i	–	0.13	0.12	0.12	0.12	0.12
	X_{BH}	–	0	0.176	0.176	0.176	0
	X_{BA}	–	0	0.142	0.142	0.142	0
Nitrogen Fractions	S_{nh}	–	0.9	0.9	2.98	0.9	0.9

Interestingly, in this present study, the model's performance during the calibration was found to be significant, influenced by the RAS and WAS, which had to be changed to reflect the reality of the studied DWWTP from low default values of 2000 and 1200 m^3/d drastically increased to 15,000 and 10,000 m^3/d before achieving the obtained good calibration. Earlier studies indicated the strong influence and contribution of WAS and RAS for successful ASP model calibration [35].

3.2. DWWTP ASP Model Validation

After the successful calibration of the plant, validation became the next task. Model validation can be defined as the excellent agreement of the model's predictions when compared with a different set of data that did not partake in the model's development within acceptable limits. The validation was achieved by considering months 2, 3, and 4 discrete average monthly effluent quality from which the calibrated model simulations results were compared with the actual as depicted in Figure 4b–d, respectively. Similarly, these validation results marched with the existing plant data well and within the acceptable limit. It is interesting to observe that all model's stoichiometry and kinetic parameters are similar and are applicable for both the calibration and validation (all month 1 to 4 simulated data). However, in each case, it requires initial characterizing of the influent stoichiometry fractions (Table 3) to depend on the influent wastewater quality parameters, which vary from month to month (Table 2). This adequately satisfies the model's assumption that all the constants model's coefficients for any influent characteristics. The model validation results show that calibrated models performed well in capturing the biological processes at the DWWTP for treating the municipal wastewater and be adequately considered to be acceptable. Thus, the performance capacity of the DWWTP was further investigated and analyzed under different operating conditions, as presented under the sections below.

3.3. Influence Parameters on DWWTP Performance and Capacity Analyses

The simulation results for DWWTP performance and capacity analyses under different steady-state operational scenarios of influencing parameters using the validated GPS-X model are presented and discussed under this section with meeting regulatory effluent discharge limits for COD, BOD_5, NO_2^- + NO_3^-, NH_3–N, and TN. Among them are some of the most influential parameters observed during the calibration process, which included wastewater quality parameters, VSS parameters biomass, and endogenous fractioning COD fractioning on activated sludge and clarifier sizing.

Table 4. Activated Sludge Tank GPS-X default or adjusted models stoichiometry and kinetic parameters which are similar for calibration and validation data.

Classification	Parameter	Unit	GPS-X Default	Calibration	Validation		
				Month 1	Month 2	Month 3	Month 4
Physical	d	m^3	29,800	29,800	29,800	29,800	29,800
	v	m	4	4	4	4	4
Composite Variable Stoichiometry							
Organic Fractions	i_{cv}	gCOD/gVSS	1.48	0.858	0.858	0.858	0.858
	f_{bod}	–	0.66	1	1	1	1
Nutrient Fractions	i_{XB}	gN/gCOD	0.068	0.236	0.236	0.236	0.236
	i_{XP}	gN/gCOD	0.068	0.288	0.288	0.288	0.288
Model Stoichiometry Parameters							
Active Heterotrophic Biomass	Y_H	gCOD/gCOD	0.666	0.666	0.666	0.666	0.666
	U_H	gCOD/gCOD	0.08	0.08	0.08	0.08	0.08
Active Autotrophic Biomass	Y_A	gCOD/gN	0.24	0.206	0.206	0.206	0.206
	U_A	gCOD/gCOD	0.08	0.0676	0.0676	0.0676	0.0676
Kinetic Parameters							
Active Heterotrophic Biomass	$\mu_{max,H}$	1/d	3.2	6.98	6.98	6.98	6.98
	$K_{S,S}$	mgCOD/L	5	0.3	0.3	0.3	0.3
	$K_{O,H}$	mgO_2/L	0.2	0.156	0.156	0.156	0.156
	$K_{O,A}$	mgO_2/L	0.2	0.237	0.237	0.237	0.237
	η_g	–	0.5	0.5	0.5	0.5	0.5
	K_{NO}	mgN/L	0.1	0.1	0.1	0.1	0.1
	K_{NH4}	mgN/L	0.05	0.229	0.229	0.229	0.229
	b_H	1/d	0.62	0.62	0.62	0.62	0.62
	$K_{ALK,H}$	$mgCaCO_3/L$	5	5	5	5	5
Active Autotrophic Biomass	$\mu_{max,A}$	1/d	0.9	4.5	4.5	4.5	4.5
	K_{NH}	mgN/L	0.7	0.109	0.109	0.109	0.109
	$K_{O,A}$	mgO_2/L	0.25	0.25	0.25	0.25	0.25
	b_A	1/d	0.17	0.0289	0.0289	0.0289	0.0289
	$K_{ALK,A}$	$mgCaCO_3/L$	25	25	25	25	25
Hydrolysis	k_h	1/d	3	12.7	12.7	12.7	12.7
	K_x	gCOD/gCOD	0.1	0.302	0.302	0.302	0.302
	η_h	–	0.6	0.192	0.192	0.192	0.192
Ammonification	k_A	m^3/gCOD/d	0.08	0.4	0.4	0.4	0.4

3.3.1. Influence of Activated Sludge and Clarifier Sizing DWWTP Capacity and Performance

The effect of reactor tanks size in terms of AST volume and CSA are provided in Figure 5a,b, respectively. As the AST volume was increased (Figure 5a), so also the concentrations of the effluent quality parameters decrease, reaching an optimal value at 30,000 m^3, which is comparable with the DWWPT AST of 29,841 m^3. Even though higher AST volume beyond this value would be inconsequential on the effluent quality for COD and BOD_5 removal, yet it was susceptible to significantly diminish the degradation of the nitrogenous compounds. Figure 5b suggests that CSA larger values above 5000 m^3 would yield better plant performance, although the operational plant clarifier area of 3810 m^2 was quite adequate. These results imply that the sizes of the plant's main reactors were optimally designed to handle the scenario during the present study. Additionally, the plant performance can be sustained even if higher inflow volumes are to be treated.

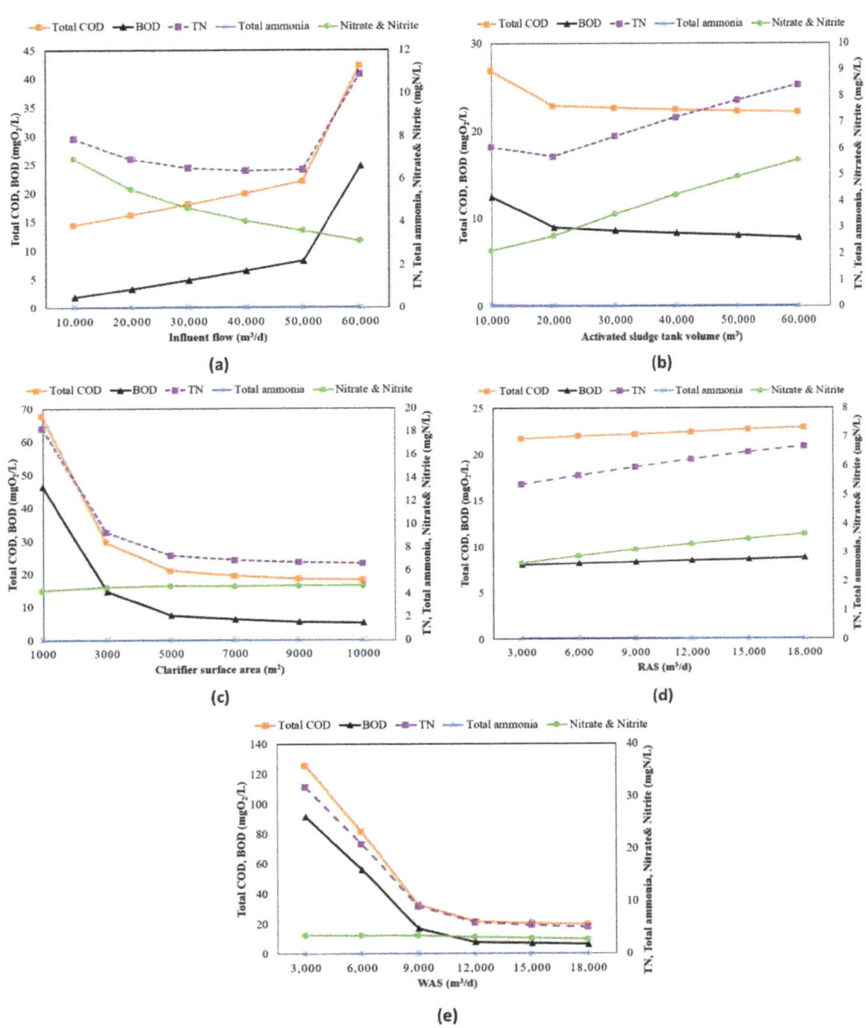

Figure 5. Influence of (**a**) influent flow (**b**) activated sludge tank (AST) volume (**c**) clarifier surface area (CSA) (**d**) recycle activated sludge (RAS) (**e**) waste activated sludge (WAS) on DWWTP capacity and performance.

3.3.2. Influence of Inflow, RAS, and WAS on DWWTP Capacity and Performance

The impact of *Q* presented in Figure 5d depicts a steady decrease in the plant's ability to remove COD and BOD_5 for increasing the *Q* from 10,000 to 50,000 m^3/d reaching maximum values of 23 and 8 mgO_2/L, respectively. Meanwhile, this inflow change resulted in a slight improvement of TN and a decline in effluent NO_3–N + NO_2–N. Afterward, higher *Q* values led to the effluent quality (TN, COD, and BOD_5) to deteriorate rapidly with the vulnerability of the plant to fail in meeting discharge limits when $Q > 60{,}000$ m^3/d. This suggests that the monthly DWWTP flow of 52,012 (±4440) m^3/d was entirely within desirable values for effective plant performance. Simultaneously, the plant can accommodate slightly higher *Q* values without undermining its performance within the acceptable effluent quality. RAS and WAS are the two vital parameters required to be set to ensure the retention of the active microbial community in ASP [35,38–40]. There existed marked differences between the influences of changes there two parameters on the DWWTP performances. This can be deduced from Figure 5d,e, which shows that the WAS has more impact on the effluent quality compared to RAS. As increasing the RAS over a considerable range, 3000 to 18,000 m^3/d, exerted no influence on the plant performance; however, there is substantial performance improvement when the WAS increased from 3000 to 9000 m^3/d. The COD and BOD_5 are dramatically decreased from 130 and 95 mgO_2/L to 40 and 20 while the TN decreased from 30 to 10 mgN/L, respectively. The insignificant effect of RAS could be attributed to the fact that the simulations were run under steady-state at the plant design while the maximum recycled ratio was 0.25 to 0.34, a typical range for conventional ASP. Invariably, the attainment of the design MLSS was only feasible at a minimum WAS of 9000 m^3/d. Thus, the respective plant operational RAS and WAS of 15,000 and 10,000 m^3/d were sufficient to sustain the plant performance for acceptable effluent quality. This is attributed to the higher influence of the WAS and the SRT which is one of the vital operational parameters of ASP as corroborated by earlier studies and discussed below. Similar to the present study, Elawwad et al. [35] reported a successful achievement optimization of a huge WWTP with great contributions of WAS flow rates.

3.3.3. Influence of MLSS, HRT, and SRT on DWWTP Capacity and Performance

As critical ASP design and operating parameters [38], the influence of MLSS, hydraulic retention time (HRT), and SRT on DWWTP performance are presented in Figure 6. For that, the treated water effluent and the sludge flow lines were employed for the assessments of the performance of the DWTTP. The MLSS trend shown in Figure 6a suggested that the plant could maintain its performance over an appreciable range of MLSS of up to 6000 mg/L. Even though above this MLSS value, the BOD_5 and nitrogen components removal will insignificantly be affected, yet there could be a dramatic increase in the COD components. This can be attributed to the sudden increased in the TN in the final effluent when the MLSS reached 6000 mg/L. While the model indicated that higher HRT resulted in better COD and TN removals (Figure 6b), in contrast, higher SRT are prone to significantly undermine the ability of the DWWTP to eliminate both COD and TN (Figure 6c) despite its effectiveness in ammonia-N removal under the various investigated operational parameters. The modeling data suggest that the condition for better performance is when the plant was set to operate at HRT > 6 h and SRT < 7 days. Results presented by Elwaad et al. [35] indicated better nitrification and denitrification when operated at SRT of 7 days, in agreement with the present study. However, other similar studies reported higher SRT values of up to 15 days [24,39,41].

These results further demonstrate that SRT—which is controlled via setting the WAS flow based on the design MLSS in the AST, and it decreases with an increase in the WAS—stands as the single vital parameter influences ASP compared to HRT [41,42]. The SRT, when operated under the lower WAS > 6000 m^3/d was expected to ensure improved plant performance for both carbon and nitrogen compounds removal (Figure 5d). Additionally, the HRT should be kept as low as possible for optimal COD removal, effective clarifier performance as well as nitrification. At the same time, higher values necessitate leading to energy expenditure due to oxygen requirements [41]. Other studies indicated that higher SRT leads to increased ASP O_2 requirements and energy expenditure, while too low is susceptible to result in partial denitrification and higher effluent nitrogen content [41].

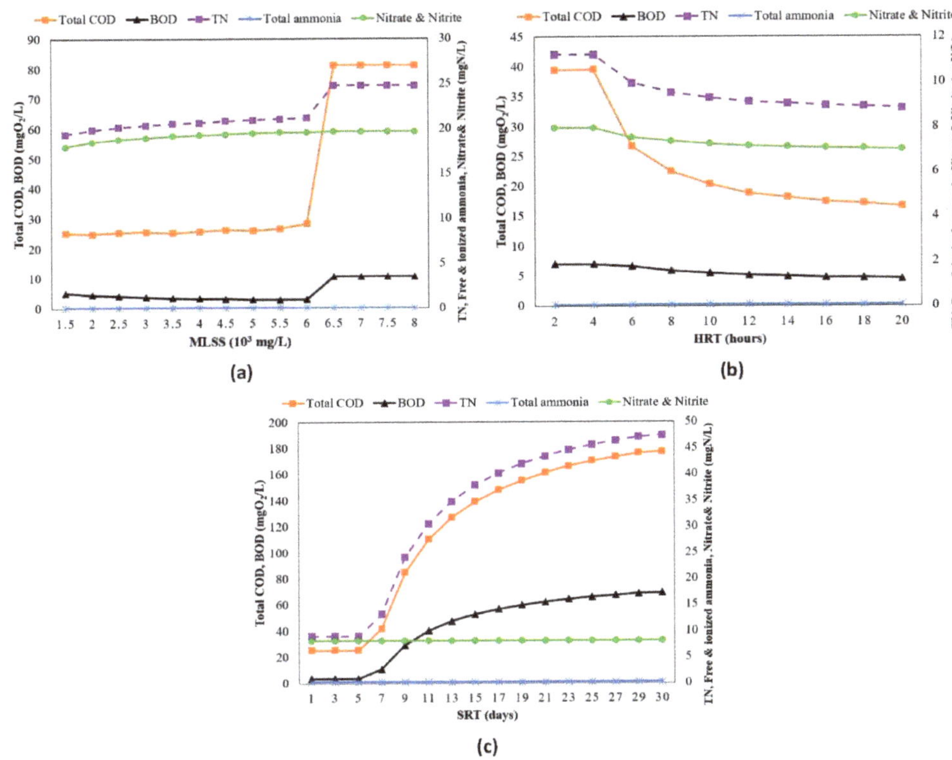

Figure 6. Influence of design (**a**) MLSS (**b**) SRT, and (**c**) hydraulic retention time (HRT) on DWWTP capacity and performance.

3.3.4. Influence of Influent Wastewater Quality Parameters on DWWTP Capacity and Performance

The results depicted in Figure 7a show that the COD_{out} and BOD_{5out} increases proportionally, with increasing COD_{in} concentration. Meanwhile, at lower COD_{in} the inhibition of effective nitrification and denitrification is manifested in the higher effluent nitrogenous parameter concentrations that decrease as the COD_{in} increases [38]. As the COD_{in} reaches beyond 250 mg/L, there was a tendency for COD_{out} and BOD_{5out} to breach the regulatory limit of 50 mg/L and 40 mg/L, respectively. This was also the case for the TN, which started to drastically increase at the 250 mg/L COD_{in} inflation point (Figure 7b). Considering that most of the average influent of the plant was far below this threshold (ranged between 129.05–215 mg/L), this implies that the plant performance was satisfactory for the handling even higher influent organic loadings. While the increase in influent total (NO_3^- + NO_2^-) insignificantly influenced both nitrogen and carbonous compounds degradation (Figure 7c), however, lower influent TN (Figure 7b) was susceptible to result in higher COD_{out} and BOD_{5out} (near the discharge limits) which could drastically reduce at higher TN values before stabilizing when the TN was beyond 14 mN/L.

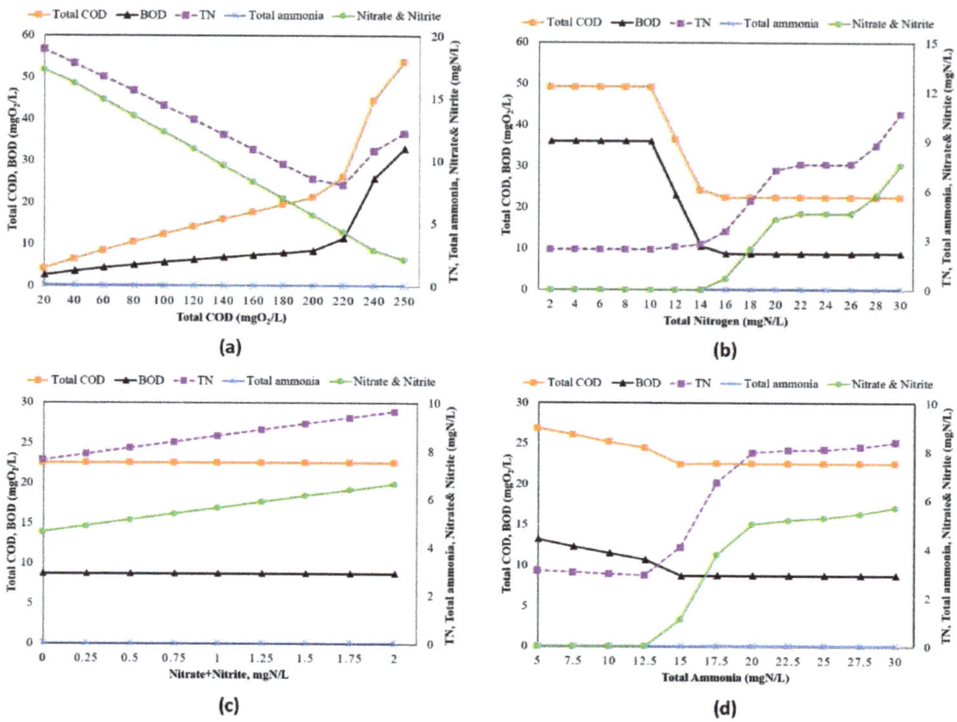

Figure 7. Influence of influent wastewater quality parameters o (**a**) COD (**b**) Total ammonia (**c**) Nitrate + Nitrite and (**d**) Total nitrogen on DWWTP capacity and performance.

Similarly, an increase in total influent NH_3−N would slightly affect COD_{out} and BOD_{5out} (Figure 7d). The plant was favorably useful for nitrogen removal when operated at lower influent NH_3−N and TN up to 12.5 and 14 mN/L, respectively. However, above these values, the nitrification and denitrification efficacy would tend to dwindle considerably. This nitrification-denitrification process, which is responsible for wastewater removal of nitrogen with approximately up to 67% of this mechanism is occurring under anoxic (i.e., anaerobic) condition [43]. However, denitrification is also known to take place under aerobic conditions by special denitrifying bacteria. Meanwhile, the influent wastewater organic nitrogen (i.e., the amine groups, $-NH_2$), is converted to ammonia via ammonification by heterotrophic bacteria as given in Equation (1) [43]. Accordingly, for organic oxidation under ASP, microbes can be either heterotrophic (mostly, bacteria or fungi and protozoa), deriving energy from organic compounds oxidation, or autotrophic, which are nitrifying bacteria getting energy from the oxidation of reduced inorganic compounds such as NH_3−N, NO_3^-, ferrous iron, and S^{2-} [38]. Thus autotrophic microbes are responsible for oxidizing NH_3−N from which energy for CO_2 uptake and growth are produced through the nitrification process via two-step oxidation: (i) oxidization of NH_3−N to NO_2^- and (ii) NO_2^- to NO_3^- by nitroso–bacteria and nitro-bacteria, in equation 2 and 3 respectively. Under anoxic conditions, NO_3^- or NO_2^- can be reduced to N_2 via the denitrification process [38]. The high rate of NH_3− N removal is manifested in the model's high ammonification rate of 0.4, which resulted in lower simulated NH_3− N for all cases investigated. This is corroborated in the higher NH_3 − N removal performance of the DWWTP lowering the NH_3−N in the effluent (average 0.261 mgN/L) compared to the influent concentration of average 18.7 mgN/L.

$$\begin{matrix}\text{Protien}\\ \text{Chitin}\\ \text{Nucleic acid}\end{matrix} \xrightarrow{\textit{Depolymerization}} \begin{matrix}\text{Amino acid}\\ \text{Urea}\\ \text{Amino sugar}\\ \text{Nucleotide}\end{matrix} \underset{\textit{Immobilization}}{\overset{\textit{Ammonification}}{\rightleftarrows}} NH_4^+ \underset{\textit{Nitrate reduction}}{\overset{\textit{Nitrification}}{\rightleftarrows}} NO_3^- \quad (1)$$

$$\begin{matrix} 2NH_4^+ + 3O_2 \longrightarrow 2NO_2^- + 4H^+ + NH_3 + 2H_2O \\ 2NO_4^- + O_2 \longrightarrow 2NO_3^- \end{matrix} \quad (2)$$

$$2NO_3^- + 2H^+ \longrightarrow N_2 + H_2O + 2.5O_2 \quad (3)$$

3.3.5. Influence of Influent Wastewater VSS Parameters on DWWTP Capacity and Performance

Figure 8a,b presents the dependencies of effluent quality parameters on influent VSS/TSS and X_{COD}/VSS ratio, respectively. Appropriate proportioning of these ratios is of paramount importance for ensuring the efficacy of WWTP biological treatment processes [38]. As shown in Figure 8a, lower *VSS/TSS* values were not favorable for the effective removal of both TN and carbonous materials in the wastewater, yet effective NH_3–N oxidization was achievable under the whole range of *VSS/TSS*. This suggests that denitrification cannot be successful under lower values of the *VSS/TSS* ratio, which needed to be maintained above 0.7, with the best performances stabilized from 0.8. Even though a similar trend is associated with X_{COD}/VSS, however, it can be observed that change in this parameter has more impact on the effluent wastewater quality compared to the *VSS/TSS* (Figure 8b). Lower X_{COD}/VSS can significantly hinder the overall plant capacity to treat the wastewater because the developed model indicated that only 12–13% of the X_{COD} accounts for the inert fraction (i.e., X_i in Table 3). Thus, to get the best DWWTP performance and ensure acceptable effluent quality, the influent wastewater X_{COD}/VSS should be maintained at a value greater than 1 with an optimal value of 1.38. This optimal value agrees with reported values of 1.41 and 1.22, as reported by Abdelsalam Elawwad et al. [35] and Nadja Hvala et al. [32], respectively. Moreover, this modeling results provided an insight into the VSS parameters considering the inconsistent data collected from the plant for the present study. This suggests that the plant *VSS/TSS* was more likely to be the range between 0.75–0.8 as obtained from the model calibration and validation.

3.3.6. Influence of Influent Wastewater COD Fractioning and Nitrogen Fractioning on DWWTP Capacity and Performance

The tCOD in wastewater organic matter is divided into biodegradable and non-biodegradable components, Equation (4), with a relative proportion of these components known to influence the efficacy of biological treatment processes [11,12,38,44,45]. However, in biological processes such as ASP, further subdivisions do exist, including active biomass (X_{BH} and X_{BA} discussed under Section 3.3.8) that forms part of the tCOD [12,24,44,45].

$$tCOD = S_s + S_i + X_s + X_i \quad (4)$$

Even though the influence of S_s on the efficacy of ASP has been reported intensively, in the literature, investigation on other components of the COD received lesser attention [45,46]. For instance, the X_S fraction and/or substrates can be a significant source of the S_s in the form of organic colloidal and/or particulate that can significantly influence the processes of COD, nitrogenous compounds removal as well as other wastewater treatment processes [46]. As shown in Figure 8c, the COD_{out} and $BOD5_{out}$ decreased from 30 and 15 mgO_2/L to about 20 and 5 mgO_2/L when the S_s was decreased from 0.1 to 0.3 before stabilizing over the considerable range (0.3–0.7). Afterward, the COD_{out} and $BOD5_{out}$ linearly increased significantly, reaching up to 90 and 70 mgO_2/L for $S_s = 1$, respectively. Despite effective nitrification and denitrification over the whole range of the S_s, which yielded TN,

NH_3–N, and NO_3–N + NO_2–N below their respective regulated limits, however, the optimal plant capacity for nitrogenous compounds removal was observed when the S_s was set at a minimum value of 0.8. However, the DWWPT model developed was at $S_s = 0.32$, which is sufficient to get all effluent quality parameters within the acceptable limit. Contrastingly, the plant capacity for removal of both nitrogenous and carbonous compounds persistently deteriorated as the inert proportion of the particulate COD (i.e., X_i) was increased (Figure 8d). The X_i should be kept below 0.2 to ensure acceptable effluent quality as corroborated by the model's adjusted values (Table 3).

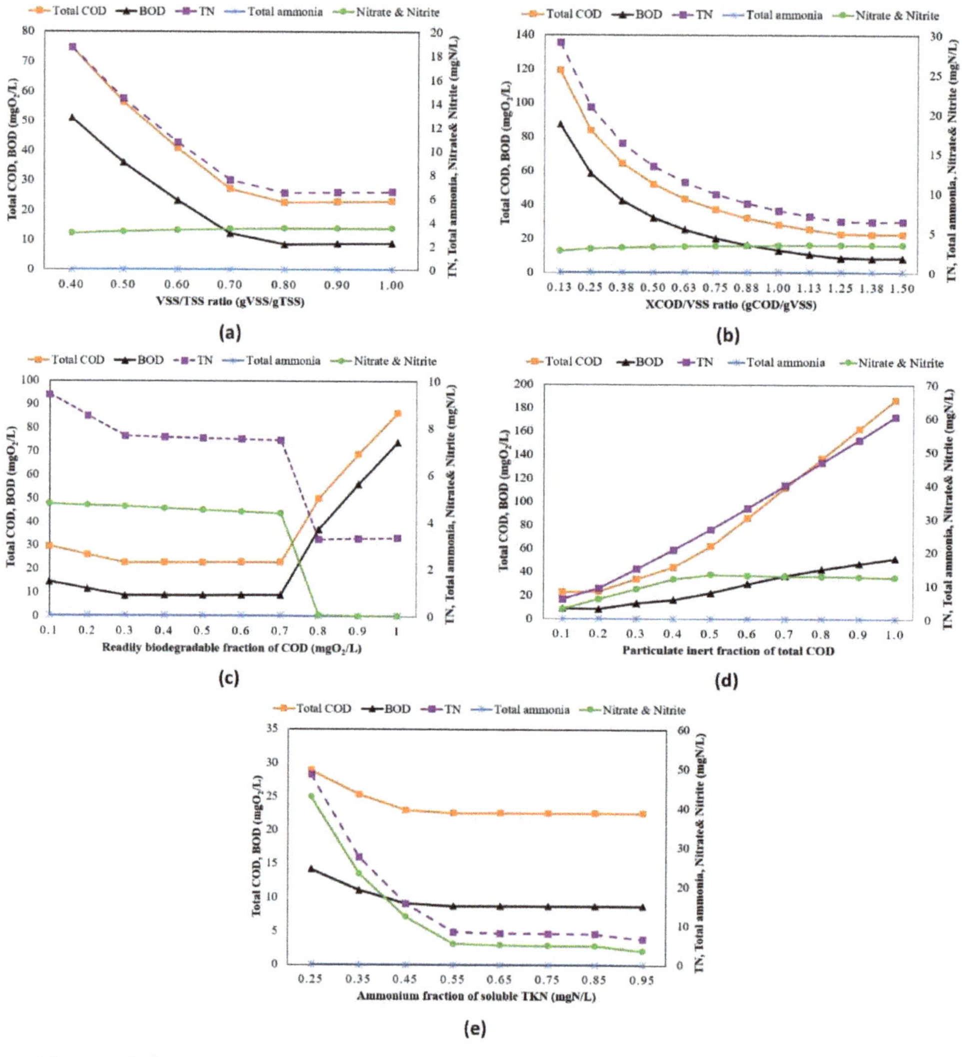

Figure 8. Influence of VSS parameters and COD fractioning and nitrogen/nutrient fractioning on DWWTP capacity and performance.

Meanwhile, as the $X_i = 0.12$ and $S_i = 0.0556$ are comparatively low, the model assumes that the higher proportion of the sbCOD (i.e., soluble biodegradable particulate and colloidal COD) effectively underwent extracellular enzymes hydrolyzation and availability for biomass assimilation [30]. On the other hand, Figure 8e shows the relevance of S_{nh} (i.e., NH_4/TKN ratio) on TN removal, which suggests that for acceptable plant performance, the S_{nh} should be kept above 0.55. Meanwhile, the S_{nh} has a lesser influence on the COD_{out} and BOD_{5out}, and increasing it improves the effluent quality, which steadied for $S_{nh} > 0.45$. For both the nitrogenous and carbonous quality parameters, the obtained simulated values of the S_{nh} for the plant data follow the GPS-X default value of 0.9. The ASM1 model adopted in this study considers both X_{ND} and X_S as transformed into S_S and S_{ND}, respectively, via hydrolysis [5,45], which is evident from the presented results.

3.3.7. Influence of COD/BOD_5, BOD_5/TKN, and COD/TKN Ratios on DWWTP Capacity and Performance

The ASPs are common, operational under carbon-limited conditions with COD/BOD_5 ratio as an acceptable index for evaluating the biological treatability efficacy of municipal wastewater [38,44,47]. The COD/BOD_5 ratio of raw domestic wastewaters has usually been reported within the range of 1.25 and 2.50 [38]. Wastewater having the COD/BOD_5 ratio below 2 can easily be biodegraded while the biodegradability of sewage is susceptible to be limited if the COD/BOD_5 > 3 [48–51]. With the mean COD/BOD_5 of 2.23 for the DWWTP, Figure 9a suggests that the DWWTP performance could be sustained for a quite appreciable range of COD/BOD_5 ratio. Moreover, the insufficient biodegradable organic carbon content concerning the nitrogen content in raw DWW influent is one of the limiting parameters for effective biological nitrogen removal. Low BOD_5/TN ratio in an influent leads to a quick carbon deficit and an uneven nitrification and denitrification process due to the competition between denitrifying bacteria and other heterotrophs substrate [20,52]. On the other hand, with a lower COD/TKN ratio, ASPs could be in more carbon-limited conditions for some domestic wastewaters, which might be inadequate to attain complete denitrification [48]. However, this is not the case for DWWTP having an average value of 7.89, which is manifested in the high denitrification of by the DWWTP.

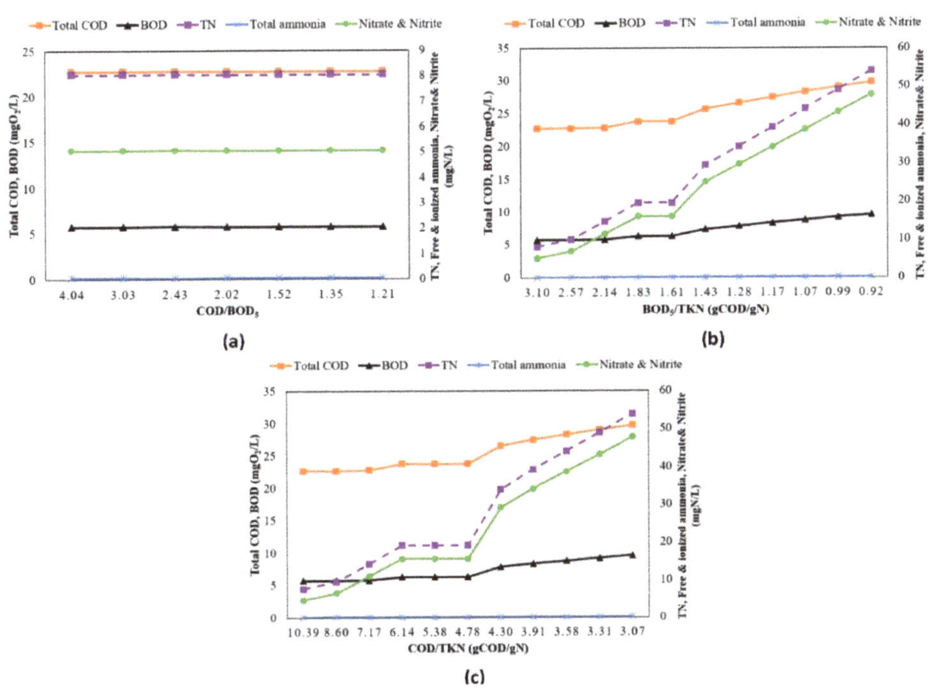

Figure 9. Influence of (**a**) COD/BOD_5 (**b**) BOD_5/TKN and (**c**) COD/TKN ratios on DWWTP capacity and performance.

Figure 9b,c reveal similar trends for BOD/TN and COD/TN ratios, which indicated better DWWTP performance when the ratios are kept comparatively at higher values. This corroborated earlier findings that showed high COD/TN > 5.5 is desirable for achieving efficient nitrogen removal [53]. Similarly, BOD_5/TN < 4 suggests nitrification occurring as a separate process for domestic wastewaters, whereas the BOD_5/TN > 4 combined approach is more likely to occur [44].

3.3.8. Influence of Influent Wastewater Biomass and Endogenous COD Fractions on DWWTP Capacity and Performance

The effects of organic biomass fractions of the COD are provided in Figure 8. The trends in Figure 10a,b are comparable for both heterotrophic (X_{BH}) and autotrophic (X_{AH}) influent of the biomass organic fraction and the final effluent quality. As both the increase of the fraction, there is a continuous decline of the effluent quality. The obtained calibrated $X_{BA} = 0.176$ and $X_{AH} = 0.142$ for the DWWTP indicated that they are quite adequate for the excellent operational performance of the plant.

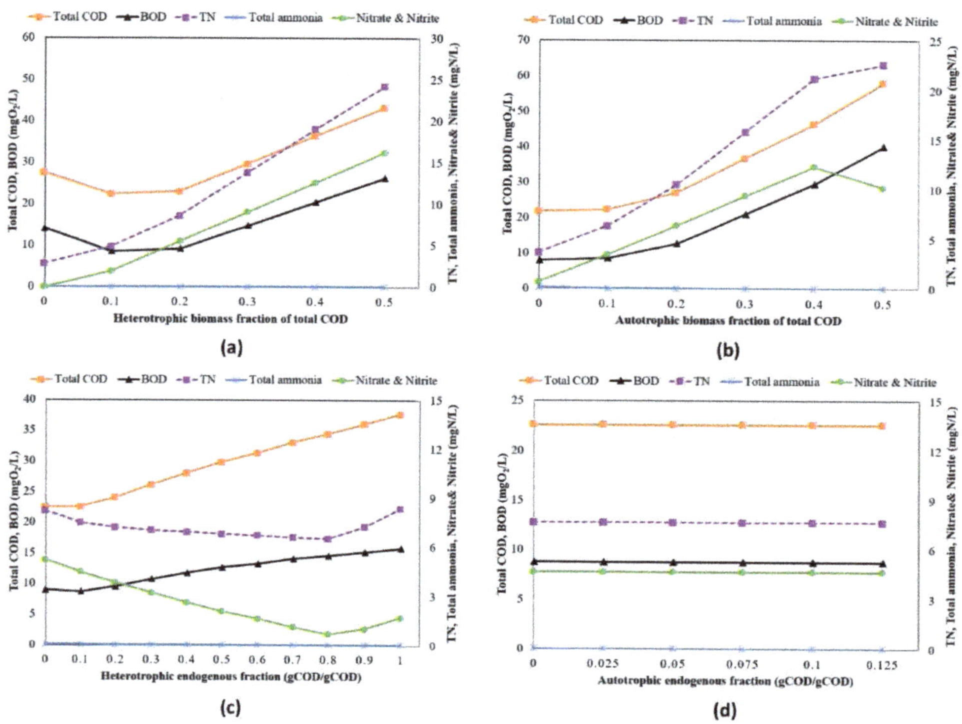

Figure 10. Influence of (**a**,**b**) biomass and (**c**,**d**) endogenous fractioning on DWWTP capacity and performance.

The influence of endogenous fraction U_H and U_A has shown in Figure 10c,d, respectively, suggests that the dynamics of U_A insignificantly influence the plant performance, which contrasted that of U_H. For the latter case, an increase in the fraction could substantially improve the quality parameters, mostly the COD. As a consequence, the model corroborates the GPS-X defaults (Table 4) [12]. The endogenous respiration phase in an ASP process considers no addition of an external substrate. The biomass is left on its own, and the substrate is generated via a decay and hydrolysis process. The process results in the release of both non-degradable matter X_i and nutrients, X_{ND} [5,7,38,45]. Thus, under this phase, the ASM1 mass balance considers only S_s, X_s, X_i, and X_{BH} [45]. In the mass balance cycle, the production of heterotrophic organisms represented by X_{BH} is lost by the decay X_{BH}, which transforms X_{BH} to X_s. Meanwhile, the generated X_s are lost via hydrolysis leading to the S_s production of which is further used by heterotrophic growth converted to produce X_{BH} at the expense of component S_o (i.e., respiration). Similarly, the biomass growth and decay follow a similar pattern for the processes involving nitrogen components (X_{ND}, S_{ND}, and S_{NH}) and nitrifying (autotrophic) organisms (X_{BA}).

4. Summary and Conclusions

In this study, a systematic approach for calibration of the ASP model for a real municipal wastewater plant activated sludge process was undertaken in GPS-X. The developed model was successfully validated while meeting the assumption of the model's constant stoichiometry and kinetic coefficients for any plant influent characteristics. The developed model was employed for the treatment of plant performance and capacity analysis. The study concludes that:

1. The sizes of the plant's main reactors were optimally designed to handle the scenario during the present study, although the plant performance can sustain higher inflow volumes.
2. Lower influent COD could inhibit the active nitrification and denitrification, while at influent COD beyond 250 mg/L, there is the tendency for effluent quality COD, BOD_5, and TN to breach the regulatory limit, implying that the plant performance can be satisfactory for handling even higher influent organic loadings.
3. The plant favorably performed well for nitrogen removal when operated at lower influent $NH_3{-}N$ and TN up to 12.5 and 14 mN/L, respectively. However, above these values, the nitrification and denitrification efficacy would tend to dwindle considerably.
4. The plant could maintain its performance over an appreciable range of MLSS of up to 6000 mg/L. While above this MLSS value, the BOD_5 and nitrogen components removal will have an insignificant effect, yet there could be a dramatic increase in the COD components attributed to a sudden increase in the effluent TN.
5. Higher HRT resulted in better COD and TN removals; in contrast, higher SRT are prone to significantly undermine the ability of the DWWTP to eliminate both COD and TN despite its effectiveness in an ammonia-N reduction under the various investigated operational parameters with better performance when the plant was set to operate at HRT > 6 h and SRT < 7 days.
6. The influence of COD/BOD_5 suggests that the DWWTP performance could be sustained for a quite appreciable range of COD/BOD_5 ratio, while the plant influent BOD_5/TN ratio of 7.89 manifested in results in high denitrification.
7. Better DWWTP performance could be achieved when the BOD_5/TN and COD/TN are kept comparatively at higher values.
8. Lower VSS/TSS values were not favorable for the effective removal of both TN and carbonous materials in the wastewater, and the ratio should be maintained above 0.7, yet effective $NH_3{-}N$ oxidization was achievable under the whole range of VSS/TSS.
9. X_{COD}/VSS has more impact on the effluent wastewater quality compared to the VSS/TSS. Lower X_{COD}/VSS can significantly hinder the overall plant capacity to treat the wastewater. To ensure the best plant performance and acceptable effluent quality, the influent wastewater X_{COD}/VSS should be maintained at a value greater than 1.
10. Increasing the Q from 10,000 to 50,000 m^3/d slightly, decreased the plant's ability to remove COD and BOD_5 while resulting in a slight improvement of TN and a decline in effluent $NO_3{-}N + NO_2{-}N$. Afterward, higher Q values led to the effluent quality (TN, COD, and BOD_5) to suddenly increased for $Q > 60{,}000$ m^3/d
11. The WAS has more impact on the effluent quality compared to RAS. Increasing the RAS over a considerable range (3000 to 18,000 m^3/d) comparatively, exerted an insignificant influence on the plant performance. However, when the WAS increased from 3000 to 9000 m^3/d, the COD

and BOD_5 noticeably decreased from 130 and 95 mgO_2/L to 40 and 20, respectively, while the TN decreased from 30 to 10 mgN/L.

12. This work further verifies the high potential modeling capabilities of ASM1 executed in GPS-X software for identifying influential and relevant kinetic and stoichiometric parameters and employing them for modeling and understating the influence of operational parameters on real WWTP when respirometry tests data are lacking

Author Contributions: O.A. conceived, designed the research idea, and secured the required data. N.D.M. performed GPS-X modeling, analyzed, and interpreted the results. I.A. process required data. All authors participated in writing the manuscript as well as its review. O.A. performed project administration and funding acquisition. All authors have read and agreed to the published version of the manuscript.

Funding: This research was funded by King Abdulaziz City for Science and Technology (KACST) Saudi Arabia through project grant number 10-WAT1336-46.

Acknowledgments: The authors acknowledge the financial support provided by King Abdulaziz City for Science and Technology (KACST) Saudi Arabia for funding this work through grant number (10-WAT1336-46) and the support of Deanship of Scientific Research (DSR) at Imam Abdulrahman bin Faisal University (IAU).

Conflicts of Interest: The authors declare no conflict of interest.

Abbreviations

ADS	Anaerobic digestion system
ASM	activated sludge model
ASP	activated sludge process
AST	Activated sludge tank
AOB	Ammonia Oxidizing Biomass
BOD	Biochemical Oxygen Demand
BOD_5	Biochemical Oxygen Demand in a 5-day incubation period in mgO_2/L
COD	Chemical Oxygen Demand in mgO_2/L
CSA	Clarifier surface area
DO	Dissolved oxygen
bCOD	Biodegradable COD
nbCOD	Non-biodegradable COD
sbCOD	Slowly Biodegradable COD in mgO_2/L
rbCOD	Readily Biodegradable COD mgO_2/L
tCOD	Total COD mgO_2/L
MLVSS	mixed liquor volatile suspended solids
TSS	Total Suspended Solids
VSS	Volatile Suspended solids
TKN	Total Kjeldahl Nitrogen
TN	Total Nitrogen
NOB	Nitrite Oxidizing Biomass
DWWTP	Dhahran north sewage wastewater treatment plant
SD	standard deviation
WWTP	wastewater treatment plant
HRT	Hydraulic retention time
RAS	recycle activated sludge (under flow)
SRT	Sludge retention time
WAS	wasted activated sludge (pumped flow)

Nomenclature

Symbol	Parameter	Unit
X_{COD}	Particulate COD	mgO_2/L
S_S	Readily biodegradable soluble fraction of COD	–
S_i	Soluble inert fraction of COD	–
X_S	Soluble Biodegradable particulate fraction of COD	–
X_i	Particulate inert fraction of COD	–
X_{BH}	Heterotrophic biomass fraction of total COD	–
X_{BA}	Autotrophic biomass fraction of total COD	–
X_P	Inert materials fraction	–
S_{nh}	ammonium fraction of soluble TKN	–
S_o	Dissolved oxygen	
NO_2^-	Nitrite	mgN/L
NO_3^-	Nitrate	mgN/L
$NH_3{-}N$	Total ammonia (free $NH_3{-}N$ and ionized $NH_4^+{-}N$ ammonia)	mgN/L
f_P	Particulate products fraction of biomass	
i_{vt}	VSS/TSS	gVSS/gTSS
i_{cv}	X_{COD}/VSS	gCOD/gVSS
f_{bod}	BOD_5/BODultimate ratio	–
i_{XB}	N content of active biomass	gN/gCOD
i_{XP}	N content of endogenous/inert mass	gN/gCOD
Y_H	heterotrophic yield	gCOD/gCOD
U_H	heterotrophic endogenous fraction	
Y_A	autotrophic yield	gCOD/gN
U_A	autotrophic endogenous fraction	
$\mu_{max,H}$	heterotrophic maximum specific growth rate	1/d
$K_{S,S}$	readily biodegradable substrate half-saturation coefficient	mgCOD/L
$K_{O,H}$	aerobic oxygen half-saturation coefficient	mgO_2/L
$K_{A,H}$	anoxic oxygen half-saturation coefficient	mgO_2/L
η_g	anoxic growth factor	–
K_{NO}	nitrate half-saturation coefficient	mgN/L
K_{NH4}	ammonia (as nutrient) half saturation coefficient	mgN/L
b_H	heterotrophic decay rate	1/d
$K_{ALK,H}$	alkalinity half-saturation coefficient for heterotrophic growth	$mgCaCO_3/L$
$\mu_{max,A}$	autotrophic maximum specific growth rate	1/d
K_{NH}	ammonia (as substrate) half-saturation coefficient	mgN/L
$K_{O,A}$	oxygen half-saturation coefficient	mgO_2/L
b_A	autotrophic decay rate	1/d
$K_{ALK,A}$	alkalinity half-saturation coefficient for autotrophic growth	$mgCaCO_3/L$
k_h	maximum specific hydrolysis rate	1/d
K_x	slowly biodegradable substrate half-saturation coefficient	gCOD/gCOD
η_h	anoxic hydrolysis factor	–
k_A	ammonification rate	m^3/gCOD/d
Q	Inflow rate	m^3/d
d	Tank depth	m
v	Volume of tank or reactor	m^3

References

1. Alagha, O.; Allazem, A.; Bukhari, A.A.; Anil, I.; Mu'azu, N.D. Suitability of SBR for wastewater treatment and reuse: Pilot-Scale reactor operated in different anoxic conditions. *Int. J. Environ. Res. Public Health* **2020**, *17*, 1617. [CrossRef]
2. Mu'azu, N.; Jarrah, N.; Zubair, M.; Alagha, O. Removal of phenolic compounds from water using sewage sludge-based activated carbon adsorption: A review. *Int. J. Environ. Res. Public Health* **2017**, *14*, 1094. [CrossRef]
3. Chan, Y.J.; Chong, M.F.; Law, C.L.; Hassell, D.G. A review on anaerobic-aerobic treatment of industrial and municipal wastewater. *Chem. Eng. J.* **2009**, *155*, 1–18. [CrossRef]
4. Sun, S.P.; Nàcher, C.P.I.; Merkey, B.; Zhou, Q.; Xia, S.Q.; Yang, D.H.; Sun, J.H.; Smets, B.F. Effective biological nitrogen removal treatment processes for domestic wastewaters with low c/n ratios: A review. *Environ. Eng. Sci.* **2010**, *27*, 111–126. [CrossRef]
5. Henze, M.; Gujer, W.; Mino, T.; van Loosedrecht, M. *Activated Sludge Models ASM1, ASM2, ASM2d and ASM3*; IWA Publishing: London, UK, 2015; Volume 5, ISBN 9781780402369.
6. Henze, M.; Gujer, W.; Mino, T.; van Loosdrecht, M.C.M. *Activated Sludge Models ASM1, ASM2, ASM2d and ASM3*; IWA Publishing: London, UK, 2000; ISBN 1900222248.
7. Nelson, M.I.; Sidhu, H.S. Analysis of the activated sludge model (number 1). *Appl. Math. Lett.* **2009**, *22*, 629–635. [CrossRef]
8. Mąkinia, J.; Zaborowska, E. Practical model applications. In *Mathematical Modelling and Computer Simulation of Activated Sludge Systems*; IWA Publishing: London, UK, 2020; pp. 599–638.
9. Al-Shahwan, A.A.; Balhaddad, A.M.S.; Al-Soudani, H.H.O.; Nuhu, D.M.; Abdel-Magid, I.M. Municipal wastewater treatment plants monitoring and evaluation: Case study Dammam Metropolitan Area. *Adv. Res. J. Multidiscip. Discov.* **2016**, *1*, 1–9.
10. Liwarska-Bizukojc, E.; Biernacki, R. Identification of the most sensitive parameters in the activated sludge model implemented in BioWin software. *Bioresour. Technol.* **2010**, *101*, 7278–7285. [CrossRef] [PubMed]
11. Liwarska-Bizukojc, E.; Olejnik, D.; Biernacki, R.; Ledakowicz, S. Calibration of a complex activated sludge model for the full-scale wastewater treatment plant. *Bioprocess Biosyst. Eng.* **2011**, *34*, 659–670. [CrossRef] [PubMed]
12. Environmental Software Solutions Inc. *Hydromantis GPS-X Technical Reference*; Environmental Software Solutions Inc.: Hamilton, ON, Canada, 2017.
13. Zeng, M.; Soric, A.; Roche, N. Modeling partial nitrification and denitrification in a hybrid biofilm reactor: Calibration by retention time distribution and respirometric tests. *Environ. Sci. Pollut. Res.* **2015**, *22*, 12849–12860. [CrossRef]
14. Jasim, N.A. The design for wastewater treatment plant (WWTP) with GPS X modelling. *Cogent Eng.* **2020**, *7*. [CrossRef]
15. Nasr, M.S.; Moustafa, M.A.E.; Seif, H.A.E.; El Kobrosy, G. Modelling and simulation of German BIOGEST/EL-AGAMY wastewater treatment plants—Egypt using GPS-X simulator. *Alex. Eng. J.* **2011**, *50*, 351–357. [CrossRef]
16. Petersen, B.; Gernaey, K.; Henze, M.; Vanrolleghem, P.A. Calibration of activated sludge models: A critical review of experimental designs. In *Biotechnology for the Environment: Wastewater Treatment and Modeling, Waste Gas Handling*; Springer: Berlin/Heidelberg, Germany, 2003; pp. 101–186.
17. Brun, R.; Kühni, M.; Siegrist, H.; Gujer, W.; Reichert, P. Practical identifiability of ASM2d parameters—Systematic selection and tuning of parameter subsets. *Water Res.* **2002**, *36*, 4113–4127. [CrossRef]
18. Makinia, J.; Rosenwinkel, K.-H.; Spering, V. Comparison of two model concepts for simulation of nitrogen removal at a full-scale biological nutrient removal pilot plant. *J. Environ. Eng.* **2006**, *132*, 476–487. [CrossRef]
19. Man, Y.; Shen, W.; Chen, X.; Long, Z.; Pons, M.N. Modeling and simulation of the industrial sequencing batch reactor wastewater treatment process for cleaner production in pulp and paper mills. *J. Clean. Prod.* **2017**, *167*, 643–652. [CrossRef]
20. Andraka, D. Reliability analysis of activated sludge process by means of biokinetic modelling and simulation results. *Water* **2020**, *12*, 291. [CrossRef]

21. Baek, S.H.; Jeon, S.K.; Pagilla, K. Mathematical modeling of aerobic membrane bioreactor (MBR) using activated sludge model no. 1 (ASM1). *J. Ind. Eng. Chem.* **2009**, *15*, 835–840. [CrossRef]
22. Kim, D.; Bowen, J.D.; Ozelkan, E.C. Optimization of wastewater treatment plant operation for greenhouse gas mitigation. *J. Environ. Manag.* **2015**, *163*, 39–48. [CrossRef]
23. Gujer, W.; Henze, M.; Mino, T.; van Loosdrecht, M. Activated sludge model No. 3. *Water Sci. Technol.* **1999**, *39*, 183–193. [CrossRef]
24. Chen, W.; Dai, H.; Han, T.; Wang, X.; Lu, X.; Yao, C. Mathematical modeling and modification of a cycle operating activated sludge process via the multi-objective optimization method. *J. Environ. Chem. Eng.* **2020**. [CrossRef]
25. Vanrolleghem, P.A.; Spanjers, H.; Petersen, B.; Ginestet, P.; Takacs, I. Estimating (combinations of) Activated Sludge Model No. 1 parameters and components by respirometry. *Water Sci. Technol.* **1999**, *39*, 195–214. [CrossRef]
26. Muoio, R.; Palli, L.; Ducci, I.; Coppini, E.; Bettazzi, E.; Daddi, D.; Fibbi, D.; Gori, R. Optimization of a large industrial wastewater treatment plant using a modeling approach: A case study. *J. Environ. Manag.* **2019**, *249*. [CrossRef]
27. Baird, R.B.; Eaton, A.D. *Standard Methods for the Examination of Water and Wastewater*, 23rd ed.; Baird, R.B., Eaton, A.D., Rice, E.W., Eds.; American Public Health Association: Washington, DC, USA; American Water Works Association: Denver, CO, USA; Water Environment Federation: Alexandria, VA, USA, 2017; ISBN 9780875532875.
28. Sharma, B.; Ahlert, R.C. Nitrification and nitrogen removal. *Water Res.* **1977**, *11*, 897–925. [CrossRef]
29. Surampalli, R.Y.; Tyagi, R.D.; Scheible, O.K.; Heidman, J.A. Nitrification, denitrification and phosphorus removal in sequential batch reactors. *Bioresour. Technol.* **1997**, *61*, 151–157. [CrossRef]
30. Pereira, S.F. *Modelling of a Wastewater Treatment Plant Using GPS-X*; Faculdade de Ciências e Tecnologia: Caparica, Portugal, 2014.
31. Vanrolleghem, P.A.; Insel, G.; Petersen, B.; Sin, G.; De Pauw, D.; Nopens, I.; Dovermann, H.; Weijers, S.; Gernaey, K. A comprehensive model calibration procedure for activated sludge models. *Proc. Water Environ. Fed.* **2003**, *2003*, 210–237. [CrossRef]
32. Hvala, N.; Vrečko, D.; Levstek, M.; Bordon, C. The use of dynamic mathematical models for improving the designs of upgraded wastewater treatment plants. *J. Sustain. Dev. Energy Water Environ. Syst.* **2017**, *5*, 15–31. [CrossRef]
33. Abdel-Kader, A.M. Studying the efficiency of grey water treatment by using rotating biological contactors system. *J. King Saud Univ. Eng. Sci.* **2013**, *25*, 89–95. [CrossRef]
34. Cox, C.D. Statistical distributions of uncertainty and variability in activated sludge model parameters. *Water Environ. Res.* **2004**, *76*, 2672–2685. [CrossRef]
35. Elawwad, A.; Matta, M.; Abo-Zaid, M.; Abdel-Halim, H. Plant-wide modeling and optimization of a large-scale WWTP using BioWin's ASDM model. *J. Water Process Eng.* **2019**, *31*. [CrossRef]
36. Mulas, M. Modelling and control of activated sludge processes. Ph.D. Thesis, University of Cagliari, Oristano, Italy, 2006.
37. Grady, C.P.L., Jr.; Daigger, G.T.; Love, N.G.; Filipe, C.D.M. *Biological Wastewater Treatment*; CRC Press: Boca Raton, FL, USA, 2011; ISBN 142000963X.
38. Metcalf, W.; Eddy, C. Wastewater engineering: Treatment and recource recovery. *Wastewater Eng. Treat. Reuse McGraw Hill. N.Y.* **2014**, *5*, 384.
39. Dey, A. Modeling simultaneous nitrification-denitrification process in an activated sludge continuous flow stirred-tank reactor: System optimization and sensitivity analysis. *Environ. Eng. Sci.* **2010**, *27*, 757–765. [CrossRef]
40. Struk-Sokoowska, J.; Rodziewicz, J.; Mielcarek, A. Effect of dairy wastewater on changes in COD fractions in technical-scale SBR type reactors. *Water Sci. Technol.* **2018**, *2017*, 156–169. [CrossRef]
41. Smith, R.; Elger, S.; Mleziva, S. Wastewater: Solids retention time control in wastewater treatment. *Filtr. Sep.* **2014**, *51*, 12–17. [CrossRef]
42. Techobanoglous, G.; Burton, F.L.; Stensel, H.D. *Wastewater Engineering: Treatment and Reuse*, 5th ed.; Metcalf and Eddy, McGraw-Hill Education: New York, NY, USA, 2014.

43. Lahiri, S.; Ghosh, D.; Sarkar, D. Biogeochemical cycling bacteria and nutrient dynamics in waste stabilization pond system. In *Wastewater Management Through Aquaculture*; Jana, B.B., Mandal, R.N., Jayasankar, P., Eds.; Springer: Singapore, 2018; pp. 29–52. ISBN 978-981-10-7248-2.
44. Orhon, D.; Ateş, E.; Sözen, S.; Çokgör, E.U. Characterization and COD fractionation of domestic wastewaters. *Environ. Pollut.* **1997**, *95*, 191–204. [CrossRef]
45. Drewnowski, J.; Szeląg, B.; Xie, L.; Lu, X.; Ganesapillai, M.; Deb, C.K.; Szulżyk-Cieplak, J.; Łagód, G. The Influence of COD fraction forms and molecules size on hydrolysis process developed by comparative our studies in activated sludge modelling. *Molecules* **2020**, *25*, 929. [CrossRef]
46. Drewnowski, J. The impact of slowly biodegradable organic compounds on the oxygen uptake rate in activated sludge systems. *Water Sci. Technol.* **2014**, *69*, 1136–1144. [CrossRef] [PubMed]
47. Cossu, R.; Lai, T.; Sandon, A. Standardization of BOD 5/COD ratio as a biological stability index for MSW. *Waste Manag.* **2012**, *32*, 1503–1508. [CrossRef] [PubMed]
48. Karef, S.; Kettab, A.; Loudyi, D.; Bruzzoniti, M.C.; Del Bubba, M.; Nouh, F.A.; Boujelben, N.; Mandi, L. Pollution parameters and identification of performance indicators for wastewater treatment plant of medea (Algeria). *Desalin. Water Treat.* **2017**, *65*, 192–198. [CrossRef]
49. Zawilski, M.; Brzeziriska, A. Variability of COD and TKN fractions of combined wastewater. *Pol. J. Environ. Stud.* **2009**, *18*, 501–505.
50. Attiogbe, F.; Glover-Amengor, M.; Nyadziehe, K. Correlating biochemical and chemical oxygen demand of effluents—A case study of selected industries in Kumasi, Ghana. *West Afr. J. Appl. Ecol.* **2009**, *11*. [CrossRef]
51. Al-Sulaiman, A.M.; Khudair, B.H. Correlation between BOD5 and COD for Al-Diwaniyah wastewater treatment plants to obtain the biodigrability indices. *Pak. J. Biotechnol.* **2018**, *15*, 423–427.
52. Khursheed, A.; Gaur, R.Z.; Sharma, M.K.; Tyagi, V.K.; Khan, A.A.; Kazmi, A.A. Dependence of enhanced biological nitrogen removal on carbon to nitrogen and rbCOD to sbCOD ratios during sewage treatment in sequencing batch reactor. *J. Clean. Prod.* **2018**, *171*, 1244–1254. [CrossRef]
53. Zhao, C.H.; Peng, Y.Z.; Wang, S.Y.; Tang, X.G. Influence of wastewater composition on biological nutrient removal in UniFed SBR process. *Water Sci. Technol.* **2008**, *58*, 803–810. [CrossRef] [PubMed]

sustainability

MDPI

Article

Application of Neural Networks and Regression Modelling to Enable Environmental Regulatory Compliance and Energy Optimisation in a Sequencing Batch Reactor

Shane Fox [1,2], James McDermott [3], Edelle Doherty [1,4], Ronan Cooney [1,4] and Eoghan Clifford [1,4,*]

1 School of Engineering, National University of Ireland Galway, H91TK33 Galway, Ireland; shane.fox@molloyprecast.com (S.F.); edelle.doherty@nuigalway.ie (E.D.); rcooney@nuigalway.ie (R.C.)
2 Molloy Environmental Systems, Clara Rd, Coleraine, R35D956 Tullamore, Ireland
3 School of Computer Science, National University of Ireland Galway, H91TK33 Galway, Ireland; james.mcdermott@nuigalway.ie
4 Ryan Institute, National University of Ireland Galway, H91TK33 Galway, Ireland
* Correspondence: eoghan.clifford@nuigalway.ie

Citation: Fox, S.; McDermott, J.; Doherty, E.; Cooney, R.; Clifford, E. Application of Neural Networks and Regression Modelling to Enable Environmental Regulatory Compliance and Energy Optimisation in a Sequencing Batch Reactor. *Sustainability* **2022**, *14*, 4098. https://doi.org/10.3390/su14074098

Academic Editors: José Luis Campos, Anuska Mosquera Corral, Ángeles Val del Río and Alba Pedrouso Fuentes

Received: 13 December 2021
Accepted: 7 March 2022
Published: 30 March 2022

Abstract: Real-time control of wastewater treatment plants (WWTPs) can have significant environmental and cost advantages. However, its application to small and decentralised WWTPs, which typically have highly varying influent characteristics, remains limited to date due to cost, reliability and technical restrictions. In this study, a methodology was developed using numerical models that can improve sustainability, in real time, by enhancing wastewater treatment whilst also optimising operational and energy efficiency. The methodology leverages neural network and regression modelling to determine a suitable soft sensor for the prediction of ammonium-nitrogen trends. This study is based on a case-study decentralised WWTP employing sequencing batch reactor (SBR) treatment and uses pH and oxidation-reduction potential sensors as proxies for ammonium-nitrogen sensors. In the proposed method, data were pre-processed into 15 input variables and analysed using multi-layer neural network (MLNN) and regression models, creating 176 soft sensors. Each soft sensor was then analysed and ranked to determine the most suitable soft sensor for the WWTP. It was determined that the most suitable soft sensor for this WWTP would achieve a 67% cycle-time saving and 51% electricity saving for each treatment cycle while meeting the criteria set for ammonium discharges. This proposed soft sensor selection methodology can be applied, in full or in part, to existing or new WWTPs, potentially increasing the adoption of real-time control technologies, thus enhancing their overall effluent quality and energy performance.

Keywords: real-time control; neural network; soft sensor; regression; sequencing batch reactor

1. Introduction

Advances in instrumentation, control and automation are aiding the development of intelligent real-time control (RTC) systems that can be used to predict, analyse and judge the real-time state of a system and self-adapt/organise based on input signals from sensors [1–5]. RTC systems can improve decision making and optimise system performance and are well suited to the control of complex and dynamic processes. However, sensors and detectors can produce large quantities of data that can be challenging to store, process and analyse. Thus, advances in analytic, decision-making, and process optimisation tools are required to enable the development of RTC systems. This has driven research into the use of numerical modelling techniques in a variety of engineering applications such as water fault detection, aquaculture and vaccine development [1,3,6–9].

An area where RTC can disruptively innovate and increase process efficiencies is in wastewater treatment. Protection of water resources and water quality is a key sustainable development goal [10], and the effective and sustainable treatment of wastewater is essential to this. Untreated wastewater results in water pollution, which affects both environmental quality and public health [11,12]. Therefore, environmental regulatory compliance in the wastewater treatment sector is vital. However, in the process of meeting regulatory compliance, wastewater treatment plants (WWTPs) can often inefficiently consume energy and be operated inefficiently due to a lack of suitable control processes.

1.1. RTC in Wastewater Treatment Facilities

Wastewater hydraulic flow rates and organic concentrations fluctuate over time; however, wastewater treatment plants are typically rigidly designed and operated to process worst-case scenarios (e.g., maximum hydraulic and design mass loading rates) [13–15]. This, in addition to stringent regulatory requirements, can result in inefficiencies in treatment capacity and energy consumption [13]. It has been noted that providing effective and efficient operation requires advanced or RTC solutions that can increase process control and efficiencies [5,16]. This is particularly true for small-scale WWTPs commonly located in towns and villages which have the additional challenges of (i) a lack of permanent operators and local expertise, (ii) relatively high energy costs, (iii) sludge handling, (iv) variable influent hydraulic or organic loads, and (v) inflexible operating regimes [17,18]. Despite these challenges, small WWTP operators are required to comply with tight regulations, which are proving difficult to meet. In Europe, the Urban Wastewater Treatment Directive (UWWTD) (91/271/EEC) specifies the standards for effluent discharged from WWTPs with population equivalents (Pes) exceeding 2000. These regulated parameters include biochemical oxygen demand (BOD), chemical oxygen demand (COD), ammonium-nitrogen (NH_4-N) and total suspended solids (TSS). In sensitive locations, additional parameters can include the monitoring of total phosphorus (TP) and total nitrogen (TN). The implementation of the Water Framework Directive (2000/60/EC) means more stringent limits can be attached to smaller WWTPs depending on status of the receiving waters.

RTC presents a viable means of advanced and targeted control which has significant potential to improve energy efficiency and environmental performance [19], this can lead to improved sustainability [20] in both large- and small-scale WWTPs. Despite considerable developments in sensor technology, real-time analysis of key parameters such as NH_4-N remains a challenge in terms of robustness, accuracy and affordability [4,20–22]. Therefore, the use of cost-efficient and reliable soft sensors as surrogates to predict certain parameters holds significant potential for disruptive innovation [23,24]. Several studies have demonstrated that sensors measuring parameters such as oxidation-reduction potential (ORP) and pH can act as surrogates for NH_4-N sensors [15,25–30] (Table 1). However, the implementation of these results at small-scale WWTPs is limited. Much of this research is limited to raw and differentiated pH and ORP sensor data as input variables. To the knowledge of the authors, no research has been conducted using a suite of pH and ORP variables (i.e., variables identified from the pH and ORP profile characteristics).

Table 1. Summary of research on advanced RTC methodologies with surrogate sensors.

Objectives	Control Methodology	Influent Type	Study Type	References
Advanced RTC Methodologies				
Strategy proposal for SBR optimisation using pH, ORP and DO profiles and fuzzy clustering algorithms for detecting critical process transitions	Fuzzy clustering with wavelet de-noising	Synthetic wastewater	Strategy examined using data collected from a pilot-scale SBR reactor	[16]
Investigation into the use of pH, ORP and DO sensors with an advanced control strategy to optimise nitrogen removal in a continuous system	Fuzzy logic	Urban wastewater with a small industrial input	Pilot-scale continuous flow plant	[31]

Table 1. *Cont.*

Objectives	Control Methodology	Influent Type	Study Type	References
Advanced RTC Methodologies				
Development of an RTC strategy using artificial NNs with ORP and pH sensors for optimised nitrogen removal and phosphorus uptake	Artificial NNs	Synthetic wastewater	Laboratory-scale continuous flow SBR reactor	[28]
Examination of using NNs for predicting biological nitrogen and phosphorus removal using ORP and pH	NNs	Synthetic wastewater	Laboratory-scale SBR reactor	[32]
Examination of the establishment of an online controlling system for nitrogen and phosphorus removal.	A primary professional intelligent control filtered noise by filtration wave and used NNs, database and deducing machine to identify each breakpoint.	Municipal Wastewater	Laboratory-scale SBR reactor	[33]
Methodology development for process monitoring and process analysis for nitrogen and phosphorus removal	Use of multi-way principal component analysis (MPCA) and clustering using historical process data	Domestic strength Synthetic wastewater	Pilot-scale SBR reactor	[34]
Validation study to assess the ability of an algorithm using networks to detect breakpoints using pH, ORP and DO sensors	NNs, de-noising was achieved using a regularisation algorithm	Municipal wastewater	Pilot-scale SBR reactor	[13]
Examination of using a software sensor for real-time estimation of nutrient concentration using pH, ORP and DO sensors	Fuzzy NN analysis	Synthetic wastewater	Bench-scale SBR reactor	[23]
Examination of using a software sensor for real-time estimation of nutrient concentration using pH, ORP and DO sensors	Genetic algorithm-based neural fuzzy system, using self-adapting fuzzy c-means clustering and genetic algorithms	Synthetic wastewater	Laboratory-scale SBR reactor	[24]
Examination of an intelligent control system to achieve advanced nitrogen removal using DO, pH and ORP sensors.	Three-layer network technology with high-performance PLCs and fuzzy control for break point identification	Municipal wastewater	Pilot-scale SBR reactor	[35]
Review article on the general use of artificial NNSAT modelling biological water and wastewater treatment processes	Artificial NNs	Several types	Several types	[36]
Examination of the use of a Gaussian-process (GP) model for the online optimisation of batch phases using pH, ORP and DO sensors.	GP regression was used to smooth the signals and GP classification was used for pattern recognition	Not specified	Laboratory-scale SBR reactor	[37]
Examination of the optimisation of a fuzzy logic controlled DO SBR system using pH and OUR trends for carbon and NH_4-N removal	Fuzzy control was used to switch on and off DO input, in order to smooth out pH and OUR profiles. The breaking point was identified using episode representation	Urban wastewater	Pilot-scale SBR reactor	[38]

Table 1. *Cont.*

Objectives	Control Methodology	Influent Type	Study Type	References
Advanced RTC Methodologies				
Examination of a methodology to develop a soft sensor monitoring of an SBR for enhanced biological phosphorus removal.	Artificial NNs	Synthetic wastewater	Laboratory-scale SBR reactor	[21]
Examination of a soft sensor for the optimisation of an SBR for biological nutrient removal	NNs	Synthetic wastewater	Laboratory-scale SBR reactor	[39]
Development of a control strategy to enhance nitrogen and phosphorus removal in an SBR reactor using pH, ORP and OUR	Use of a data acquisition system with curve fitting and characteristic point detection	Municipal wastewater	Semi industrial pilot SBR reactor	[40]
Development of a reliable RTC and supervision tool for DO control	Fuzzy NNs	Industrial wastewater	Aerated submerged biofilm wastewater treatment process	[41]
Development of a soft computing method to predict sludge volume index (SVI) values in a real WWTP	Recurrent self-organising NN	Municipal WWTP	Model based on SBR WWTP	[42]
Examination applies a self-organising cascade neural network (SCNN) with random weights to a non-linear system	Cascade NNs	Municipal WWTP	Model based on municipal WWTP	[43]
Proposal using a model-free learning control (MFLC) system to control advanced oxidation in the treatment of industrial wastewaters	Reinforcement learning	Phenol wastewater	Laboratory pilot plant	[44]
Development of a model for predicting TSS and chemical oxygen demand removal	Fuzzy inference system with principal control analysis	Papermill process wastewater	Papermill WWTP with an anaerobic digester and submerged biofilm biological reactor	[45]
Identifying model to predict effluent nitrogen concentrations and assessment of controller efficiency in terms of economic and environmental performances	Recurrent NNs for model identification and dynamic matrix control as predictive control (PC) algorithm and Benchmark Simulation Model 1 to test these PC configurations	Biological wastewater	Activated sludge process of a municipal WWTP	[46]
Development of soft sensor to predict effluent concentrations such as COD, TSS and TN content	NN with principal component analysis	Biological wastewater	Activated sludge process of large-scale municipal WWTP	[30]

RTC using surrogate sensors requires developing relationships between the primary variable(s) of interest and the surrogate variables being measured. For example, an operator may wish to employ the following rule for controlling a wastewater treatment plant: "when $y < t$, stop processing", where y is the concentration of the chemical of interest and t is a threshold for safe discharge. When using surrogate sensors, the task then reduces to a non-linear modelling problem since "y" is not measured directly. Instead, a number of variables (x_n) are analysed to develop functions, whereby $y = f(x_1, x_2, \ldots x_n)$. Several authors have taken this type of approach (Table 1), focusing particularly on fuzzy modelling and advanced neural network (NN) approaches, including recurrent networks [23], cascade networks [43], self-organising network structures [42,43] and fuzzy-neural network hybrids [24,41,45]. There has also been work in developing NN-based soft sensors, using principal component analysis (PCA) to select the optimal number of input vectors [30,47]. These PCA-based NNs were applied to a large-scale municipal wastewater plant, where they predicted concentrations of COD, TN and TSS (among others) using measurements of oxygen and nitrogen concentrations with influent flow rate and alkalinity. However, to the authors' knowledge, no work has been reported on using a standard feed-forward NN for regression. Standard feed-forward NNs often perform well in non-linear system modelling, so this is an important research gap.

The current study proposes a range of soft sensors, which can be selected according to weights assigned to criteria that might vary with site-specific requirements. There is an abundance of labelled data collected in real-world conditions (which reflect the application of the methodology in practice); hence, there is no need for a self-organising structure. The appropriate network structure can be investigated by comparing the performance of alternative structures directly.

Finally, this study takes a different approach to dealing with non-linear time-varying system dynamics, by using a recurrent or other dynamic network for this aspect. The data are pre-processed to produce a large selection of input variables, which encode information about time-varying aspects of the data. This approach makes the choice of input variables crucial. To address this, this study compares several variable sets (combinations of input variables)—each of which is assessed using a set of criteria describing key, usable features for performance optimisation. In contrast to [45] this study employs regularisation for feature reduction where needed, and leverages manually investigated feature subsets, rather than using PCA. This study presents a methodology capable of identifying the most suitable soft sensor, utilising surrogate probes and inferential estimating models, for RTC of small and decentralised WWTPs. This methodology can cater for the dynamic nature of small and decentralised WWTPs as well as ensuring key onsite goals which can be prioritised in soft sensor selection.

1.2. Numerical Modelling Methods

Regression is the task of modelling a real dependent variable y as a function of independent variables $f(x_n)$, minimising the errors between y and $f(x_n)$. A training set, a dataset of known values for x_n and y, is required to develop the model with the goal of accurate out-of-sample prediction, which is typically measured using a hold-out or test set. A common regression technique is multiple linear regression (MLR), a linear least-squares approximation of the data. MLR provides equations linking a number of input variables (x_n) to a target variable (y) using Equation (1) [48].

$$y = w_0 + w_1 x_1 + \cdots + w_n x_n \quad (1)$$

where w_0 is the intercept, w_n is a coefficient (or slope) for x_n and n is the number of input variables. Out-of-sample accuracy can be improved by using regularisation methods which add a penalty term to the model input variables, shrinking the freedom of the input variable during learning [48]. A popular regularisation method is the least absolute shrinkage and selection operator (LASSO) [22,49].

In contrast, NNs are non-linear models with many more degrees of freedom, hence they can be used to model more complex systems. They do not require a priori knowledge about the systems' structure. They are trained using various gradient descent algorithms [32,50]. A typical NN structure can have one input layer, one or more hidden layers, and one output layer, as illustrated in Figure 1 [39]. Each layer has several nodes. Within a layer, the jth node computes a linear combination of its input variables ($x_1, x_2, x_3, \ldots, x_n$), coming from the previous layer, with each signal having an associated weight ($w_{1j}, w_{2j}, w_{3j}, \ldots, w_{nj}$) [51]. A second input to the node is the bias (b_j), a constant that governs the node's net input. Weights are multiplied by corresponding inputs to create a weighted input using Equation (2).

$$y_j = b_j + \sum_{i=1}^{n} w_{ij} \times x_i \tag{2}$$

where i represents the inputs and j represents each node.

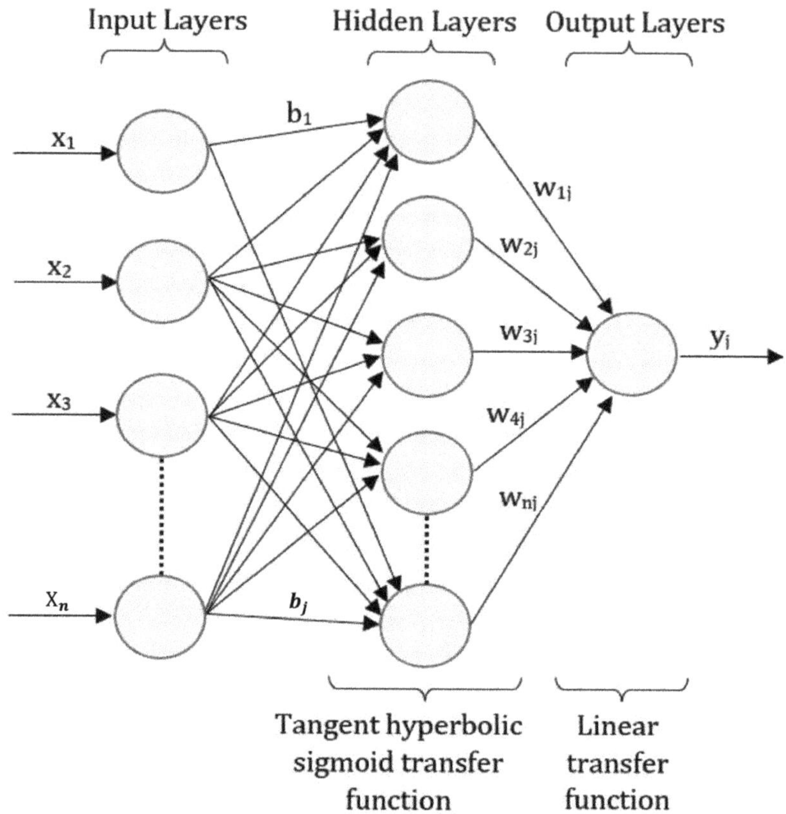

Figure 1. Typical NN structure with n inputs, j nodes in the hidden layer, a hyperbolic tangent sigmoid transfer function, and a single output layer with a linear transfer function.

The node then applies a transfer function to give its output. Several transfer functions are commonly used including logistic sigmoid, hyperbolic tangent sigmoid and linear functions.

Beginning with the independent variables, values are fed into each successive layer, with outputs from one layer becoming inputs to the next. At the output layer, a single value is output, which is the predicted value of y for the current inputs x_n. Training proceeds by adjusting weights and biases using gradient descent algorithms, such as Levenberg–Marquardt back-propagation [52–56] and Levenberg–Marquardt back-propagation with Bayesian regularisation [57–60], to minimise error at the output.

The specific goal, in this study, was to create a model to accurately predict current NH_4-N concentration (output) given current and previous ORP and pH values (inputs). This study investigated two types of regression methods, (i) multiple linear regression (MLR) (R_{lin}) and (ii) MLR with LASSO regularisation (R_{reg}), and two types of NN training algorithms, (i) Levenberg–Marquardt back-propagation (NN_{lm}) and (ii) Levenberg–Marquardt back-propagation with Bayesian regularisation (NN_{br}). Results were analysed in two ways,

(i) prediction of the general NH_4-N trend and (ii) performance when predicting a specific NH_4-N concentration—for example a regulatory discharge limit (performance was assessed in terms of accuracy of prediction, and time and energy savings achieved in the treatment cycle). Furthermore, a weighting and ranking system was used to determine the overall best setup that can enable optimal operational, environmental and energy performance.

2. Materials and Methods

The case-study site comprised a sequencing batch reactor (SBR), receiving wastewater from a residential development. The influent wastewater to the SBR comprised domestic wastewater that had undergone primary clarification. The SBR comprised a two-chamber precast concrete tank (a primary settlement chamber and a reaction chamber), with working volumes of 2.42 m^3 (hydraulic retention time (HRT) of 4 days) and 1.56 m^3 (HRT of 2.6 days), respectively (Figure 2). Influent raw wastewater fed into the primary tank using a pump. This pump was operated using a programme that mimicked the typical diurnal domestic house flow pattern (Table 2) according to the European Standards for evaluation of domestic wastewater treatment systems (CEN 12566-3 2006) [61]. The system was aerated mechanically as required.

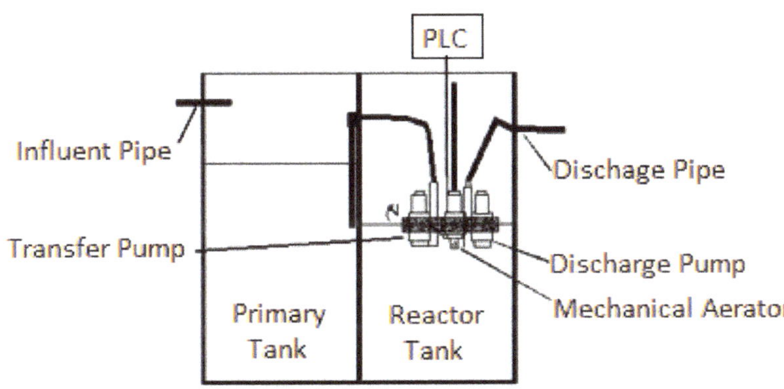

Figure 2. Schematic of pilot SBR unit.

Table 2. Diurnal flow pattern used to feed the primary chamber of the SBR pilot unit (CEN, 2006).

Time of Day	% of Total Volume	Volume (Litres)	Time of Day	% of Total Volume	Volume (Litres)
0:00–6:00	0	0	15:00–16:00	0	0
6:00–7:00	10	60	16:00–17:00	0	0
7:00–8:00	10	60	17:00–18:00	0	0
8:00–9:00	10	60	18:00–19:00	20	120
9:00–10:00	5	30	19:00–20:00	20	120
10:00–11:00	5	30	20:00–21:00	5	30
11:00–12:00	5	30	21:00–22:00	5	30
12:00–13:00	0	0	22:00–23:00	5	30
13:00–14:00	0	0	23:00–0:00	0	0
14:00–15:00	0	0			

2.1. Cycle Control

A Siemens LOGO! PLC controlled a 464 min cycle comprising the following phases: 2 min fill phase, 400 min aeration phase, 60 min settlement phase and 2 min discharge phase

(Figure 3). The aerated phase comprised 20 min blocks, each of which had a 5 min period during which the aeration system was turned on, followed by a 15 min quiescent period.

Figure 3. Illustration of cycle sequence (the on-off aeration pattern is demonstrated using the grey and white sequence in the aeration period).

A feed pump installed in the reactor chamber (switched on for 5 s, to create a siphon) moved liquid from the primary settlement chamber into the reaction chamber as required. Siphoning was terminated when the liquid level in the primary chamber went below (i) the inlet level of the feed pipe, (ii) the liquid level, or (iii) once the two chambers had equalised. As only the volume available over the feed pipe was transferred for treatment, this technique resulted in a dynamic feed volume. Table 3 details the operations in each phase.

Table 3. Overview of the SBR treatment cycle.

Phase (Step)	Operation	Description	Illustration
Fill (1)	Pump: A-On	The pump was switched on for 5 s, subsequently creating a siphon that moved liquid from the primary chamber into the reaction chamber. Siphoning terminated when the liquid level in the primary chamber went below the inlet level of the feed pipe or the liquid level or once the two chambers had equalised.	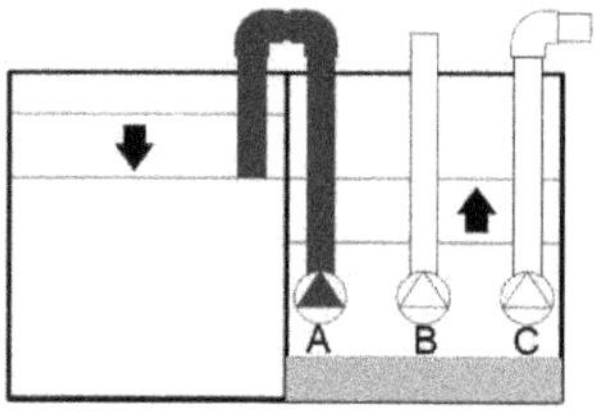
Aerobic—Repeated for 400 min (2)	(a) Aeration: B-On	The aeration period consisted of a"repetitive sequence of (a) aeration on for 5 min and (b) off for 15 min.	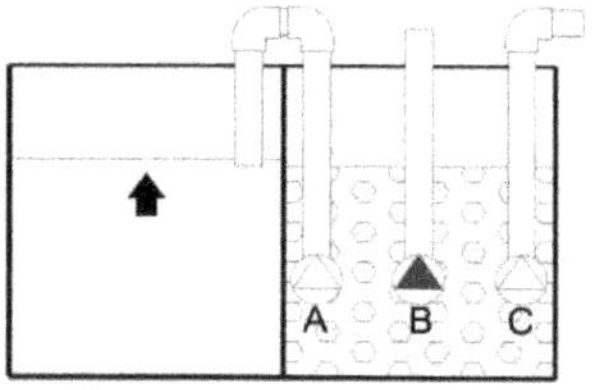
	(b) Rest		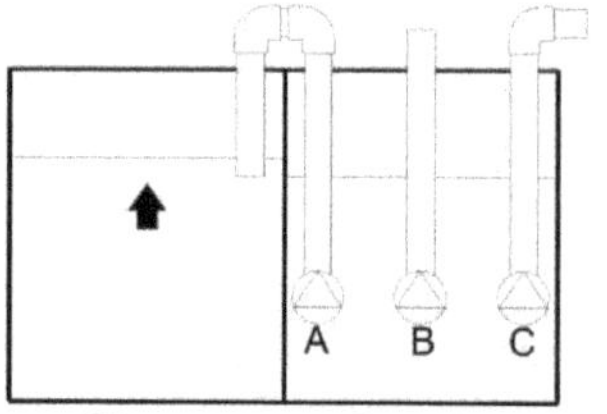

Table 3. *Cont.*

Phase (Step)	Operation	Description	Illustration
(3)	Settle	A settle time allowed an activated sludge settle prior to discharge creating an upper layer of clarified treated wastewater.	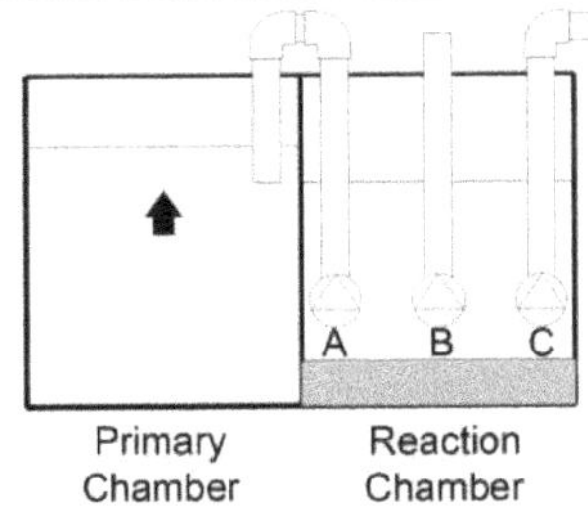
(4)	Discharge: C-On	The discharge pump I is used to remove the clarified treated wastewater from the upper portion of the reactor tank.	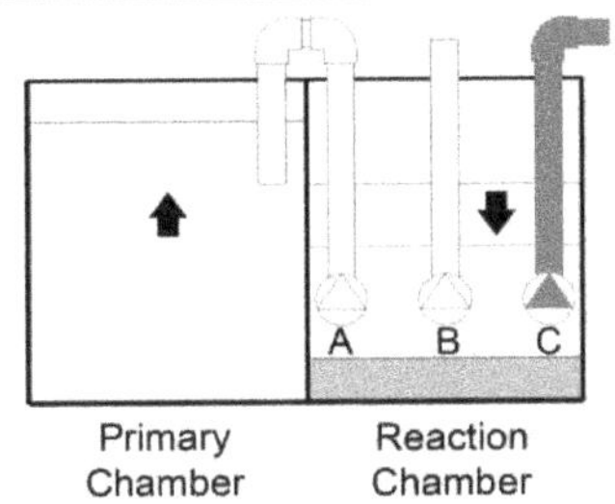
Symbol Definition		Pump On ; Pump Off	
Legend		**A**—transfer pump, **B**—mechanical aerator, **C**—discharge pump	

2.2. Monitoring

Influent and effluent wastewater samples were taken from the primary tank and from a collection vessel placed on the discharge line of the SBR, respectively. Filtered COD and TSS were tested in accordance with standard methods [62] whereby samples were passed through 1.2 μm Whatman GF/C microfiber filters. Total nitrogen (TN) was measured using a Biotector TOC TN TP Analyser (BioTector Analytical Limited, Cork, Ireland). Filtered NH_4-N and NO_3-N were measured using a Thermo Clinical Labsystem, Konelab 20 Nutrient Analyser (Fisher Scientific, Waltham, MA, USA). Hach sc1000 multi-meters monitored data collected from pH, ORP and NH_4-N sensors, in the reactor chamber. pH and ORP were measured at 1 min intervals while NH_4-N was measured at 5 min intervals on a 24 h basis (to match the pH and ORP data, NH_4-N data were linearly interpolated to create a data point every 1 min). All sensors were fitted approximately 500 mm below the lowest liquid level within the reaction chamber and above any potential sludge blanket that might be formed during settlement. All instruments were calibrated, maintained and operated in accordance with manufacturers' instructions.

2.3. Overview of NH_4-N, pH and ORP Profiles

A typical profile for NH_4-N saw an increase in concentrations as influent was mixed with the treated wastewater remaining in the reactor from the previous cycle. NH_4-N concentrations peaked soon after the fill phase. The time and magnitude of this peak varied depending on influent hydraulic volumes, organic carbon and NH_4-N concentrations. Following this peak, NH_4-N concentrations decreased due to organic carbon oxidation and subsequent nitrification. At approximately 225 min, the rate of decrease in NH_4-N concentrations reduced/levelled off and continued thus for the remainder of the cycle.

A cyclical rise and fall in both pH′ (Figure 4a) and ORP (Figure 4c) profiles during the aeration phase occurred, as the aerator switched on and off, resulting in a peak (or apex) and trough (nadir) in each aeration period in both pH (Figure 4b) and ORP (Figure 4d) profiles. The increase in pH, corresponding to the aeration-on period, was likely, in this case, to be due to CO_2 stripping [28]. The decreases in pH and ORP profiles during the 15 min quiescent period were likely due to a reduction in microbial activity over the course of the aerobic phase [63]. pH reduction was greatest and tailed off following the apex before a subsequent nadir was reached. A similar pattern was observed in the ORP profile. In general, pH decreases as alkalinity is consumed during the nitrification progresses [25]. The trend in pH decreased in response to aeration-on periods as a result of CO_2 stripping (Figure 4b). ORP generally increased during aeration; on completion of nitrification, ORP change accelerated; this acceleration was caused by an abundance of DO [64].

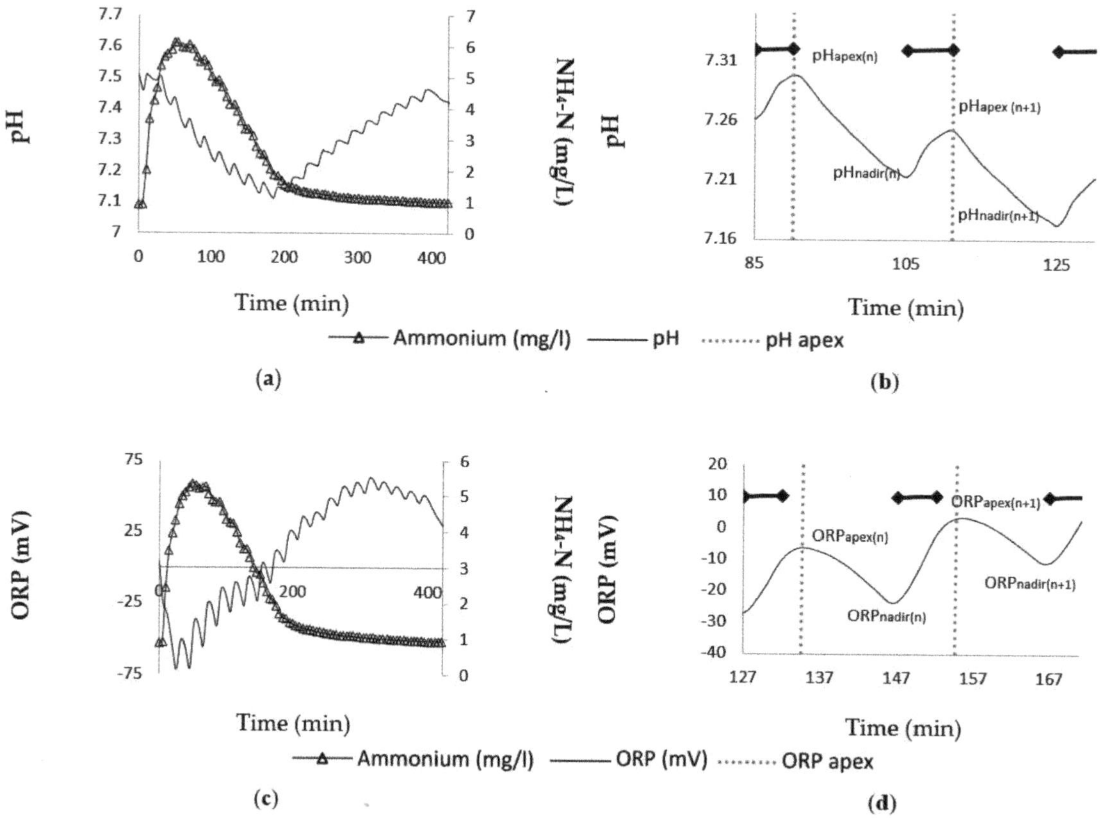

Figure 4. (**a**) pH and NH_4-N plotted against time for a sample cycle. (**b**) Example of a pH profile within three aeration periods plotted against time for a sample cycle. The black lines indicate "aeration-on" periods. (**c**) ORP and NH_4-N plotted against time for a sample cycle. (**d**) Example of an ORP profile with three aeration periods plotted against time for a sample cycle. The black lines indicate "aeration-on" periods.

3. Application

The methodology consisted of four main steps, namely, (i) data collection and pre-processing, (ii) experimental setup, (iii) soft sensor analyses and (iv) weighting and ranking application (Figure 5).

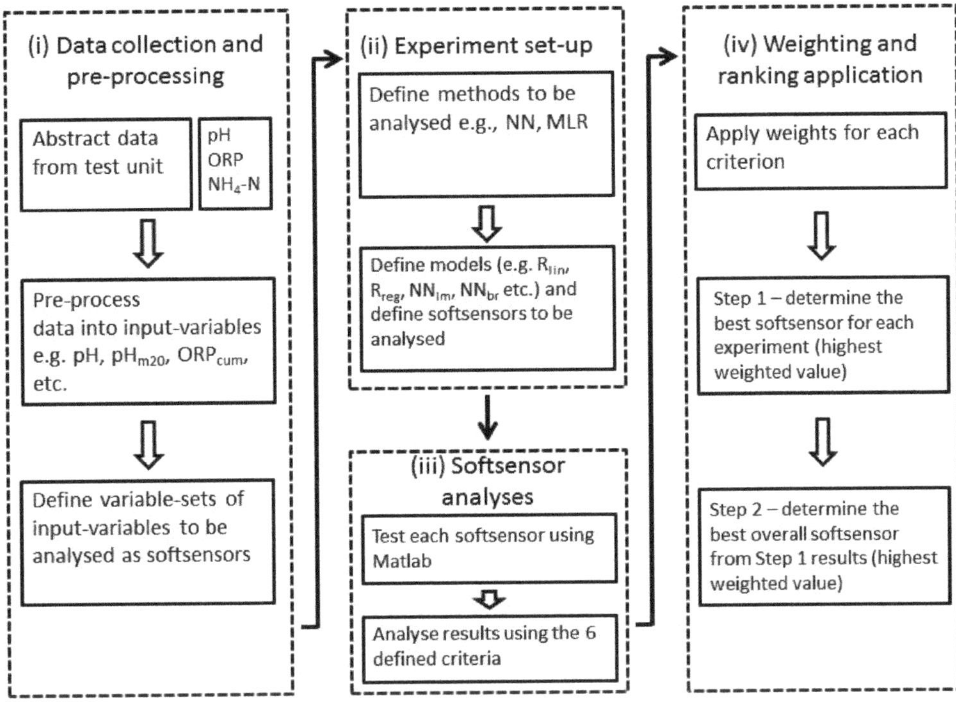

Figure 5. Summary of overall procedure.

3.1. Assessed Input Variables

A number of unprocessed (pH and ORP) and processed input variables were constructed and added to the set of independent variables (Table 4). The selected processed input variables were constructed using the profile features identified in Section 2.3. For example, the change in pH_{apex} values ($pH_{\Delta apex}$) was observed to decrease with NH_4-N reduction and was considered useful in identifying the end of NH_4-N removal. The set of independent variables was then analysed in 22 variable sets encompassing a broad range of combinations. Each variable set included a unique collection of input variables (Table 5).

Within each 464 min cycle, data collected between 0 and 45 min and 402 and 464 min were excluded to eliminate the effects of filling and settlement periods (as these phases were not part of the biological reaction phases of the treatment cycle). Between 0 and 45 min, the effects of the filling stage were still apparent in terms of raw influent mixing with existing wastewater in the system. The settlement and discharge phase was between 402 and 464 min. Data from 41 treatment cycles (each 464 min in duration) were collected, 12 of which (approximately 30%) were randomly separated for use as a test dataset, and the remainder were used as a training dataset.

Table 4. pH and ORP processed input variables.

Input Variable	Description
pH	Raw pH data
pH_{ma20}	Moving average of pH over the previous 20 min of data (i.e., 1 aeration block; Section 2.1)
pH_{cum}	Cumulative sum of pH data over the duration of the cycle
pH_{apex}	pH apex values during each aeration period
$pH_{\Delta apex}$	Change in sequential pH apex values over a treatment cycle

Table 4. *Cont.*

Input Variable	Description
pH_{nadir}	pH nadir values during each aeration period
$pH_{nadir\text{-}apex}$	pH nadir value minus pH apex value for each aeration period
ORP	Raw ORP data
ORP_{ma20}	Moving average of ORP over the previous 20 min of data
ORP_{cum}	Cumulative sum of ORP data over the duration of the cycle
ORP_{apex}	ORP apex values during each aeration period
$ORP_{\Delta apex}$	Change in sequential ORP apex values over a treatment cycle
ORP_{nadir}	ORP nadir values during each aeration period
$ORP_{nadir\text{-}apex}$	ORP nadir value minus ORP apex value for each aeration period
$pH_{ma20}XORP_{ma20}$	pH_{ma20} input variable multiplied by the ORP_{ma20} input variable

Table 5. Input variables to each variable set.

Variable Sets	Input Variables														
	pH	pH_{ma20}	pH_{cum}	pH_{apex}	$pH_{\Delta apex}$	pH_{nadir}	$pH_{nadir\text{-}apex}$	ORP	ORP_{ma20}	ORP_{cum}	ORP_{apex}	$ORP_{\Delta apex}$	ORP_{nadir}	$ORP_{nadir\text{-}apex}$	$pH_{ma20}.ORP_{ma20}$
A	X	X	X	X	X	X	X	X	X	X	X	X	X	X	X
B	X	X		X	X	X	X	X	X		X	X	X	X	X
C		X		X	X	X	X		X		X	X	X	X	X
D		X		X	X	X	X		X		X	X	X	X	
E		X		X	X		X		X		X	X		X	
F		X	X	X	X		X		X		X	X		X	
G		X	X	X	X		X								
H			X		X		X								
I		X	X		X		X					X		X	
J									X	X	X	X		X	
K			X				X								
L			X		X										
M										X				X	
N										X		X			
O										X		X		X	
P		X		X	X		X								
Q		X		X					X		X				
R				X	X						X	X			
S		X			X				X			X			
T							X							X	
U		X					X		X					X	
V		X					X								

3.2. Models

Two types of inferential estimation models were examined, namely regression and NNs. Two regression models were assessed, MLR without regularisation (R_{lin}) and MLR with LASSO regularisation (R_{reg}). Levenberg–Marquardt back-propagation (NN_{lm}) and Levenberg–Marquardt back-propagation with Bayesian regularisation (NN_{br}) were the two NN training models used. Within the NN training models, a hyperbolic tangent sigmoid hidden layer transfer function and a linear output layer transfer function were used. Each model contained one hidden layer of X neurons, notated as $NN_{lm[X]}$ and $NN_{br[X]}$ (X being the number of input variables in the variable set under investigation). Additional NN_{lm} and NN_{br} models were created by adjusting the number of neurons in the hidden layer to half the number of input variables, i.e., X/2 ($NN_{lm[0.5X]}$ and $NN_{br[0.5X]}$) and twice the number of input variables, i.e., 2X ($NN_{lm[2X]}$ and $NN_{br[2X]}$).

The feed-forward neural network architecture we have chosen is suitable for non-linear system modelling. As the input data are structured, not spatial, we do not need weight-sharing schemes such as convolution. Since we aim to produce an instantaneous soft sensor (i.e., its output reflects the current state of the system), we do not need a stateful network such as a recurrent network. Our choices for (i) transfer function and regularisation, (ii) the number of hidden nodes tested as a hyperparameter and (iii) values chosen, relative to the number of input variables ($\leq$15), are long-standing best practice [58,65]. The main advantages of our design are that it is simple, robust, easy to train, and not demanding to run even on low-power devices in the field. More sophisticated designs are possible and could have potential performance advantages but were considered out of scope.

In total, 176 soft sensors (i.e., a model applied to a variable set) were analysed. These soft sensors consisted of eight models with 22 identified variable sets using 15 input variables (Table 5, Figure 6). MATLAB was used as the computing environment to apply each of the models.

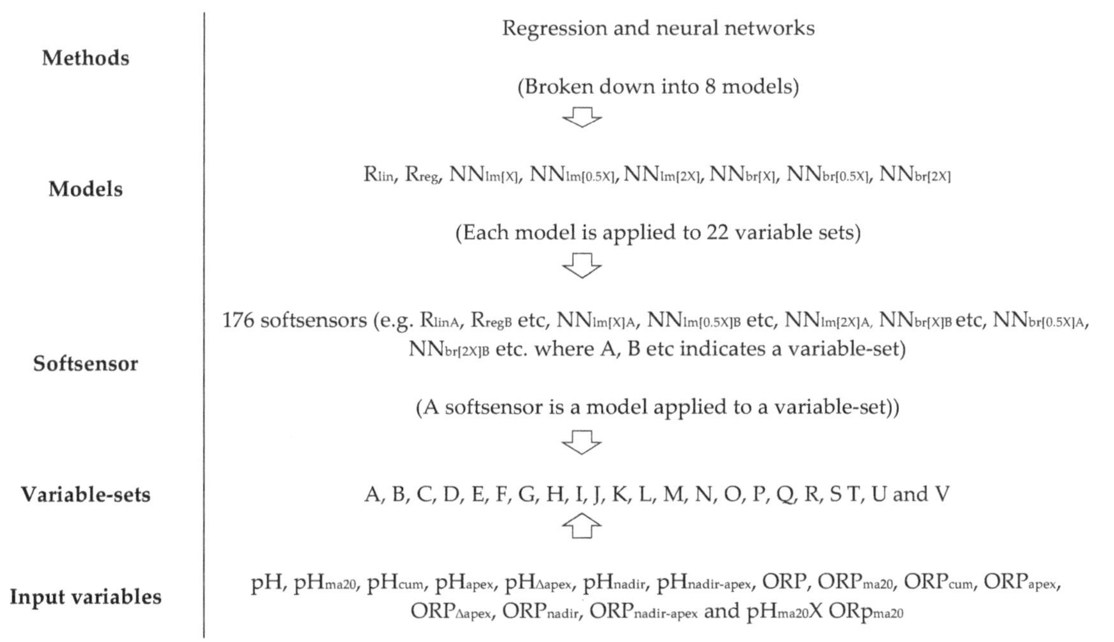

Figure 6. Breakdown of methods, models, variable sets, soft sensors and input variables.

3.3. Analyses

The effectiveness of the models was assessed across six criteria, split between two categories. Category A assessed the accuracy of the general NH_4-N trend prediction; and Category B the accuracy of the predicted trend at a selected NH_4-N concentration, known as the "cut-off threshold value". This value was set at 2 mg NH_4-N/l for the purposes of this study; site-specific values can vary due to local regulations. The assessment criteria are listed in Table 6.

Table 6. Analyses criteria.

Criterion	Description	Practical Application
Category A		
Criterion 1A: R^2	Referred to as the coefficient of determination, it is an indicator of the strength of the relationship between variables. 0 indicates a poor relationship, while 1 indicates a very close relationship.	Measures the strength of the relationship between predicted NH_4-N trend and actual NH_4-N trend
Criterion 2A: RMSE	Root mean square error (RMSE) is a standard statistical metric to measure model performance; it measures the difference between sample and predictor values and is a good measure of accuracy. The lower the RMSE value the more accurate the prediction.	Measures the average accuracy of the predicted NH_4-N trend against the actual NH_4-N trend
Category B		
Criterion 1B: Percentage of NH_4-N removal (NH_{4rem}(%))	This criterion returns the percentage NH_4-N removal from the peak NH_4-N ($NH_{4\,peak}$) concentration (during any given cycle) from a model controlled cycleto the actual NH_4-N concentration achieved on-site in a full (non-controlled) treatment cycle ($NH_{4\,final}$). The higher the NH_{4rem} value the better the soft sensor. $NH_{4\,rem} = \left(\frac{NH_{4\,thres} - NH_{4\,final}}{NH_{4\,peak} - NH_{4\,final}}\right) \times 100\%$ where NH_{4rem} is the percentage of potential NH_4-N removal achieved, $NH_{4\,thres}$ is the actual NH_4-N concentration where the cycle was terminated by the selected cut-off threshold (mg NH_4-N/l), $NH_{4\,final}$ is the final NH_4-N concentration at the end of a full cycle (mg NH_4-N/l) and $NH_{4\,peak}$ is the highest NH_4-N concentration. $NH_{4\,thres}$ could be related to an ammonium discharge limit at a given site.	Provides a comparison of the NH_4-N concentration at which the cycle would have been ended by the model during a controlled cycle and the actual final NH_4-N concentration at the end of a non-controlled cycle
Criterion 2B: Percentage of time saved (T_{save})	This criterion returns the time saved (as a percentage of a non-controlled cycle) by the soft sensor in question, at the selected cut-off threshold value, when compared to the full treatment cycle (and expressed as a percentage). The higher the T_{save} value, the better the soft sensor. $T_{save} = (1 - \frac{T_{thres}}{T_{fixed}}) \times 100$ where T_{save} is the time saving (%), T_{thres} is the time at which the cycle would be ended by the model in a controlled scenario and T_{fixed} is the fixed time cycle length (min) set in an uncontrolled scenario.	Indicates the time saved with the selected cut-off threshold value. For example, the model might be asked to terminate the treatment cycle when NH_4-N concentrations are predicted to reach a certain concentration (e.g., a discharge limit concentration). In general, the greater the time saved, the better, as in practice it increases system capacity
Criterion 3B: Number of successful cycles (SC)	During the application of the soft sensors, it was noticed that some soft sensors may end a treatment cycle very early due to the addition and subsequent mixing of influent at the start of a treatment cycle. This can influence pH and ORP trends temporarily and cause cycles to be ended at an early stage (often prior to the new influent beign completely missed with existing wastewater in the system). Where a cycle was ended before $NH_{4\,peak}$ occurred, a soft sensor was deemed unsuccessful for that cycle.	Allows for elimination of soft sensors that would end cycles too early
Criterion 4B: Absolute error (Ab_{error})	This criterion assessed the accuracy of the soft sensor in meeting a specific threshold concentration for effluent NH_4-N discharges.	Indicates the accuracy of each soft sensor at the cut-off threshold value

3.4. Ranking System

A ranking and weighting system was developed to compare the overall impact of each soft sensor. This was necessary as soft sensors may differ in their impact on the overall performance and efficiency of the SBR. For example, a soft sensor may achieve good R^2 performance, but also return a poor RMSE result. This example scenario would produce results in line with the actual NH_4-N trend but not necessarily close to the actual concentration, thus the overall result would not be acceptable. In consultation with WWTP operators, weights were applied to each of the criteria (Table 7). In general, the overriding concern in WWTPs is to meet environmental regulation, thus Ab_{error} would be considered vital. For indicative purposes, the weights outlined in Table 7 were applied to this study. It should be noted that weightings may vary depending on site-specific requirements and demands. In addition, these weights can be adjusted to promote site-specific goals. For example, increasing T_{save} would promote the selection of a soft sensor with good energy saving characteristics, but this may result in poor effluent quality.

Table 7. Applied weights.

Criterion	Weight	Comments
Ab_{error}	10	Ab_{error} indicates the accuracy of the soft sensor at the selected cut-off threshold value. Important as facilities must achieve regulatory compliance
RMSE	5	RMSE indicates the accuracy of the soft sensor when estimating the concentration over a cycle
NH_{4rem}	4	Provides an indication of the NH_4-N removal performance of the soft sensor
R^2	3	Indicates how well the predicted NH_4-N trend matches the actual trend
T_{save}	2	Indicates the time saving and energy savings of the soft sensor
SC	1	Least important as low SC values indicate more cycles will finish earlier than they should

Soft sensor results were ranked against each other for each criterion, with better results receiving a higher rank value (ranked values are 1 to n, where n is the number of soft sensors in question). The ranked value was then multiplied by the corresponding criterion weight to acquire the weighted value. Weighted values were then added together and compared to determine the most appropriate soft sensor as follows:

Step 1, determine the best soft sensor (highest weighted value) for each model using the system described above (Equation (3));

Step 2, determine the best soft sensor (highest weighted value) (and thus the overall best soft sensor) from Step 1 results using the system described above.

$$Weighted\ Value_{Softsensor} = \sum_{n=1}^{n=i} (Rank_n \times Weight_n) \quad (3)$$

where n = each criterion detailed in Table 7.

3.5. Further Analyses

Although determining the best soft sensor was the main objective of this study, a number of other studies, using the same criteria and weights, were also executed including (i) whether MLR and NN regularisation improved results, (ii) a comparison between MLR and NN methods, (iii) how adjusting the number of neurons in the NN hidden layers affected results, and (iv) an examination of which variable sets, which variables and which models were best. It should be noted that the model, variable set, etc., identified for the best soft sensor may differ from that for the best identified model, variable set, etc. The aim of this study was not just to identify the best soft sensor (combination of model and variable set), but also the best overall model and variable set.

4. Results

The overall influent and effluent results for the SBR are summarised in Table 8.

Table 8. Average influent and effluent results (average daily hydraulic volume = 0.9 m^3).

Parameter	Average Influent	Influent st.dev.	Average Effluent	Influent st.dev.	% Removal	*n* Inf/Eff
	mg/L	mg/L	mg/L	mg/L		
COD_f	405	126	120	85	70.3	9/14 14
TN	87.4	36	16.2	7.9	81.5	12/18
NH_4-N	49.6	20	1.1	1.2	97.8	17/28
NO_3-N	-	-	2.5	4.3	-	-/27

n is number of samples; Inf—influent; Eff—effluent.

4.1. Regression Results

Two regression models were assessed, R_{lin} and R_{reg}. Detailed results for each model are displayed in Tables A1 and A2, respectively. For NH_{4rem}, results varied between 20% and 97% for R_{lin} (average value of 66%) and between 75% and 93% for R_{reg} (average value of 84%). Average T_{save} and ab_{error} results were 51% and 0.98 mg NH_4-N/l for R_{lin} and 51% and 0.73 mg NH_4-N/l R_{reg}. An overview of these results shows that R_{reg} was better than R_{lin}, as it, on average, achieved better NH_{4rem} and ab_{error} results while maintaining a similar T_{save} result, thus resulting in better and more reliable effluent concentration predictions.

4.2. Neural Network Results

NNs were assessed using two algorithms, namely NN_{lm} and NN_{br}. Overall results for $NN_{lm[X]}$ are displayed in Table A3. The average NH_{4rem} result for $NN_{lm[X]}$ was 65% with corresponding T_{save} and ab_{error} results of 60% and 1.52 mg NH_4-N/l, respectively. The application of $NN_{lm[0.5X]}$ (Table A4) returned an average NH_{4rem} result of 72% and average T_{save} and ab_{error} results of 59% and 0.84 mg NH_4-N/l, respectively. Average results for $NN_{lm[2X]}$ (Table A5) were 59%, 65% and 1.52 mg NH_4-N/l for NH_{4rem}, T_{save} and ab_{error}, respectively.

$NN_{br[X]}$ returned average T_{save}, ab_{error} and NH_{4rem} results of 60%, 1.35 mg NH_4-N/l and 67%, respectively (Table A6). $NN_{br[X]}$ was further assessed using $NN_{br[0.5X]}$ and $NN_{br[2X]}$. $NN_{br[0.5X]}$ returning average T_{save}, ab_{error} and NH_{4rem} results of 60%, 1.03 mg NH_4-N/l and 69%, respectively, while $NN_{br[2X]}$ results for T_{save}, ab_{error} and NH_{4rem} were 64%, 1.33 mg NH_4-N/l and 61%, respectively. Overall results for $NN_{br[0.5X]}$ and $NN_{br[2X]}$ are displayed in Tables A7 and A8, respectively.

$NN_{lm[0.5X]}$ was the best soft sensor in terms of NH_{4rem} and ab_{error} results, while $NN_{lm[2X]}$ had the best T_{save} result. It should be noted that these average results are only indicative of the overall performance of the soft sensor and do not represent the ability of individual soft sensors at predicting NH_4-N trends during the cycle itself.

4.3. Weighting and Ranking Results

To decide the best soft sensor a weighting and ranking system was applied. Table 9 summarises the overall results from this study (full details are available in Table A9). The first step determined the best variable set (i.e., combination of independent input variables) for each model and the second step determined the best soft sensor.

Overall, $NN_{br[2X]U}$ was determined to be the most efficient soft sensor based on the weighting system. Variable set U used a combination of moving averages with nadir-apex values for both pH and ORP. This soft sensor achieved an average NH_{4rem} result of 88% over the 12 test cycles with corresponding T_{save} and ab_{error} results of 67%, 0.57 mg NH_4-N/l, respectively (Figure 7). This equated to a 51% reduction in electricity costs for the SB system due to the time savings during the treatment cycle (which in commercial settings may reduce aeration costs).

Table 9. Step 1 ranking results and Step 2 ranking.

Soft Sensor	Comparison of the Best Soft Sensor for Each Model against Each Criterion (Step 1)						Overall Ranking (Step 2)
	Average R^2 in Last 200 Min of the Cycle	Average RMSE in Last 200 Min of the Cycle	Averages at 2 mg NH_4-N/l Trigger				Ranking
			NH_{4rem} (%)	T_{save} (%)	ab_{error} (mg/L)	SC	
R_{linH}	0.553	0.5	86	53	0.58	12	6
R_{regL}	0.646	0.479	85	55	0.58	12	4
$NN_{lm[X]O}$	0.675	0.464	91	37	0.69	12	9
$NN_{lm[0.5X]M}$	0.634	0.457	92	36	0.59	12	7
$NN_{lm[X]R}$	0.465	0.457	77	63	0.48	11	3
$NN_{lm[X]K}$	0.653	0.441	64	60	0.80	12	8
$NN_{br[X]T}$	0.584	0.346	73	56	0.60	11	4
$NN_{br[0.5X]V}$	0.723	0.402	75	59	0.58	11	2
$NN_{br[2X]U}$	0.769	0.196	88	67	0.57	11	1

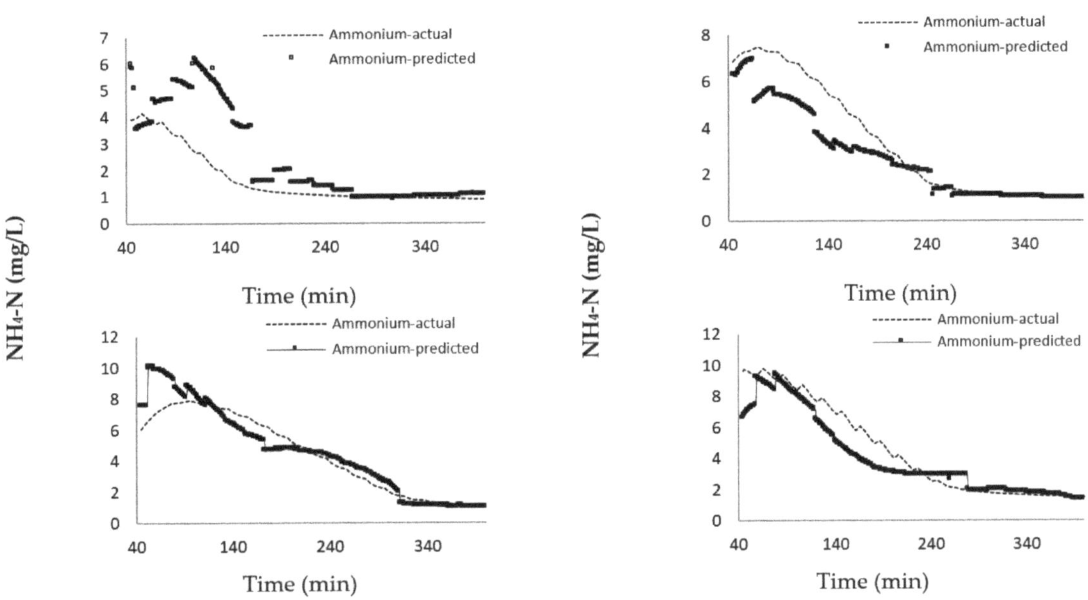

Figure 7. Comparison of modelled and measured NH_4-N concentrations for 4 of the 12 test cycles with the application of $NN_{br[2X]U}$ soft sensor.

4.4. Comparison between Methodologies Applied

Using the weighting and ranking method and comparing R_{lin} to R_{reg} for each variable set, it was observed that R_{lin} was marginally better than R_{reg} (in this comparison R_{lin} performed better for 54.5% of the model/variable set combinations). A similar comparison was carried out comparing individual variable sets for the three sets of hidden layer neuron models for NN_{lm} ($NN_{lm[X]}$, $NN_{lm[0.5X]}$ and $NN_{lm[2X]}$) and NN_{br} ($NN_{br[X]}$, $NN_{br[0.5X]}$ and $NN_{br[2X]}$). For both NN_{lm} (77.3% of total number of variable sets) and NN_{br} (45.5%), 0.5X was most efficient, while 2X was least efficient (performed best for only 4.5% and 18.2% model/variable set combinations), for NN_{lm} and NN_{br}, respectively. Bearing this in mind, and comparing NN_{lm} against NN_{br} for 0.5X, the non-regularised model, NN_{lm} (68.2%), was the better performing NN model. A further comparison was carried out to compare the leading NN ($NN_{lm[0.5X]}$) and regression (R_{lin}) models for individual input variables.

This showed that R_{lin} performed better in 54.5% of variable sets. Alternatively, a study of the final ranked results (Table 9) shows that three of the top four ranked soft sensors use the NN_{br} model; therefore, for future applications, it may be possible to use this model only. This result suggests that regularisation has indeed helped to avoid some over-fitting suffered by the unregularised NN_lm models. Table 10 compares each variable set for each soft sensor. The aggregate of variable set rank gives an indication of overall variable set performance (when compared to other models).

Table 10. Ranking results for each variable set model for each soft sensor.

Soft Sensor	R_{lin}	R_{reg}	$NN_{lin[X]}$	$NN_{lin[0.5X]}$	$NN_{lin[2X]}$	$NN_{br[X]}$	$NN_{br[0.5X]}$	$NN_{br[2X]}$
A	3	1	5	2	6	7	4	8
B	1	3	5	2	8	7	4	6
C	1	2	4	2	8	6	7	5
D	4	3	7	1	5	6	2	8
E	4	2	7	1	4	3	6	8
F	2	4	6	1	5	3	7	8
G	2	3	7	1	4	6	5	8
H	4	5	7	1	2	8	3	6
I	4	2	7	1	3	6	8	5
J	4	1	8	2	6	5	3	7
K	2	3	8	5	1	7	3	5
L	7	2	8	1	3	4	6	5
M	8	6	3	1	2	5	4	7
N	7	4	5	2	6	7	1	3
O	8	2	1	3	7	4	5	6
P	1	3	7	2	3	5	6	8
Q	1	4	5	2	6	3	7	8
R	6	3	5	1	7	2	4	8
S	3	1	6	5	7	4	2	8
T	8	3	6	4	6	1	2	5
U	3	6	2	4	8	5	7	1
V	4	6	5	3	6	8	2	1
Sum	87	692	124	47	113	112	98	64
Rank	3	2	7	1	6	5	4	8

A similar study comparing variable sets (Table A9) identified the top three variable sets as T ($pH_{nadir\text{-}apex}$ and $ORP_{nadir\text{-}apex}$), V (pH_{ma20} and $pH_{nadir\text{-}apex}$) and M (ORP_{cum} and $ORP_{nadir\text{-}apex}$)—each of these used only two input variables, suggesting that simpler variable sets can lead to better models. The nadir-apex input variable seems particularly useful, and more generally the processed input variables were clearly providing added value to the numerical modelling.

5. Discussion

As detailed in the results, soft sensors selected using NNs and regression models, in this case the $NN_{br[2X]U}$ soft sensor, have the potential to generate large operational savings such as reduced treatment cycle duration and reduced electricity usage, whilst also meeting discharge requirements. This study was conducted in a small-scale WWTP, using a suite of pH and ORP variables (i.e., variables identified from both pH and ORP profile characteristics in the SBR). Several studies have demonstrated that ORP and pH sensors can act as surrogates for NH_4-N sensors [15,25–29,31]; however, the implementation of these results at small-scale WWTPs is limited, and many of these studies did not look at pH and ORP sensors in a combined manner.

For the task at hand, the use of the NN training (optimisation) method was quite standard. The main advantage of the linear regression model was interpretability. The effect of each variable on the output of the model was easy to understand. Neural network models are often able to fit data better at the cost of interpretability. However, neural network models can be interrogated and visualised to give a good understanding of their effect.

The motivation for using Bayesian regularisation was to help avoid over-fitting. Over-fitting is the scenario where the model fits the training data well but fails to generalise to unseen data. Regularisation pushes the model towards a simpler form which may fit the training data slightly less but is more likely to generalise.

Wastewater pollutant concentration datasets are suitable for application in NNs as they have a large number of inputs, each of which can vary significantly. In addition, given the 24/7 nature of wastewater treatment, large datasets can be collected from wastewater sensors, which can improve NN suitability even further. However, as discussed in Section 3 of this paper, NNs must be carefully designed and trained to ensure that the outputs are suitable for use in real-time control applications. Given the black box nature of NNs, careful attention is required when assessing input variables, selecting models and assign rankings.

The methodology proposed in this paper creates an opportunity for WWTPs utilising SBRs (and indeed any WWTP utilising other batch treatment processes) to select their own custom soft sensor to optimise on-site treatment processes. In addition, the methodology can be repeated over time in WWTPs to adapt to any significant on-site changes such as, substantial changes in influent wastewater constitution due to the connection of new wastewater sources, etc. However, it can be labour intensive to apply the methodology in a new site, particularly if it is difficult to source the database of parameters required to train the model. To assist with this, further research on this topic would include the application of the best sensor across a larger number of site-based systems, and further adaptation to enable control of biological nitrogen and phosphorous removal where required. Recent work investigated the prediction of N and P removal in municipal wastewater using microalgae modelling response surface methodology, multilayer perceptron artificial neural network and support vector regression [66]. However, despite this and other recent work there is a need to focus on robust methods for system control.

RTC using soft sensors offers many benefits from a managerial perspective. Improved treatment efficacy (in terms of discharge compliance) can be achieved in a more consistent manner without the need for manual intervention by WWTP operatives, whilst electrical energy savings can ease the burden in terms of financial management and assist with meeting targets such as the EU Energy Efficiency Directive (EED). As the equipment required for this methodology is economical, readily available, and easy to use, highly skilled operators are not required to apply the technology, the capital and operating costs are low which enhances sustainability of the technology in smaller WWTPs.

RTC may also be particularly advantageous in WWTPs which are subject to changing loadings due to seasonal changes in tourism, which can lead to seasonal, weekly or daily fluctuations, both hydraulically and organically, which can be difficult to manage. The technology could also be used to extend the duration of treatment cycles to ensure discharge compliance in the instance where a WWTP may be over-loaded in terms of pollutant load (dependent on site-specific conditions such as upstream wastewater storage provisions and other operational considerations allowing for extended cycle times), or reduce the treatment cycle duration to the minimum time required to meet discharge regulations, which can allow a WWTP to treat additional hydraulic load, if required.

6. Conclusions and Outlook

This research presents a methodology for enabling real-time control of NH_4-N removal in wastewater treatment systems. The methodology was developed using a case-study SBR system treating residential wastewater. MLR and NN techniques were used and compared to develop suitable soft sensors that could enable RTC of wastewater treatment systems. This study also presented a method for selecting the optimal soft sensor based on the specific outcomes required at any site.

The estimating models' studies included linear regression (R_{lin}) and regularised linear regression (R_{reg}) and NN models leveraging Levenberg–Marquardt back-propagation (NN_{lm}) and Levenberg–Marquardt back-propagation with Bayesian regularisation (NN_{br}). The impact of neuron numbers in each NN model was also analysed. It was determined that for a typical treatment cycle, the best preforming soft sensor, using the site-specific criteria at this site (which heavily weighted accuracy in effluent NH_4-N concentration prediction) used Bayesian regularisation and would achieve an average treatment time saving of 67%, resulting in an average energy saving of 51% of electricity costs. The controlled treatment cycle would achieve 88% NH_4-N removal when compared to the fixed time treatment cycle but, significantly, ensured discharges remained within the threshold discharge concentration set. These results highlight how the methodology can provide a level of targeted control, which can significantly improve the sustainability of wastewater treatment by balancing the needs of safe discharge and efficient energy usage.

The methodology proposed to determine the most efficient soft sensor for any given site can allow a more targeted approach to enable a site to adapt as on-site considerations change. The models studied can be implemented on basic programmable logic controllers typically used for small-scale SBR systems, making the methodology suitable even in small WWTPs with limited resources. The methodology also has the potential to be applied to existing SBRs, making it a cost-effective option for process upgrade works in existing WWTPs.

One limitation of this research is that the methodology is focused specifically on SBRs. There is additional potential for the procedure to be modified to suit other technologies; in particular, systems that treat wastewater in batches. Further research on this topic would include the application of the best sensor across a larger number of site-based systems and further adaptation to enable control of biological nitrogen and phosphorous removal where required.

Author Contributions: Conceptualisation, E.C. and S.F.; methodology, E.C., S.F., J.M., E.D. and R.C.; formal analysis, S.F., E.D., R.C., E.C. and J.M.; investigation, S.F., R.C., E.D., J.M. and E.C.; data curation, S.F., R.C. and E.D.; writing—original draft preparation, S.F., E.D., R.C., E.C. and J.M.; supervision, E.C. and J.M.; funding acquisition, E.C. All authors have read and agreed to the published version of the manuscript.

Funding: The authors wish to acknowledge the support of Molloy Environmental Systems; the Energy Systems Integration Partnership Programme is financially supported by Science Foundation Ireland under the SFI Strategic Partnership Programme (Grant Number SFI/15/SPP/E3125) and the Irish Research Council.

Institutional Review Board Statement: Not applicable.

Informed Consent Statement: Not applicable.

Data Availability Statement: Additional data have been provided (Tables A1–A9). Other raw data are available on request.

Conflicts of Interest: The authors declare no conflict of interest.

Appendix A

Table A1. R_{lin} results.

Soft Sensor	Average R^2	Average R^2 in Last 200 Min	Average RMSE	Average RMSE in Last 200 Min	Averages at 2 mg NH_4-N/l Trigger			
					NH_{4rem} (%)	T_{save} (%)	ab_{error} (mg/L)	SC
R_{linA}	0.744	0.639	1.229	0.62	81	59	0.66	12
R_{linB}	0.779	0.515	1.273	0.506	86	58	0.64	12
R_{linC}	0.777	0.466	1.315	0.498	85	53	0.66	12
R_{linD}	0.777	0.465	1.313	0.496	82	57	0.99	11
R_{linE}	0.647	0.709	1.246	0.614	76	66	0.81	10
R_{linF}	0.725	0.554	1.234	0.55	77	60	0.62	11

Table A1. *Cont.*

Soft Sensor	Average R^2	Average R^2 in Last 200 Min	Average RMSE	Average RMSE in Last 200 Min	Averages at 2 mg NH_4-N/l Trigger			
					NH_{4rem} (%)	T_{save} (%)	ab_{error} (mg/L)	SC
R_{linG}	0.639	0.746	1.28	0.584	80	61	0.31	11
R_{linH}	0.773	0.553	1.32	0.5	86	53	0.58	12
R_{linI}	0.721	0.582	1.349	0.63	77	50	0.88	11
R_{linJ}	0.598	0.641	1.515	0.907	87	27	1.41	12
R_{linK}	0.765	0.867	1.498	0.627	93	40	0.82	12
R_{linL}	0.766	0.594	1.337	0.568	79	56	1.19	11
R_{linM}	0.364	0.397	1.634	0.949	97	14	1.31	12
R_{linN}	0.334	0.699	1.664	0.98	97	14	1.39	12
R_{linO}	0.351	0.683	1.776	0.913	20	86	3.67	5
R_{linP}	0.639	0.746	1.25	0.584	80	61	0.31	11
R_{linQ}	0.696	0.834	1.509	0.676	93	36	0.78	12
R_{linR}	0.779	0.635	1.32	0.544	80	55	1.17	11
R_{linS}	0.779	0.628	1.339	0.581	88	50	0.69	12
R_{linT}	0.493	0.742	1.455	0.604	62	69	1.17	10
R_{linU}	0.739	0.866	1.463	0.577	89	47	0.65	12
R_{linV}	0.78	0.868	1.399	0.528	91	43	0.75	12

Table A2. R_{reg} results.

Soft Sensor	Average R^2	Average R^2 in Last 200 Min	Average RMSE	Average RMSE in Last 200 Min	Averages at 2 mg NH_4-N/l Trigger			
					NH_{4rem} (%)	T_{save} (%)	ab_{error} (mg/L)	SC
R_{regA}	0.745	0.633	1.198	0.581	82	59	61	12
R_{regB}	0.714	0.502	1.233	0.554	75	61	72	11
R_{regC}	0.72	0.492	1.238	0.559	77	60	68	11
R_{regD}	0.727	0.507	1.232	0.547	77	60	66	11
R_{regE}	0.729	0.544	1.232	0.546	77	59	65	11
R_{regF}	0.762	0.661	1.195	0.553	78	61	100	11
R_{regG}	0.748	0.521	1.254	0.504	83	55	50	12
R_{regH}	0.79	0.665	1.226	0.448	78	59	97	11
R_{regI}	0.727	0.507	1.232	0.547	78	60	61	11
R_{regJ}	0.732	0.872	1.471	0.664	91	36	82	12
R_{regK}	0.789	0.837	1.315	0.523	88	50	84	12
R_{regL}	0.782	0.646	1.234	0.479	85	55	58	12
R_{regM}	0.698	0.853	1.495	0.64	92	37	78	12
R_{regN}	0.67	0.854	1.533	0.665	91	38	86	12
R_{regO}	0.693	0.853	1.497	0.642	92	37	77	12
R_{regP}	0.748	0.521	1.255	0.505	79	56	49	11
R_{regQ}	0.791	0.843	1.358	0.693	88	43	84	12
R_{regR}	0.727	0.534	1.219	0.563	84	55	56	12
R_{regS}	0.741	0.567	1.263	0.546	85	52	58	12
R_{regT}	0.772	0.866	1.405	0.524	93	43	74	12
R_{regU}	0.822	0.853	1.321	0.558	88	46	84	12
R_{regV}	0.817	0.869	1.336	0.551	87	47	87	12

Table A3. $NN_{lm[X]}$ results.

Soft Sensor	Average R^2	Average R^2 in Last 200 Min	Average RMSE	Average RMSE in Last 200 Min	Averages at 2 mg NH_4-N/l Trigger			
					NH_{4rem} (%)	T_{save} (%)	ab_{error} (mg/L)	SC
$NN_{lm[X]A}$	0.552	0.568	1.255	0.547	56	63	1.02	11
$NN_{lm[X]B}$	0.546	0.441	1.347	0.314	50	64	1.62	10
$NN_{lm[X]C}$	0.53	0.44	1.139	0.336	55	69	0.86	10
$NN_{lm[X]D}$	0.489	0.273	1.201	0.429	55	65	1.55	11
$NN_{lm[X]E}$	0.265	0.301	1.05	0.288	60	67	1.29	11
$NN_{lm[X]F}$	0.512	0.451	1.055	0.481	54	62	1.80	11
$NN_{lm[X]G}$	0.626	0.422	1.039	0.396	60	67	1.13	10
$NN_{lm[X]H}$	0.639	0.555	1.027	0.372	56	70	2.56	10
$NN_{lm[X]I}$	0.47	0.431	1.33	0.7	49	67	1.87	10
$NN_{lm[X]J}$	0.42	0.436	1.698	0.845	53	60	2.85	11
$NN_{lm[X]K}$	0.711	0.589	1.214	0.481	66	65	2.31	10
$NN_{lm[X]L}$	0.649	0.642	0.142	0.548	65	65	6.49	9
$NN_{lm[X]M}$	0.732	0.669	1.438	0.494	91	38	0.66	12
$NN_{lm[X]N}$	0.705	0.764	1.379	0.661	92	34	0.95	12
$NN_{lm[X]O}$	0.658	0.675	1.302	0.464	91	37	0.69	12
$NN_{lm[X]P}$	0.431	0.509	1.165	0.439	52	62	1.13	11
$NN_{lm[X]Q}$	0.514	0.516	1.165	0.552	68	56	1.08	11
$NN_{lm[X]R}$	0.528	0.565	1.094	0.517	67	63	0.78	11
$NN_{lm[X]S}$	0.67	0.248	1.147	0.583	67	65	0.78	12
$NN_{lm[X]T}$	0.721	0.663	1.368	0.488	77	56	0.89	12
$NN_{lm[X]U}$	0.539	0.401	1.084	0.374	63	61	0.63	12
$NN_{lm[X]V}$	0.619	0.44	1.327	0.532	75	56	0.55	11

Table A4. $NN_{lm[0.5X]}$ results.

Soft Sensor	Average R^2	Average R^2 in Last 200 Min	Average RMSE	Average RMSE in Last 200 Min	Averages at 2 mg NH_4-N/l Trigger			
					NH_{4rem} (%)	T_{save} (%)	ab_{error} (mg/L)	SC
$NN_{lm[0.5X]A}$	0.623	0.615	1.135	0.532	58	56	1.72	11.00
$NN_{lm[0.5X]B}$	0.426	0.328	0.947	0.39	52	67	1.88	11.00
$NN_{lm[0.5X]C}$	0.635	0.393	0.1093	0.39	55	67	1.56	11.00
$NN_{lm[0.5X]D}$	0.57	0.439	0.8965	0.294	45	68	1.26	10.00
$NN_{lm[0.5X]E}$	0.671	0.327	1.05	0.374	59	67	1.03	12.00
$NN_{lm[0.5X]F}$	0.643	0.577	1.309	0.523	52	67	1.57	11.00
$NN_{lm[0.5X]G}$	0.662	0.442	1.142	0.428	67	65	0.85	11.00
$NN_{lm[0.5X]H}$	0.818	0.717	1.107	0.379	65	66	0.86	11.00
$NN_{lm[0.5X]I}$	0.68	0.444	1.025	0.476	54	70	0.99	11.00
$NN_{lm[0.5X]J}$	0.71	0.779	1.445	0.727	54	73	2.63	12.00
$NN_{lm[0.5X]K}$	0.752	0.617	1.087	0.494	64	60	0.80	12.00
$NN_{lm[0.5X]L}$	0.79	0.68	1.233	0.489	66	66	0.69	11.00
$NN_{lm[0.5X]M}$	0.708	0.634	1.392	0.457	91	40	0.71	12.00
$NN_{lm[0.5X]N}$	0.775	0.681	1.406	0.538	74	55	1.03	11.00
$NN_{lm[0.5X]O}$	0.772	0.764	1.498	0.653	54	67	6.36	10.00
$NN_{lm[0.5X]P}$	0.692	0.404	1.2	0.453	56	69	1.11	11.00
$NN_{lm[0.5X]Q}$	0.526	0.283	1.269	0.523	37	74	1.80	8.00
$NN_{lm[0.5X]R}$	0.672	0.465	1.044	0.457	51	68	1.73	10.00

Table A4. *Cont.*

Soft Sensor	Average R^2	Average R^2 in Last 200 Min	Average RMSE	Average RMSE in Last 200 Min	Averages at 2 mg NH_4-N/l Trigger			
					NH_{4rem} (%)	T_{save} (%)	ab_{error} (mg/L)	SC
$NN_{lm[0.5X]S}$	0.58	0.26	1.129	0.619	53	71	1.58	10.00
$NN_{lm[0.5X]T}$	0.72	0.752	1.377	0.496	76	56	0.97	12.00
$NN_{lm[0.5X]U}$	0.6	0.547	1.096	0.43	61	69	1.20	11.00
$NN_{lm[0.5X]V}$	0.775	0.88	1.113	0.488	52	63	1.15	11.00

Table A5. $NN_{lm[2X]}$ results.

Soft Sensor	Average R^2	Average R^2 in Last 200 Min	Average RMSE	Average RMSE in Last 200 Min	Averages at 2 mg NH_4-N/l Trigger			
					NH_{4rem} (%)	T_{save} (%)	ab_{error} (mg/L)	SC
$NN_{lm[2X]A}$	0.557	0.438	1.497	0.877	58	56	1.72	11.00
$NN_{lm[2X]B}$	0.485	0.145	1.613	0.904	52	67	1.88	11.00
$NN_{lm[2X]C}$	0.442	0.445	1.448	0.613	55	67	1.56	11.00
$NN_{lm[2X]D}$	0.447	0.245	1.369	0.542	45	68	1.26	10.00
$NN_{lm[2X]E}$	0.609	0.376	1.301	0.499	59	67	1.03	12.00
$NN_{lm[2X]F}$	0.509	0.385	1.252	0.55	52	67	1.57	11.00
$NN_{lm[2X]G}$	0.571	0.491	1.145	0.446	67	65	0.85	11.00
$NN_{lm[2X]H}$	0.64	0.555	1.008	0.416	65	66	0.86	11.00
$NN_{lm[2X]I}$	0.671	0.387	1.016	0.478	54	70	0.99	11.00
$NN_{lm[2X]J}$	0.435	0.38	2.182	1.3	54	73	2.63	12.00
$NN_{lm[2X]K}$	0.601	0.653	1.102	0.441	64	60	0.80	12.00
$NN_{lm[2X]L}$	0.64	0.433	1.062	0.47	66	66	0.69	11.00
$NN_{lm[2X]M}$	0.71	0.702	1.361	0.62	91	40	0.71	12.00
$NN_{lm[2X]N}$	0.62	0.631	2.083	1.254	74	55	1.03	11.00
$NN_{lm[2X]O}$	0.409	0.594	4.09	3.405	54	67	6.36	10.00
$NN_{lm[2X]P}$	0.528	0.42	1.126	0.381	56	69	1.11	11.00
$NN_{lm[2X]Q}$	0.466	0.409	1.4	0.576	37	74	1.80	8.00
$NN_{lm[2X]R}$	0.508	0.313	1.403	0.815	51	68	1.73	10.00
$NN_{lm[2X]S}$	0.493	0.342	1.308	0.703	53	71	1.58	10.00
$NN_{lm[2X]T}$	0.646	0.32	1.337	0.478	76	56	0.97	12.00
$NN_{lm[2X]U}$	0.607	0.649	1.181	0.498	61	69	1.20	11.00
$NN_{lm[2X]V}$	0.455	0.465	1.222	0.469	52	63	1.15	11.00

Table A6. $NN_{br[X]}$ results.

Soft Sensor	Average R^2	Average R^2 in Last 200 Min	Average RMSE	Average RMSE in Last 200 Min	Averages at 2 mg NH_4-N/l Trigger			
					NH_{4rem} (%)	T_{save} (%)	ab_{error} (mg/L)	SC
$NN_{br[X]A}$	0.46	0.288	1.368	0.996	50	67	1.32	12
$NN_{br[X]B}$	0.514	0.515	1.265	0.431	54	64	2.40	11
$NN_{br[X]C}$	0.59	0.378	1.078	0.341	61	68	1.26	11
$NN_{br[X]D}$	0.651	0.395	1.29	0.417	57	58	1.29	10
$NN_{br[X]E}$	0.529	0.329	0.999	0.268	61	62	0.86	11
$NN_{br[X]F}$	0.559	0.421	0.942	0.424	55	69	0.93	10
$NN_{br[X]G}$	0.665	0.667	1.063	0.407	63	66	1.19	11
$NN_{br[X]H}$	0.698	0.457	1.363	0.465	61	68	7.58	10

Table A6. *Cont.*

Soft Sensor	Average R^2	Average R^2 in Last 200 Min	Average RMSE	Average RMSE in Last 200 Min	Averages at 2 mg NH_4-N/l Trigger			
					NH_{4rem} (%)	T_{save} (%)	ab_{error} (mg/L)	SC
$NN_{br[X]I}$	0.588	0.547	1.089	0.522	61	64	1.96	10
$NN_{br[X]J}$	0.548	0.556	1.833	0.796	79	51	1.44	12
$NN_{br[X]K}$	0.65	0.679	1.044	0.443	64	66	0.92	11
$NN_{br[X]L}$	0.622	0.661	1.183	0.502	75	61	0.74	11
$NN_{br[X]M}$	0.762	0.721	1.392	0.461	90	44	0.79	12
$NN_{br[X]N}$	0.702	0.598	1.452	0.702	92	39	1.03	12
$NN_{br[X]O}$	0.607	0.286	1.365	0.578	89	46	0.71	12
$NN_{br[X]P}$	0.559	0.469	0.992	0.389	60	59	0.76	12
$NN_{br[X]Q}$	0.499	0.462	1.182	0.541	70	57	0.92	12
$NN_{br[X]R}$	0.52	0.484	1.18	0.5	57	67	0.74	11
$NN_{br[X]S}$	0.594	0.276	1.057	0.57	75	64	0.64	11
$NN_{br[X]T}$	0.715	0.584	1.327	0.346	73	56	0.60	11
$NN_{br[X]U}$	0.533	0.469	0.935	0.358	63	57	0.67	11
$NN_{br[X]V}$	0.72	0.638	1.024	0.395	68	66	1.00	11

Table A7. $NN_{br[0.5X]}$ results.

Soft Sensor	Average R^2	Average R^2 in Last 200 Min	Average RMSE	Average RMSE in Last 200 Min	Averages at 2 mg NH_4-N/l Trigger			
					NH_{4rem} (%)	T_{save} (%)	ab_{error} (mg/L)	SC
$NN_{br[0.5X]A}$	0.54	0.477	0.975	0.4	54	64	0.68	10
$NN_{br[0.5X]B}$	0.542	0.424	1.009	0.225	47	73	1.17	9
$NN_{br[0.5X]C}$	0.595	0.539	1.188	0.466	58	68	1.48	11
$NN_{br[0.5X]D}$	0.682	0.374	1.042	0.396	72	65	0.84	11
$NN_{br[0.5X]E}$	0.562	0.638	0.946	0.269	58	69	1.04	11
$NN_{br[0.5X]F}$	0.59	0.551	1.07	0.448	50	69	1.85	10
$NN_{br[0.5X]G}$	0.642	0.559	1.064	0.431	71	63	1.01	11
$NN_{br[0.5X]H}$	0.815	0.675	1.124	0.382	69	66	0.85	11
$NN_{br[0.5X]I}$	0.593	0.451	1.244	0.637	55	65	2.03	10
$NN_{br[0.5X]J}$	0.713	0.636	1.523	0.722	94	38	1.04	12
$NN_{br[0.5X]K}$	0.778	0.554	1.092	0.418	79	59	0.86	12
$NN_{br[0.5X]L}$	0.732	0.638	1.198	0.588	68	64	1.18	11
$NN_{br[0.5X]M}$	0.717	0.692	1.367	0.52	92	37	0.75	12
$NN_{br[0.5X]N}$	0.787	0.843	1.356	0.589	93	41	0.79	12
$NN_{br[0.5X]O}$	0.775	0.77	1.45	0.662	90	41	0.94	12
$NN_{br[0.5X]P}$	0.622	0.679	1.033	0.437	64	61	1.02	11
$NN_{br[0.5X]Q}$	0.486	0.326	1.224	0.585	65	66	1.34	12
$NN_{br[0.5X]R}$	0.568	0.409	1.115	0.501	61	66	0.77	11
$NN_{br[0.5X]S}$	0.546	0.315	1.149	0.564	66	65	0.60	11
$NN_{br[0.5X]T}$	0.704	0.482	1.383	0.463	82	52	0.62	12
$NN_{br[0.5X]U}$	0.686	0.668	1.089	0.397	63	67	1.17	12
$NN_{br[0.5X]V}$	0.739	0.723	1.094	0.402	75	59	0.58	11

Table A8. $NN_{br[2X]}$ results.

Soft Sensor	Average R^2	Average R^2 in Last 200 Min	Average RMSE	Average RMSE in Last 200 Min	Averages at 2 mg NH_4-N/l Trigger			
					NH_{4rem} (%)	T_{save} (%)	ab_{error} (mg/L)	SC
$NN_{br[2X]A}$	0.458	0.513	2.083	1.202	61	61	2.08	10
$NN_{br[2X]B}$	0.537	0.274	1.275	0.502	56	65	1.53	10
$NN_{br[2X]C}$	0.54	0.339	1.099	0.46	61	61	1.08	11
$NN_{br[2X]D}$	0.543	0.3	1.261	0.447	44	69	1.77	10
$NN_{br[2X]E}$	0.511	0.435	1.041	0.347	52	66	1.81	10
$NN_{br[2X]F}$	0.487	0.502	1.555	0.993	53	64	1.28	10
$NN_{br[2X]G}$	0.506	0.496	1.036	0.395	47	73	1.34	10
$NN_{br[2X]H}$	0.633	0.482	1.086	0.364	64	67	1.88	11
$NN_{br[2X]I}$	0.573	0.459	1.209	0.496	54	71	1.04	11
$NN_{br[2X]J}$	0.335	0.551	2.084	1.01	42	67	2.36	10
$NN_{br[2X]K}$	0.662	0.557	1.195	0.475	67	66	0.88	11
$NN_{br[2X]L}$	0.667	0.551	1.216	0.49	72	62	0.71	11
$NN_{br[2X]M}$	0.71	0.59	1.513	0.631	92	42	1.10	12
$NN_{br[2X]N}$	0.64	0.546	1.302	0.603	90	44	0.79	12
$NN_{br[2X]O}$	0.526	0.442	1.431	0.562	67	53	1.10	10
$NN_{br[2X]P}$	0.465	0.486	1.131	0.425	44	74	1.53	9
$NN_{br[2X]Q}$	0.403	0.409	1.605	0.923	49	70	1.64	10
$NN_{br[2X]R}$	0.462	0.395	1.437	0.512	45	75	2.64	9
$NN_{br[2X]S}$	0.49	0.307	1.154	0.644	52	69	0.84	10
$NN_{br[2X]T}$	0.629	0.309	1.374	0.406	65	58	0.86	11
$NN_{br[2X]U}$	0.581	0.769	0.942	0.196	88	67	0.57	11
$NN_{br[2X]V}$	0.643	0.538	0.981	0.342	70	59	0.49	11

Table A9. Step 1 ranking results for each model.

Soft Sensor	R_{lin}	R_{reg}	$NN_{lin[X]}$	$NN_{lin[0.5X]}$	$NN_{lin[2X]}$	$NN_{br[X]}$	$NN_{br[0.5X]}$	$NN_{br[2X]}$	Sum	Rank
A	13	16	11	6	7	1	15	3	72	16
B	17	5	6	2	1	3	7	6	47	20
C	16	4	16	5	10	10	4	14	79	15
D	10	11	5	4	8	5	17	5	65	18
E	9	15	13	20	13	14	13	8	105	11
F	15	1	4	14	9	7	2	7	59	19
G	19	21	12	16	20	11	12	13	124	4
H	22	6	8	19	19	2	21	11	108	9
I	5	14	2	10	15	4	1	15	66	17
J	3	8	1	7	4	6	6	1	36	22
K	11	13	6	8	22	17	16	19	112	8
L	6	22	3	18	21	19	5	20	114	6
M	2	10	21	21	18	20	19	16	127	3
N	4	3	15	13	12	8	18	18	91	12
O	1	12	22	9	3	13	10	10	80	14
P	19	20	9	17	16	15	8	9	113	7
Q	12	2	10	1	5	9	3	4	46	21
R	8	17	16	21	2	12	10	2	88	13
S	14	18	14	12	6	16	14	12	106	10
T	7	19	18	11	17	22	20	17	131	1
U	21	9	20	3	14	21	9	22	119	5
V	18	7	19	15	11	18	22	21	131	1

References

1. Baudart, G.; Hirzel, M.; Kate, K.; Ram, P.; Shinnar, A. Lale: Consistent automated machine learning. *arXiv* **2020**, arXiv:2007.01977.
2. Brasil, J.; Macedo, M.; Lago, C.; Oliveira, T.; Júnior, M.; Oliveira, T.; Mendiondo, E. Nature-based solutions and real-time control: Challenges and opportunities. *Water* **2021**, *13*, 651. [CrossRef]
3. Hashim, H.; Ryan, P.; Clifford, E. A statistically based fault detection and diagnosis approach for non-residential building water distribution systems. *Adv. Eng. Inform.* **2020**, *46*, 101187. [CrossRef]
4. Icke, O.; van Es, D.M.; de Koning, M.F.; Wuister, J.J.; Ng, J.; Phua, K.M.; Koh, Y.K.; Chan, W.J.; Tao, G. Performance improvement of wastewater treatment processes by application of machine learning. *Water Sci. Technol.* **2020**, *82*, 2671–2680. [CrossRef] [PubMed]
5. Yang, C.; Zhang, Y.; Huang, M.; Liu, H. Adaptive dynamic prediction of effluent quality in wastewater treatment processes using partial least squares embedded with relevance vector machine. *J. Clean. Prod.* **2021**, *314*, 128076. [CrossRef]
6. Keshavarzi Arshadi, A.; Webb, J.; Salem, M.; Cruz, E.; Calad-Thomson, S.; Ghadirian, N.; Collins, J.; Diez-Cecilia, E.; Kelly, B.; Goodarzi, H.; et al. Artificial Intelligence for COVID-19 Drug Discovery and Vaccine Development. *Front. Artif. Intell.* **2020**, *3*, 65. [CrossRef]
7. Lapuschkin, S.; Wäldchen, S.; Binder, A.; Montavon, G.; Samek, W.; Müller, K.-R. Unmasking Clever Hans predictors and assessing what machines really learn. *Nat. Commun.* **2019**, *10*, 1096. [CrossRef]
8. Ong, E.; Wong, M.U.; Huffman, A.; He, Y. COVID-19 Coronavirus Vaccine Design Using Reverse Vaccinology and Machine Learning. *Front. Immunol.* **2020**, *11*, 1581. [CrossRef]
9. Chen, C.L.P.; Zhang, C.-Y. Data-intensive applications, challenges, techniques and technologies: A survey on Big Data. *Inf. Sci.* **2014**, *275*, 314–347. [CrossRef]
10. Ridzuan, N.H.A.M.; Marwan, N.F.; Khalid, N.; Ali, M.H.; Tseng, M.L. Effects of Agriculture, Renewable Energy, and Economic Growth on Carbon Dioxide Emissions: Evidence of the Environmental Kuznets Curve. *Resour. Conserv. Recycl.* **2020**, *160*, 104879. [CrossRef]
11. Khan, S.A.; Ponce, P.; Yu, Z.; Golpîra, H.; Mathew, M. Environmental Technology and Wastewater Treatment: Strategies to Achieve Environmental Sustainability. *Chemosphere* **2022**, *286*, 131532. [CrossRef] [PubMed]
12. Zuccarello, P.; Manganelli, M.; Conti, G.O.; Copat, C.; Grasso, A.; Cristaldi, A.; De Angelis, G.; Testai, E.; Stefanelli, M.; Vichi, S.; et al. Water Quality and Human Health: A Simple Monitoring Model of Toxic Cyanobacteria Growth in Highly Variable Mediterranean Hot Dry Environments. *Environ. Res.* **2021**, *192*, 110291. [CrossRef]
13. Luccarini, L.; Bragadin, G.L.; Colombini, G.; Mancini, M.; Mello, P.; Montali, M.; Sottara, D. Formal verification of wastewater treatment processes using events detected from continuous signals by means of artificial neural networks. Case study: SBR plant. *Environ. Model. Softw. Environ. Data News* **2010**, *25*, 648–660. [CrossRef]
14. Torregrossa, D.; Castellet-Viciano, L.; Hernández-Sancho, F. A data analysis approach to evaluate the impact of the capacity utilization on the energy consumption of wastewater treatment plants. *Sustain. Cities Soc.* **2019**, *45*, 307–313. [CrossRef]
15. Vives Fàbregas, M.T. *SBR Technology for Wastewater Treatment: Suitable Operational Conditions for a Nutrient Removal*; Universitat de Girona: Girona, Spain, 2004.
16. Bernardelli, A.; Marsili-Libelli, S.; Manzini, A.; Stancari, S.; Tardini, G.; Montanari, D.; Anceschi, G.; Gelli, P.; Venier, S. Real-time model predictive control of a wastewater treatment plant based on machine learning. *Water Sci. Technol.* **2020**, *81*, 2391–2400. [CrossRef]
17. Fox, S.; Cahill, M.; O'Reilly, E.; Clifford, E. Decentralized wastewater treatment using 'pumped flow biofilm reactor' (PFBR) technology. *Water Pract. Technol.* **2016**, *11*, 93–103. [CrossRef]
18. Norton, J.W. Decentralized Systems. *Water Environ. Res.* **2009**, *81*, 1440–1450. [CrossRef]
19. Di Cicco, M.R.; Masiello, A.; Spagnuolo, A.; Vetromile, C.; Borea, L.; Giannella, G.; Iovinella, M.; Lubritto, C. Real-Time Monitoring and Static Data Analysis to Assess Energetic and Environmental Performances in the Wastewater Sector: A Case Study. *Energies* **2021**, *14*, 6948. [CrossRef]
20. Ashagre, B.B.; Fu, G.; Butler, D. Automation and real-time control of urban wastewater systems: A review of the move towards sustainability: Journal of Water Supply. *Res. Technol.-Aqua* **2020**, *69*, 751–768.
21. Aguado, D.; Ribes, J.; Montoya, T.; Ferrer, J.; Seco, A. A methodology for sequencing batch reactor identification with artificial neural networks: A case study. *Comput. Chem. Eng.* **2009**, *33*, 465–472. [CrossRef]
22. Haimi, H.; Mulas, M.; Corona, F.; Vahala, R. Data-derived soft-sensors for biological wastewater treatment plants: An overview. *Environ. Model. Softw. Environ. Data News* **2013**, *47*, 88–107. [CrossRef]
23. Huang, M.; Ma, Y.; Wan, J.; Chen, X. A sensor-software based on a genetic algorithm-based neural fuzzy system for modeling and simulating a wastewater treatment process. *Appl. Soft Comput.* **2015**, *27*, 1–10. [CrossRef]
24. Huang, M.-z.; Wan, J.-q.; Ma, Y.-w.; Li, W.-j.; Sun, X.-f.; Wang, Y. A fast predicting neural fuzzy model for on-line estimation of nutrient dynamics in an anoxic/oxic process. *Bioresour. Technol.* **2010**, *101*, 1642–1651. [CrossRef] [PubMed]
25. Akın, B.S.; Ugurlu, A. Monitoring and control of biological nutrient removal in a Sequencing Batch Reactor. *Process Biochem.* **2005**, *40*, 2873–2878. [CrossRef]
26. Ga, C.H.; Ra, C.S. Real-time control of oxic phase using pH (mV)-time profile in swine wastewater treatment. *J. Hazard. Mater.* **2009**, *172*, 61–67. [CrossRef]

27. Guo, H.J.; Peng, Y.Z.; Wang, S.Y.; Zheng, Y.N.; Huang, H.J.; Ge, S.J. Effective and robust partial nitrification to nitrite by real-time aeration duration control in an SBR treating domestic wastewater. *Process Biochem.* **2009**, *44*, 979–985. [CrossRef]
28. Tanwar, P.; Nandy, T.; Ukey, P.; Manekar, P. Correlating on-line monitoring parameters, pH, DO and ORP with nutrient removal in an intermittent cyclic process bioreactor system. *Bioresour. Technol.* **2008**, *99*, 7630–7635. [CrossRef]
29. Won, S.G.; Ra, C.S. Biological nitrogen removal with a real-time control strategy using moving slope changes of pH(mV)- and ORP-time profiles. *Water Res.* **2011**, *45*, 171–178. [CrossRef]
30. De Canete, J.F.; del Saz-Orozco, P.; Baratti, R.; Mulas, M.; Ruano, A.; Garcia-Cerezo, A. Soft-sensing estimation of plant effluent concentrations in a biological wastewater treatment plant using an optimal neural network. *Expert Syst. Appl.* **2016**, *63*, 8–19. [CrossRef]
31. Kim, H.; Hao, O.J. pH and Oxidation-Reduction Potential Control Strategy for Optimization of Nitrogen Removal in an Alternating Aerobic-Anoxic System. *Water Environ. Res.* **2001**, *73*, 95–102. [CrossRef]
32. Luccarini, L.; Porrà, E.; Spagni, A.; Ratini, P.; Grilli, S.; Longhi, S.; Bortone, G. Soft sensors for control of nitrogen and phosphorus removal from wastewaters by neural networks. *Water Sci. Technol.* **2002**, *45*, 101–107. [CrossRef] [PubMed]
33. Li, J.; Ni, Y.; Peng, Y.; Guowei, G.; Jingen, L.; Su, W.; Guobiao, C.; Changjin, O. On-line controlling system for nitrogen and phosphorus removal of municipal wastewater in a sequencing batch reactor (SBR). *Front. Environ. Sci. Eng.* **2008**, *2*, 99–102. [CrossRef]
34. Villez, K.; Sin, G.; Vanrolleghem, P.A.; Ruiz, M.; Colomer, J.; Rosén, C.; Vanrolleghem, P.A. Combining multiway principal component analysis (MPCA) and clustering for efficient data mining of historical data sets of SBR processes. *Water Sci Technol.* **2008**, *57*, 1659–1666. [CrossRef] [PubMed]
35. Yang, Q.; Gu, S.; Peng, Y.; Wang, S.; Liu, X. Progress in the Development of Control Strategies for the SBR Process. *Clean Soil Air Water* **2010**, *38*, 732–749. [CrossRef]
36. Khataee, A.; Kasiri, M. Modeling of Biological Water and Wastewater Treatment Processes Using Artificial Neural Networks. *CLEAN–Soil Air Water* **2011**, *39*, 742–749. [CrossRef]
37. Kocijan, J.; Hvala, N. Sequencing batch-reactor control using Gaussian-process models. *Bioresour. Technol.* **2013**, *137*, 340–348. [CrossRef]
38. Puig, S.; Corominas, L.; Adama, T.; Colomer; Balaguer, M.; Colprim, J. An On-line Optimisation of a SBR Cycle for Carbon and Nitrogen Removal Based on On-line pH and OUR: The Role of Dissolved Oxygen Control. *Water Sci. Technol.* **2006**, *53*, 171–178. [CrossRef]
39. Hong, S.H.; Lee, M.W.; Lee, D.S.; Park, J.M. Monitoring of sequencing batch reactor for nitrogen and phosphorus removal using neural networks. *Biochem. Eng. J.* **2007**, *35*, 365–370. [CrossRef]
40. Casellas, M.; Dagot, C.; Baudu, M. Set up and assessment of a control strategy in a SBR in order to enhance nitrogen and phosphorus removal. *Process Biochem.* **2006**, *41*, 1994–2001. [CrossRef]
41. Mingzhi, H.; Jinquan, W.; Yongwen, M.; Yan, W.; Weijiang, L.; Xiaofei, S. Control rules of aeration in a submerged biofilm wastewater treatment process using fuzzy neural networks. *Expert Syst. Appl.* **2009**, *36*, 10428–10437. [CrossRef]
42. Han, H.-G.; Li, Y.; Guo, Y.-N.; Qiao, J.-F. A soft computing method to predict sludge volume index based on a recurrent self-organizing neural network. *Appl. Soft Comput.* **2016**, *38*, 477–486. [CrossRef]
43. Li, F.; Qiao, J.; Han, H.; Yang, C. A self-organizing cascade neural network with random weights for nonlinear system modeling. *Appl. Soft Comput.* **2016**, *42*, 184–193. [CrossRef]
44. Syafiie, S.; Tadeo, F.; Martinez, E.; Alvarez, T. Model-free control based on reinforcement learning for a wastewater treatment problem. *Appl. Soft Comput.* **2011**, *11*, 73–82. [CrossRef]
45. Wan, J.; Huang, M.; Ma, Y.; Guo, W.; Wang, Y.; Zhang, H.; Li, W.; Sun, X. Prediction of effluent quality of a paper mill wastewater treatment using an adaptive network-based fuzzy inference system. *Appl. Soft Comput.* **2011**, *11*, 3238–3246. [CrossRef]
46. Foscoliano, C.; Del Vigo, S.; Mulas, M.; Tronci, S. Predictive control of an activated sludge process for long term operation. *Chem. Eng. J.* **2006**, *304*, 1031–1044. [CrossRef]
47. Mulas, M.; Corona, F.; Sirviö, J.; Hyvönen, S.; Vahala, R. Full-scale implementation of an advanced control system on a biological wastewater treatment plant. *IFAC-Pap.* **2016**, *49*, 1163–1168. [CrossRef]
48. James, G.; Witten, D.; Hastie, T.; Tibshirani, R. *An Introduction to Statistical Learning*; Springer: New York, NY, USA, 2013; Volume 112.
49. Souza, F.A.A.; Araújo, R.; Mendes, J. Review of soft sensor methods for regression applications. *Chemom. Intell. Lab. Syst.* **2016**, *152*, 69–79. [CrossRef]
50. Abyaneh, H.Z. Evaluation of multivariate linear regression and artificial neural networks in prediction of water quality parameters. *J. Environ. Health Sci. Eng.* **2014**, *12*, 40. [CrossRef]
51. Nasr, M.S.; Moustafa, M.A.E.; Seif, H.A.E.; El Kobrosy, G. Application of Artificial Neural Network (ANN) for the prediction of EL-AGAMY wastewater treatment plant performance-EGYPT. *Alex. Eng. J.* **2012**, *51*, 37–43. [CrossRef]
52. Arslan, O.; Yetik, O. ANN based optimization of supercritical ORC-Binary geothermal power plant: Simav case study. *Appl. Therm. Eng.* **2011**, *31*, 3922–3928. [CrossRef]
53. Hagan, M.T.; Menhaj, M.B. Training feedforward networks with the Marquardt algorithm. *IEEE Trans. Neural Netw.* **1994**, *5*, 989–993. [CrossRef] [PubMed]

54. Nawi, N.M.; Khan, A.; Rehman, M.Z. CSLM: Levenberg marquardt based back propagation algorithm optimized with cuckoo search. *J. ICT Res. Appl.* **2013**, *7*, 103–116. [CrossRef]
55. Rahimi-Ajdadi, F.; Abbaspour-Gilandeh, Y. Artificial Neural Network and stepwise multiple range regression methods for prediction of tractor fuel consumption. *Meas. J. Int. Meas. Confed.* **2011**, *44*, 2104–2111. [CrossRef]
56. Wagh, V.; Panaskar, D.; Muley, A.; Mukate, S.; Gaikwad, S. Neural network modelling for nitrate concentration in groundwater of Kadava River basin, Nashik, Maharashtra, India. *Groundw. Sustain. Dev.* **2018**, *7*, 436–445. [CrossRef]
57. Dan Foresee, F.; Hagan, M.T. Gauss-Newton approximation to Bayesian learning. In Proceedings of the International Conference on Neural Networks (ICNN'97), Houston, TX, USA, 12 June 1997; Volume 3, pp. 1930–1935.
58. MacKay, D.J.C. Bayesian interpolation. *Neural Comput.* **1992**, *4*, 415–447. [CrossRef]
59. Sharma, A.K.; Sharma, R.K.; Kasana, H.S. Prediction of first lactation 305-day milk yield in Karan Fries dairy cattle using ANN modeling. *Appl. Soft Comput.* **2007**, *7*, 1112–1120. [CrossRef]
60. Tien Bui, D.; Pradhan, B.; Lofman, O.; Revhaug, I.; Dick, O.B. Landslide susceptibility assessment in the Hoa Binh province of Vietnam: A comparison of the Levenberg–Marquardt and Bayesian regularized neural networks. *Geomorphology* **2012**, *171–172*, 12–29. [CrossRef]
61. *CEN 12566-3*; Small Wastewater Treatment Systems for up to 50 PT—Part 3: Packaged and/or Site Assembled Domestic Wastewater Treatment Plants. National Standards Authority of Ireland: Dublin, Ireland, 2006; pp. 1–34.
62. Federation, Water Environmental; Aph Association. *Standard Methods for the Examination of Water and Wastewater*; Port City Press: Baltimore, MD, USA, 2005.
63. Chang, C.H.; Hao, O.J. Sequencing Batch Reactor System for Nutrient Removal: ORP and pH Profiles. *J. Chem. Technol. Biotechnol. Int. Res. Process Environ. AND Clean Technol.* **1996**, *67*, 27–38. [CrossRef]
64. Holman, J.B. The Application of pH and ORP Process Control Parameters within the Aerobic Denitrification Process. Ph.D. Thesis, University of Canterbury, Christchurch, New Zealand, 2004.
65. Bishop, C.M. *Neural Networks for Pattern Recognition*; Oxford University Press: Oxford, UK, 1995; ISBN 978-0198538646.
66. Hossain, S.Z.; Sultana, N.; Jassim, M.S.; Coskuner, G.; Hazin, L.M.; Razzak, S.A.; Hossain, M.M. Soft-Computing Modeling and Multiresponse Optimization for Nutrient Removal Process from Municipal Wastewater Using Microalgae. *J. Water Process Eng.* **2022**, *45*, 102490. [CrossRef]

sustainability MDPI

Article

Recovery of Polyhydroxyalkanoates from Cooked Mussel Processing Wastewater at High Salinity and Acidic Conditions

Alba Pedrouso *, Andrea Fra-Vazquez, Angeles Val del Rio and Anuska Mosquera-Corral

CRETUS Institute, Department of Chemical Engineering, Universidade de Santiago de Compostela, Rua Lope Gomez de Marzoa s/n, E-15782 Santiago de Compostela, Spain; andreafravazquez@gmail.com (A.F.-V.); mangeles.val@usc.es (A.V.d.R.); anuska.mosquera@usc.es (A.M.-C.)

* Correspondence: alba.pedrouso@usc.es

Received: 14 November 2020; Accepted: 8 December 2020; Published: 11 December 2020

Abstract: Polyhydroxyalkanoates (PHA) are biodegradable polymers that can be intracellularly produced by microorganisms valorizing organic-rich wastes. In the present study, a PHA production system was fed with mussel cooker wastewater after acidogenic fermentation. Besides low pH (4.0 ± 0.3) and high salt (21.7 ± 2.9 g NaCl/L) concentrations, this wastewater also contained nitrogen concentrations (0.8 ± 0.1 g N/L), which were previously reported to be a challenge to the PHA accumulating bacteria enrichment. Bacteria with a PHA storage capacity were selected in an enrichment sequencing batch reactor (SBR) after 60 days of operation. The enriched mixed microbial culture (MMC) was mainly formed by microorganisms from phylum *Bacteroidetes*, and genera *Azoarcus, Comamonas* and *Thauera* from phylum *Proteobacteria*. The MMC was able to accumulate up to 25 wt% of PHA that was mainly limited by the wastewater nitrogen content, which promoted biomass growth instead of PHA accumulation. Indeed, when the presence of nutrient was limited, PHA stored in the accumulation reactor increased to up to 40.9 wt%. This work demonstrated the feasibility of the enrichment of a MMC with a PHA storage ability valorizing the fish-canning industrial wastewater at low pH, which is generally difficult to treat in wastewater treatment plants.

Keywords: fish-canning industry; industrial wastewater valorization; low pH; mixed microbial culture; PHA; value-added products

1. Introduction

Fish and seafood processing industries consume vast volumes of water and, consequently, generate large amounts of wastewater with high organic matter, nutrients and salt concentrations. Depending on the processed fish product (tuna, mussels, etc.) and seasonal variations, the generated waste streams show different compositions [1,2]. In fact, effluents with distinct qualities and flows may be produced even in the same facility, depending on the processing steps, challenging its adequate treatment [3,4]. The high loaded wastewaters generated in these facilities often contain chemical oxygen demand (COD) concentrations up to 42 g COD/L characterized by its high complexity with not only easily-biodegradable carbohydrates but also relevant protein fractions (15–20% of wet weight), and lipids (0.1–44 g/L) [2,5].

Different biological treatment technologies have been developed to treat these effluents, focusing on the achievement of more efficiency, less energy consumption and fewer surface requirements for implantation [5]. Méndez et al. [6] achieved, in an anaerobic digester, a COD removal efficiency of 80% treating effluents generated in a factory processing different fish products with high salinity (up to 15 g Cl^-/L). Picos–Benítez et al. [7] reported that the biogas yield decreases by 64% when the salt concentration increases from 0 to 20 g NaCl/L while treating wastewater from the evisceration process in the fish processing industry. At fish and seafood canning facilities, the wastewater treatment plant (WWTP) configuration highly varies depending on the wastewater characteristics and the local discharge limits. The most common technologies are a dissolved air floatation (DAF) pretreatment to remove grease and oil followed by a biological treatment that might use anaerobic (in large industries) or aerobic systems to remove organic matter and, eventually, nitrogen. Indeed, there are fish and seafood canning facilities that treat their effluents by only applying physical-chemical treatment steps to remove grease and solids and then, the wastewater is discharged to the municipal WWTP to be polished. In the last few decades, in the frame of the circular economy concept, other added-value products were proposed as an alternative to biogas to obtain more sustainable materials. For example, fish and seafood industry wastes (as organic-rich streams) can be valorized to produce volatile fatty acids (VFA) suitable for use as a chemical platform to obtain other compounds such as biopolymers [8,9].

In search of new materials, polyhydroxyalkanoates (PHA) have appeared as an alternative to fossil fuel-based plastics due to their similar properties [8,10]. PHA constitute a group of biodegradable polymers that can be produced by microorganisms as carbon and energy reserves [11]. PHA production by pure microbial cultures are the most widely employed strategies at industrial scale due to the high accumulation capacities of single strains. However, these strategies are limited by the specific substrates and sterile conditions required, which increase the operational costs and, therefore, the final product price. Thus, mixed microbial cultures (MMC) have appeared as a promising alternative to pure cultures due to the expected reduction of PHA production costs, which is associated with the low price of the substrates (wastes) and the fact that sterile conditions are not required. The typical PHA production system using MMC comprises three stages [12]: (1) acidogenic fermentation of the raw feedstock to produce a VFA-rich effluent; (2) selection of PHA-accumulating microorganisms in an enrichment reactor; and, finally, (3) accumulation assays to maximize the PHA production.

Wastewater streams produced in the food processing industry containing high organic content appear as suitable substrates for PHA production, such as those from sugar factories [13], the brewery industry [14] and cheese whey from cheese production [15]. Therefore, PHA production using MMC represents an opportunity to recover organic carbon from wastewater. However, PHA production from complex waste streams must take into consideration certain characteristics of the wastewater, which might decrease the storage capacity of the system [8,16]. In the case of the cooked mussel processing wastewater, the presence of high NaCl concentrations has been traditionally considered as being inhibitory for aerobic [17] and anaerobic biological processes [18]. Passanha et al. [19] studied the effect of salts on a pure culture of *Cupriavidus necator* and obtained the highest PHA production with the addition of 9 g NaCl/L, which yielded 30% higher PHA than the control without salt. The effect of saline conditions over non-adapted MMC was evaluated by Palmeiro-Sánchez et al. [20] and observed that the salt provoked a decrease of the PHA production rate, with a half inhibitory concentration (IC_{50}) value close to 6 g NaCl/L. pH value is another crucial operational parameter since microbial activities are critically dependent on the pH homeostasis because most proteins and enzymes have different optimal pH range values to operate. For example, bacteria can grow at external pH ranges from 5.5 to 9.0, but they generally maintain their cytoplasmic pH in a narrow range of 7.5–7.7 [21]. Therefore, most of the microorganisms have strategies to maintain a significantly more alkaline cytoplasmic pH relative to the outside pH value.

The present study evaluates the potential of pre-fermented VFA-rich wastewater, coming from the cookers of a mussel-processing factory [9], to be used as a substrate to produce PHA. An enrichment reactor was operated with the ability to use high salt concentrations and a low pH. Main PHA-accumulating microorganisms selected in the MMC were identified. Furthermore, several fed-batch assays were carried out to test the maximum storage capacity in PHA of the microbial community selected.

2. Materials and Methods

2.1. PHA Production System

The PHA production system comprising two stages was operated. In the first stage, PHA-accumulating microorganisms were selected in an enrichment sequencing batch reactor (SBR) operated under the feast-famine regime. Then, the maximum PHA accumulation capacity of the enriched MMC was promoted in fed-batch assays.

2.1.1. Enrichment Reactor

A double jacket tubular SBR with a working volume of 2 L was operated under non-sterile and fully aerobic conditions to select a MMC enriched in PHA-accumulating bacteria. Since many microbial populations naturally produce PHA, activated sludge from a municipal WWTP was used as inoculum with an initial volatile suspended solids (VSS) concentration of 2.4 g VSS/L. The complete medium mixture was achieved by using the supply of air (6 L/min), which was introduced through a ceramic air diffuser located at the bottom of the reactor. The temperature was controlled at 30 °C by using a thermostatic bath (Techne Inc., Burlington, NJ, USA). The dissolved oxygen (DO) concentration was measured with an oxygen pocket meter provided with a membrane sensor (Hach-Lange, Loveland, CO, USA). The DO concentration and the pH value in the reactor media were not controlled. The SBR cycles lasted for 12 h and were divided into 3 phases: feeding (15 min), aerobic reaction (690 min) and withdrawal (15 min). Cycles were controlled by a programmable logic controller (PLC, Siemens S7-224 CPU). Continuous aeration (no settling stage) resulted in equal sludge retention times (SRT) and hydraulic retention times (HRT) of 1 day.

Cooked mussel processing wastewater was fermented at 35 °C to produce a VFA-rich stream (details provided in Fra-Vázquez, Pedrouso, Val del Rio and Mosquera-Corral [9]). Then, after solids separation by centrifugation, wastewater was fed to the system. The mixture of VFA in this wastewater was composed of acetic acid (HAc), propionic acid (HPr), butyriç acid (HBu) and valeric acid (HVa). The wastewater (described in Table 1) was five times diluted with tap water to achieve an optimal organic loading rate of 2.5 g COD/(L·d) to feed the PHA production system. Fluctuations in the composition of the raw wastewater fed to the previous acidified system resulted in variable acidified effluents [9]. Hence, the feeding of the enrichment reactor also showed these fluctuations. Moreover, 10 mg/L of allylthiourea (ATU) were added to the feeding to prevent nitrification.

Table 1. Average composition of the acidified mussel cookers wastewater.

Parameter	Value
pH	3.99 ± 0.30
sCOD (g/L)	13.01 ± 3.25
VFA (g COD_{VFA}/L)	4.06 ± 1.27
% HAc (as COD)	42.06 ± 7.05
% HPr (as COD)	16.43 ± 9.67
% HBu (as COD)	36.73 ± 11.69
% HVa (as COD)	6.83 ± 3.74
Carbohydrates (g/L)	0.04 ± 0.02
Proteins (g/L)	2.34 ± 0.67
Total Nitrogen (g N/L)	0.82 ± 0.14
Ammonium (g NH_4^+-N/L)	0.21 ± 0.06
NaCl (g/L)	21.69 ± 2.92

2.1.2. Accumulation Assays

Biomass samples were collected from the enrichment SBR at the end of the operational cycle to perform the accumulation assays in a 2-L fed-batch reactor (FB-R). The accumulatio experiments were also carried out at 30 °C, controlled by using a thermostatic bath (Techne Inc., Burlington, NJ, USA). The DO concentration was monitored but neither the DO concentration nor the pH value were controlled. Two types of carbon sources were tested: the acidified cooked mussel processing wastewater without dilution (Table 1) and a mixture of VFA. The latter was prepared to mimic the composition of the acidified wastewater in terms of VFA (43:7:42:8, as HAc:HPr:HBu:HVa in Cmmol fraction percentages), to serve as a basis to evaluate the effects of the complex matrix of the industrial wastewater on the accumulation experiment. Both substrates were manually added in pulses every time that an increase in the DO concentration was observed, which coincided with the complete depletion of the carbon source added in the previous pulse.

2.2. *Identification of Microbial Populations*

Fresh biomass samples were collected from the enrichment SBR. The fluorescence in situ hybridisation (FISH) technique was carried to identify the microbial populations. Biomass samples were fixed with paraformaldehyde 4% (*wt/vol*) solution according to the procedure described by Amann et al. [22]. Hybridization was performed at 46 °C for 90 min, adjusting the percentages of formamide to each probe. Bacterial cells were hybridized with several FISH probes.

General FISH probes used were: EUB338mix, for all *Bacteria*; a mix of CFB562 and CF319ab for phylum *Bacteroidetes*; and ALF1b, BET42a and GAM42a probes for classes *Alpha-*, *Beta-* and *Gamma-*, respectively, from phylum *Proteobacteria*. More specific probes were: PAE997, PAR1244, Cte, Zra23a, MZ1 and Azo644 for genera *Pseudomonas, Paracoccus, Comamonas, Zoogloea, Thauera* and *Azoarcus*, respectively. All probes were 5′ labeled by using fluorochromes FITC (Fluorescein-5-isocyanate) or Cy3 (Carbocyanine 3). DAPI (4, 6-diamidino-2-phenylindole) was used as a universal dye for the detection of all DNA in the samples. Fluorescence signals were captured using an acquisition system (Coolsnap, Roper Scientific Photometrics) coupled with an epifluorescence microscope (Axioskop 2, Zeiss, Obercochen, Germany). The semi-quantitative counting of the bacterial populations, based on the biovolume fraction, was performed with DAIME software [23].

2.3. *Analytical Methods*

Total Suspended Solids (TSS) and VSS were analyzed according to *Standard Methods for the Examination of Water and Wastewater* [24]. Liquid samples were filtered through a cellulose-ester filter of 0.45 μm of pore size (Advantec, Japan) for the quantification of total organic carbon (TOC), total nitrogen (TN), ammonium (NH_4^+), soluble COD (sCOD) [24], ions (e.g., Na^+ and Cl^-), proteins, carbohydrates and VFA concentrations. TOC and TN concentrations were determined by catalytic combustion in a TOC-L CSN analyzer (Shimadzu, Kyoto, Japan). Ammonium was determined following the methodology described by Bower and Holm–Hansen [25]. Ion chromatography (861 Advanced Compact IC system, Methrom, Herisau, Switzerland) was used to determine the concentration of Na^+ and Cl^-, among other ions, to calculate the salt concentration. VFA concentration was determined by gas chromatography (GC) (Hewlett Packard 5890 A, Palo Alto, CA, USA). Protein content was measured according to Lowry et al. [26] using bovine serum albumin (BSA, Sigma) as the standard. Carbohydrate concentration was measured following the methodology described by Loewus [27] and quantified as glucose (Sigma, St. Louis, MO, USA) equivalents.

For the quantification of the PHA content inside cells, fresh biomass samples were collected, and formaldehyde was added to stop the microbial activity. Then, samples were immediately centrifuged, frozen and freeze-dried to obtain a solid phase. The method proposed by Smolders et al. [28] was applied. The PHA sample content was analyzed by GC (6850 Series II, Agilent Tehnologies) equipped with the HP-INNOWAX detection column (Agilent, Santa Clara, CA, USA). PHA quantification was done using a commercial PHA standard (Sigma) containing 88% of hydroxybutyrate (HB) and 12% of hydroxyvalerate (HV). HB and HV are distinguished by the different retention times in the obtained chromatograph.

2.4. Calculations

Detailed calculations can be found in the Supplementary Material. The amount of PHA accumulated inside the cells was determined, on dry weight basis, as the percentage (wt%) of the measured volatile solids (VS). The HB:HV ratio was calculated as Cmmol. The active biomass (X) was obtained by subtracting the mass of the stored compounds from the VSS mass. The elemental composition of the active biomass was assumed to be $CH_{1.8}O_{0.5}N_{0.2}$ [29].

The specific VFA consumption rates (q_{VFA}, $Cmmol_{VFA}/(Cmmol_X \cdot h)$) and PHA production rates (q_{PHA}, $Cmmol_{PHA}/(Cmmol_X \cdot h)$) were determined from the maximum slopes of the curves representing the obtained experimental data, divided by the active biomass. Yields (Y) of biopolymers (PHA or separate HB and HV) on substrates ($Cmmol_{PHA}/Cmmol_{VFA}$) were obtained by dividing the corresponding production rate ($Cmmol_{PHA}/h$) by the VFA consumption rate ($Cmmol_{VFA}/h$). A similar procedure is applied for the calculation of the active biomass yield ($Cmmol_X/Cmmol_{VFA}$). The HB:HV ratio was calculated as the amount of each homopolymer divided by the total amount of PHA.

The concentration of proteins and carbohydrates as COD was calculated using the following factors: 1.5 g $COD_{protein}$/g protein and 1.1 g $COD_{carbohydrate}$/g carbohydrate [30]. The nitrogen content in proteins was assumed to be 15% of the weight.

3. Results and Discussion

3.1. Selection of PHA-Accumulating Microorganisms

3.1.1. Enrichment of the Mixed Microbial Culture

The enrichment SBR was operated for 180 days to obtain a MMC with PHA-accumulating capacity treating cooked mussel processing wastewater. This substrate showed a complex composition, which mainly consisted of proteins, carbohydrates and salt. The C/N of the feeding during the operation presented an average value of 6.5 ± 0.9.

The performance of the enrichment SBR was monitored by the feast phase length (Figure 1), which was measured by the change in the DO concentration profile. The feast phase corresponds to the period with low DO concentrations and the famine phase to the opposite situation. The shorter the feast length, the more enriched the system was. In the present study, whereas the DO concentration profile randomly varied during the first cycles, after one week of operation a clear feast-famine profile was observed, with feast length values of approximately 6 h (approximately 50% of the cycle length). Then, from day 60 of operation onwards, the feast phase length remained on average at a value of 2.9 ± 0.4 h, which was of 22% of the cycle length on day 180. Dionisi et al. [31] found that the selection of microorganisms with storage response takes place when the feast phase length is lower than approximately 20% of the overall length of the cycle. The increase of the degree of enrichment of the PHA-accumulating culture coincided with the rise of its storage capacity. The average PHA accumulation was of 4.5 ± 2.2 wt% during the first weeks and increased up to 12.8 ± 0.8 wt% from day 60 of operation onwards (Figure 1).

The obtained results indicated that the enrichment of the system was accomplished after two months of operation. However, the complex composition of the wastewater used as a substrate could prevent the further decrease of the feast length and the achievement of higher PHA accumulation values. Furthermore, the presence of different carbon sources (not only VFA but also proteins were present) probably allowed both accumulating and non-accumulating bacterial groups to coexist in the mixed culture.

The main bacterial populations present in the enrichment SBR were identified by the FISH technique. The activated sludge used as inoculum was characterized by the presence of a wide range of microbial populations but in a low relative abundance. However, once the MMC was enriched, the diversity decreased and the dominance of a few groups increased. The PHA-accumulating mixed culture was composed of two phyla: *Bacteroidetes* and *Proteobacteria*. Members of this latter were identified mainly as class *Betaproteobacteria* and in a low abundance as class *Gammaproteobacteria*. Microorganisms from genera *Azoarcus* and *Thauera* were observed as the most abundant ones (Figure 2a,b), followed by genera *Comamonas*, all of them from class *Betaproteobacteria* (Figure 2c). The presence of these microbial populations is ubiquitous in PHA-producing systems from wastewater and residual streams. Carvalho et al. [32], for example, observed that the PHA-producing community fed with fermented molasses was dominated by a combination of genera *Azoarcus*, *Thauera* and *Paraccocus*. Microorganisms affiliated to phylum *Bacteroidetes* were also identified (Figure 2d), which have been demonstrated to be able to store PHA using mixed cultures fed with different substrates [33].

The complex composition of the substrate contributed to select microbial populations that adapted to unfavorable operational parameters. Salt concentrations present in the acidified cooked mussel processing wastewater may select certain microorganisms that, despite being non-halophilic with optimal growth in less than 11 g NaCl/L, are able to tolerate high NaCl concentrations and are defined as halotolerant [34]. For example, bacteria belonging to phylum *Bacteroidetes* and *Proteobacteria* were found to be the dominant groups in an acetate-enriched MMC of a PHA production system using estuarine sediments as inoculum [35]. Certain members from class *Gammaproteobacteria* such as genera *Pseudomonas* and *Halomonas*, have been identified as halotolerant but also PHA-storing bacteria [34]. Therefore, the selection of members from phylum *Bacteroidetes* and class *Gammaproteobacteria* in the mixed culture operated with acidified cooked mussel processing wastewater correlated with the operational conditions in terms of high salt concentrations in the enrichment SBR.

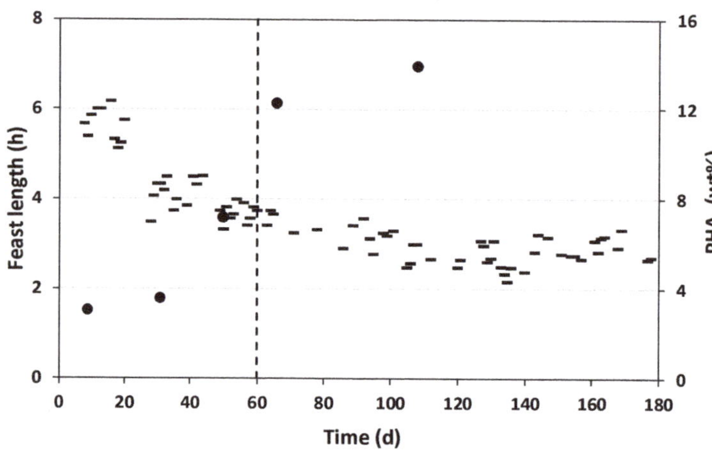

Figure 1. Evolution of the feast length during the operation of the enrichment sequencing batch reactor (SBR) (-) and maximum Polyhydroxyalkanoates (PHA) percentage accumulated at the end of the feast phase (•). The discontinuous vertical line indicates the beginning of the enrichment of the mixed culture.

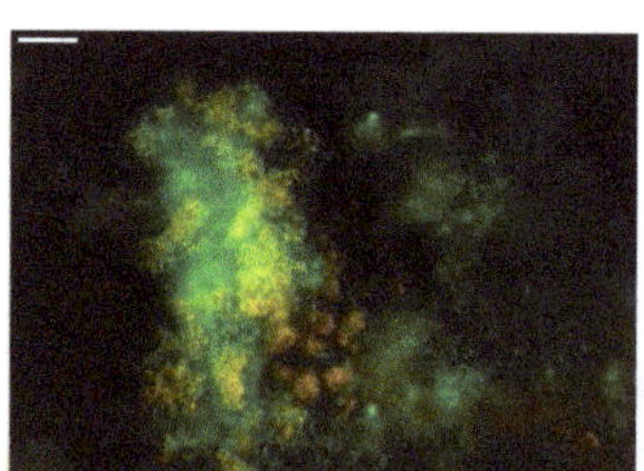

(a)

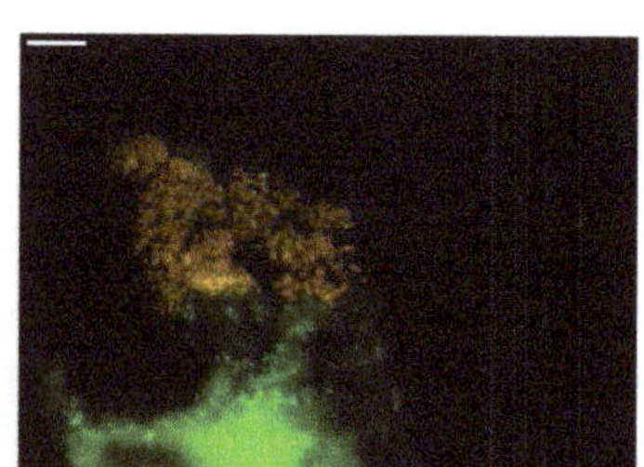

(b)

Figure 2. *Cont.*

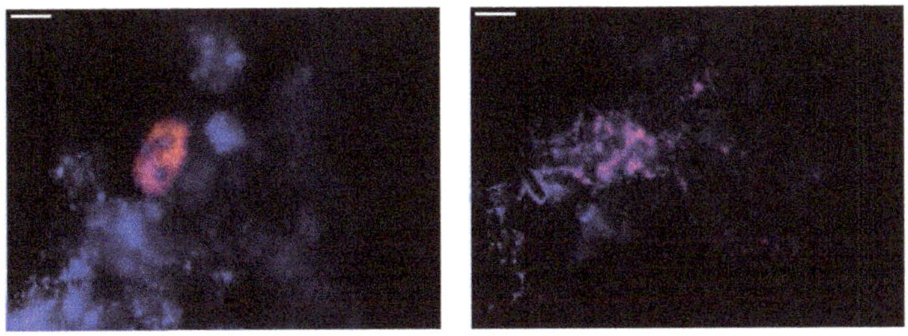

(c) (d)

Figure 2. FISH images of the mixed microbial culture (MMC) enriched with acidified cooked mussel processing wastewater. (**a**) *Azoarcus* (Cy3-red) and *Bacteria* (FITC-green). (**b**) *Thauera* (Cy3-red) and *Bacteria* (FITC-green). (**c**) *Comamonas* (Cy3-red) and all DNA (DAPI-blue). (**d**) *Bacteroidetes* (Cy3-red) and all DNA (DAPI-blue). In A and B, the orange color indicates both Cy3 and FITC-labeled probes hybridized; In (**c**,**d**), the pink color indicates both Cy3 and DAPI-labeled probes that are hybridized. The bar represents 10 μm.

3.1.2. Characterization of Enrichment Cycles

The mixed culture performance was characterized by monitoring several enrichment cycles throughout the operation of the SBR. Data from a representative operational cycle on day 66 once the steady-state conditions were achieved showed the correlation between the end of the feast phase, after approximately 2.8 h of the cycle, and the complete depletion of the VFA and the maximum PHA production (as the sum of HB and HV) of 12.3 wt% (Figure 3a). The composition of the produced PHA was 90:10 as the HB and HV ratio. Even after the complete depletion of the VFA, the soluble COD was still present in the liquid media and was mainly associated with the incomplete degradation of the proteins (Figure 3b). This enrichment cycle was carried out at a concentration of approximately 5 g/L of NaCl.

The observed yields of PHA over VFA during the feast phase on day 66 of operation were 0.21 $Cmmol_{HB}/Cmmol_{VFA}$ and 0.03 $Cmmol_{HV}/Cmmol_{VFA}$. The observed values were in the range of those obtained by other authors working with PHA production systems from complex substrates. Oliveira et al. [15] obtained a maximum PHA accumulation of 17% during the enrichment of a MMC fed with fermented cheese whey, with a production yield of 0.18 $Cmol_{PHA}/Cmol_{substrate}$. Tamis et al. [36] observed a PHA yield value of 0.37 g COD_{PHA}/g $COD_{substrate}$ using fermented wastewater from a candy bar factory, whereas Korkakaki et al. [37] employed raw leachate from the process of hydrolysis of the organic fraction of municipal solid waste and estimated a yield of 0.3 g COD_{PHA}/g COD_{VFA}.

Approximately 60% of the TN was consumed during the enrichment cycle, which was initially composed by ammonium (40%) and proteins (60%) (Figure 3b). Since ATU was added to inhibit the nitrifying activity, nitrogen was only consumed during the enrichment cycle for biomass growth. During the feast phase, the increase of the active biomass and the decrease of the TN concentration indicated that VFA were used as a carbon source both for growth and PHA storage (Figure 3). Then, during the famine phase, the ammonium concentration increased due to the degradation of proteins. Part of that ammonium, generated by the protein hydrolysis, was consumed for the microrganisms growth, as shown by the TN consumption during this phase. Approximately 40% of the TN remained in the effluent, which corresponded to 30 mg N/L of protein. Biomass concentration at the end of the cycle doubled the initial value.

Finally, the evolution of the feast/famine regime also correlated with the pH profile (Figure 3a). As no pH control was used in the enrichment SBR, the pH sharply dropped from 7.9 to 4.6 at the beginning of the enrichment cycle, after the addition of the acidified wastewater as a substrate. The reactor operated under acidic conditions during the feast phase, but the pH value increased along with the VFA consumption up to 8. Then, the pH remained in this value during the famine phase until the end of the cycle. These results indicated the successful enrichment, in PHA-accumulating bacteria, of a MMC operated under acidifying conditions during the feast phase. Few authors have previously reported on studies of PHA enrichment systems that operated at different pH values in the neutral or basic range: 7 and 8 [38], 7.5, 8.5 and 9.5 [39], between 8 and 9 [40]. Moreover, Montiel–Jarillo, et al. [41] studied the MMC enrichment at a controlled pH of 7.5 or with uncontrolled pH that was established at 9 using pure acetate as carbon source. These authors found slightly higher PHA accumulation when pH was not controlled. There is no previous information on the enrichment of a PHA-storing mixed culture developed under acidic conditions, as in the case of the present study, performed without pH control.

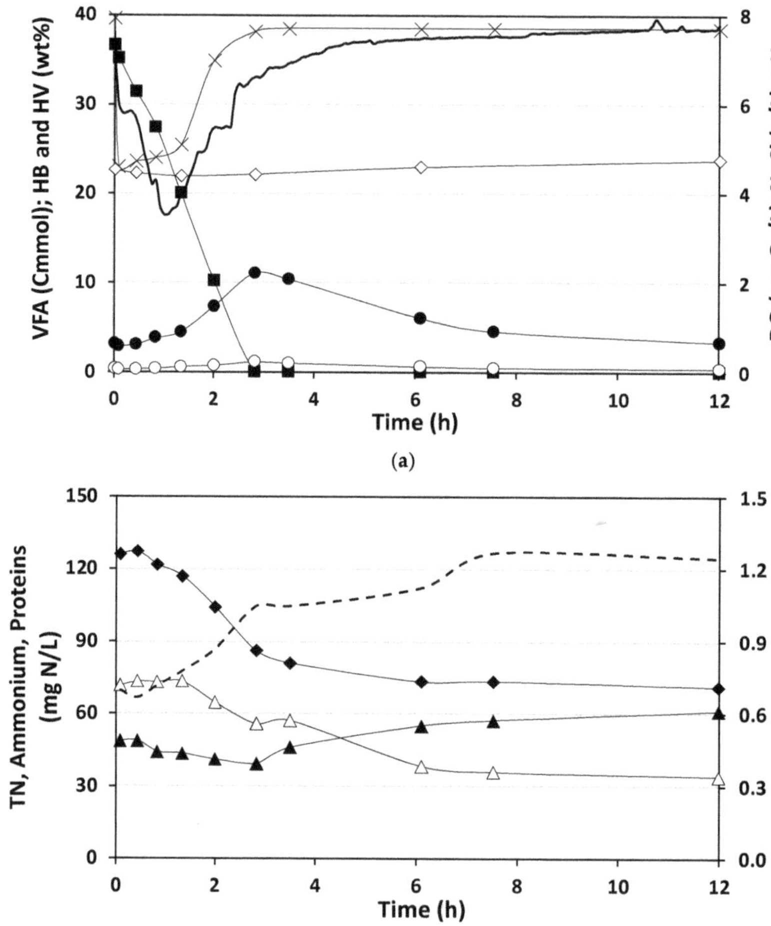

(b)

Figure 3. Characterization of the operational cycle of the enrichment SBR measured on day 66. (**a**) VFA (■), DO (-) and NaCl (◇) concentrations; HB (•) and HV (○) percentages; and pH value (×). (**b**) TN (♦), ammonium (▲), proteins (△) and active biomass (- - -) concentrations.

3.2. PHA Accumulation Capacity of the Enriched MMC

3.2.1. PHA Accumulation with Acidified Cooked Mussel Processing Wastewater

PHA fed-batch accumulation assays were conducted with biomass harvested from the enrichment SBR at the end of operational cycles. At this point, the biomass concentration was approximately 1 g VSS/L. The acidified mussel cookers processing wastewater was fed, as carbon source, and added in pulses to prevent substrate inhibition. Considering the link between the DO concentration and the external carbon source consumption, substrate pulses were added each time the DO concentration increased. The DO concentration increased indicated that the previously added VFA were completely depleted.

The maximum PHA content obtained in accumulation assays (day 120) had a 24.95 wt%, which corresponded to a volumetric productivity of 278.3 mg PHA/(L·h) (Figure 4). The presence of nitrogen as ammonium and proteins in the acidified wastewater had a negative effect on the maximum PHA content and promoted the biomass growth instead of the PHA storage during the batch accumulation assays. Carbon balance calculations showed that only approximately 20% of carbon contributed to PHA accumulation ($Y_{PHA/TOC}$ = 0.24), and that the 50% was used for biomass growth ($Y_{X/TOC}$ = 0.52). Indeed, Oliveira, et al. [42] already reported that it is impossible to obtain a culture enriched in PHA-accumulating bacteria capable of using proteins as the sole nitrogen source requiring the presence of ammonium to growth.

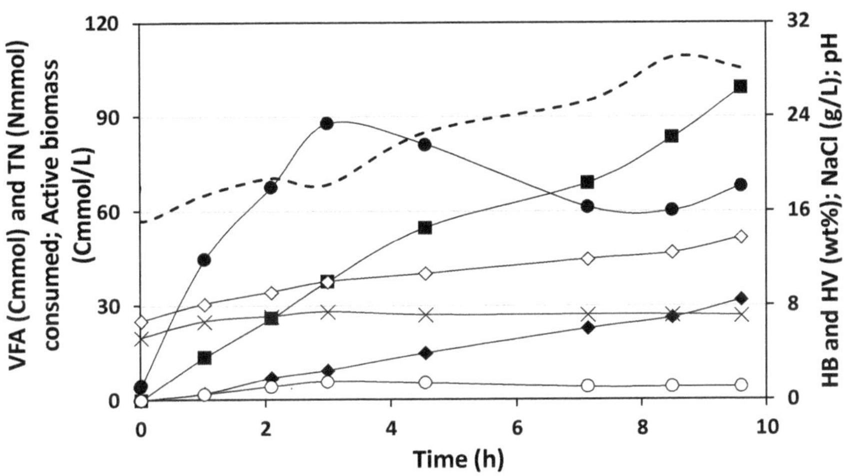

Figure 4. Characterization of an accumulation assay on day 120 of operation using the acidified mussel cookers wastewater as a substrate. Cumulative amount of consumed VFA (■) and TN (♦); active biomass (- - -) and NaCl (◇) concentrations; HB (•) and HV (○) accumulated percentages; and pH value (×). The time of pulse addition is the same as the data points since pulses were supplied just after the sample collection.

The biomass growth, associated with the addition of nitrogen in each pulse, was clearly observed in the active biomass concentration profile during the fed-batch accumulation experiment. The solids concentration at the end of the experiment almost doubled the initial value (Figure 4). Previous studies have demonstrated that the presence of nutrient and carbon sources in the accumulation assays promote cell growth instead of the intracellular accumulation of PHA [43]. However, the biomass production rate was lower during the first 3 h of the experiment, when 25% of the total biomass was produced, which correlated with the 25% of the nitrogen consumed during the same period. Once the maximum accumulation was achieved, after 3 h of the experiment, the percentage of PHA decreased due to cell growth, which leads to a decrease of PHA-accumulation percentage by the formation of new biomass. The HB:HV average ratio in all the accumulation fed-batch experiments was of 83:17. Small variations in the composition of the biopolymer were due to changes in the mixture of the VFA present in the acidified effluent. The highest HB production (HB:HV ratio of 95:5) corresponded to the lowest concentrations of valeric and propionic acids (precursors of HV synthesis) in the substrate.

Moreover, the addition of the substrate in different pulses generated the increase of salt concentration in the reactor throughout the time of the experiment (Figure 4). The initial salt concentration corresponded to that present in the effluent of the enrichment SBR, which was approximately 5 g NaCl/L (the wastewater fed was 5 times diluted). However, this value doubled up to 13.74 g NaCl/L after subsequent addition of substrate pulses (added as acidified wastewater without dilution). Although the biomass continued growing according to the nitrogen consumption and solids measurements, the PHA accumulation decreased after 4 h when the salt concentration was above 10 g NaCl/L. Palmeiro–Sánchez et al. [20] evaluated the effects of NaCl over PHA accumulation in a non-adapted mixed culture. They observed a decrease in the PHA accumulation from 34.6 to 17.4 wt% when the salt concentration increased from 7 to 13 g NaCl/L, respectively. They also observed the degradation of the accumulated polymer, probably as a reaction of the microorganisms to overcome the stress produced by the high salt concentration.

3.2.2. Maximum Accumulation Capacity Evaluated with Mimicked VFA Mixture

As previously discussed, the presence of nitrogen in the acidified effluent used as substrate limited the maximum PHA accumulated by the enriched MMC and promoted the biomass growth. For this reason, to determine the maximum accumulating capacity of the culture, an experiment with a mimicked media was carried out. A mixture of VFA in the same proportion as in the composition of the acidified cooked mussel processing wastewater was used as a substrate in fed-batch assays and no nitrogen source was added.

When the industrial wastewater was replaced by the mimicked VFA mixture, maximum PHA storage of 16.6 wt% was achieved in 6 h, with a yield of 0.30 $Cmmol_{PHA}/Cmmol_{VFA}$. However, the pH of the liquid media after the addition of the VFA-pulse gradually decreased and finally reached a value of 3 after 8 h of the experiment (Figure 5a). Most bacteria can grow at pH values of 5.5–9.0, and maintain their cytoplasmatic pH in a narrow range of 7.5–7.7 [21]. Although the MMC was enriched at acidic conditions, these results indicated that the PHA production was inhibited when the pH dropped below 4, which was the average pH value during the feast phase. The active biomass remained stable at the beginning of the experiment but later, it increased from 60 to 90 Cmmol/L. This growth correlates with the decrease of PHA accumulated (after 6 h of experiment) inside the cells. Residual ammonium coming with the seeding sludge collected at the end of the enrichment cycle (70 mg N/L) was used for growing. At the end of the cycle, ammonium concentration was 6 mg N/L, while protein concentration variation was not observed.

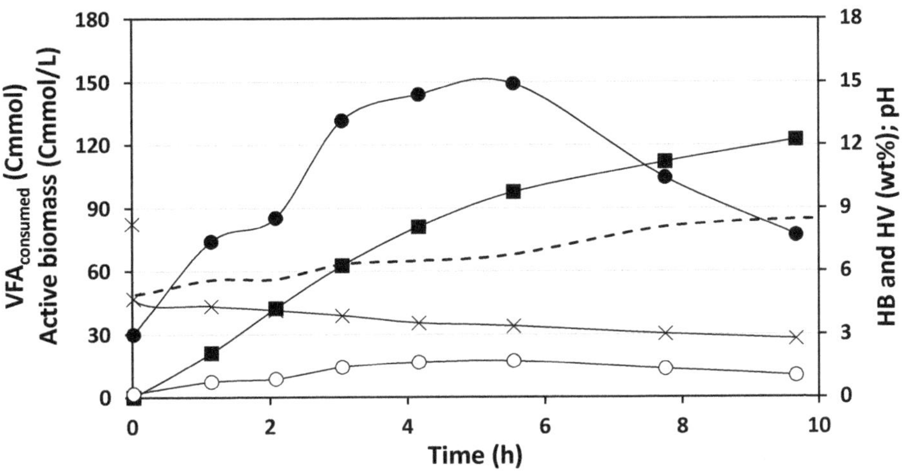

(a)

Figure 5. *Cont.*

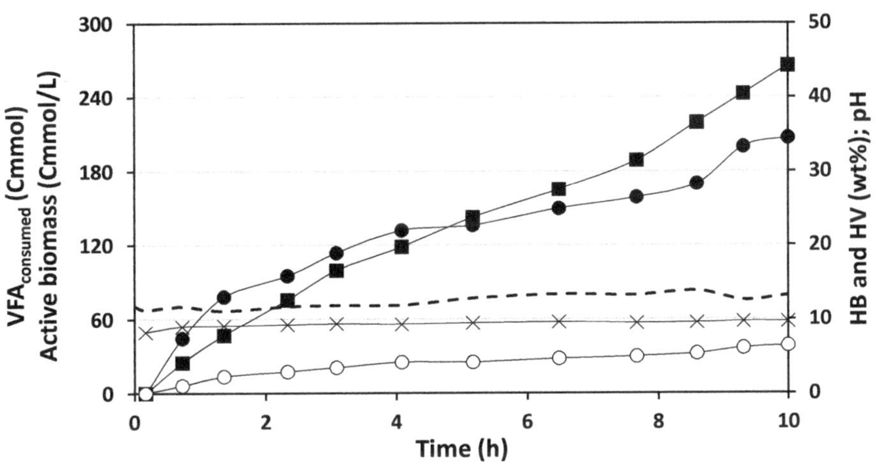

(b)

Figure 5. Performance of accumulation assays using a mixture of VFA as a substrate. (**a**) Without pH adjusted. (**b**) With the pH adjusted to approximately 7.0. VFA consumed (■), pH value (×), active biomass (- - -), HB (•) and HV (○) accumulated percentages. The time of pulse addition is the same as for the data points, since pulses were supplied just after the sample collection.

Since the PHA storage was improved by limiting the nitrogen source but inhibited by the acidic pH, another experiment was carried out by adding NaOH to the VFA mixture to buffer the influent mixture at pH 7 (Figure 5b). In this way, the acidification of the medium after the addition of the carbon pulses was avoided. The pH value only slightly increased due to the CO_2 stripping, and the pH value was maintained at an average value of 9.3 ± 0.2 during the whole experiment. The maximum accumulation of the total PHA increased up to 40.9 wt% after 10 h and it might potentially be higher, since a plateau on the VFA consumption rate was not observed. The PHA production yield was of 0.48 $Cmmol_{PHA}/Cmmol_{VFA}$, which was higher than in the experiment without the pH control and no variations on active biomass concentration were observed. Moreover, the PHA productivity was of 0.2 g PHA/(L·h), three times higher than in the previous experiment. Therefore, the pH control in the substrate promoted the maintenance of the pH value inside the reactor, and eventually the increase of the PHA storage. The HB:HV ratio in both experiments was similar to that obtained in the fed-batch experiments with acidified cooked mussel processing cooker wastewater. Since the mixture of VFA was prepared in the same proportion as the acidified mussel cookers wastewater, the enriched MMC showed the same response in terms of the type of PHA produced.

3.3. Global PHA Production System

The performance of the complete PHA production system from cooked mussel processing wastewater, including acidification (data published in previous work [9]), enrichment and accumulation units, corresponded to a global yield of 0.09 kg COD_{PHA}/kg COD_{fed} (Figure 6). Approximately 10% of the soluble COD present in the raw wastewater from the mussel cookers factory is recovered as PHA. This value was calculated considering the values obtained during the operation of the enrichment SBR under steady-state conditions and during the accumulations assay with the acidified mussel cookers wastewater.

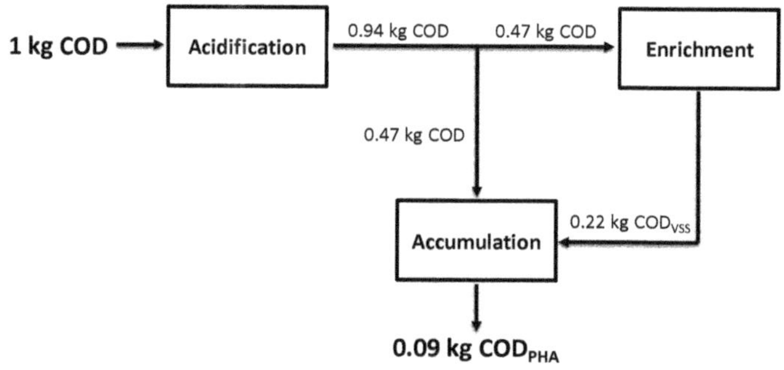

Figure 6. Overall efficiency of the PHA production process with the acidified mussel cookers wastewater, using the conversion of 1 kg of soluble COD from the influent wastewater into PHA as a basis.

This approximately 10% recovery of COD as PHA is much lower than the 34% reported by Tamis, et al. [44] fed with paper mill wastewater. Nevertheless, the wastewater used in the present study had higher complexity due to the presence of salt and proteins.

One important factor that should be noted is the absence of pH control in the whole PHA production system. Even though the enrichment SBR was operated under a low pH during the feast phase, due to the acidic nature of the acidified mussel cookers processing wastewater, the MMC has demonstrated to be enriched in PHA-accumulating microorganisms. This is important for further implementation at large scale since the reduction in the use of chemicals is economically advantageous. However, operational conditions need to be optimized to reduce the large pH decrease. A possible alternative is to use alternative alkalinity sources such as mussel shells that are also generated in the factory.

The effluent generated in the accumulation unit still contained a high nitrogen concentration due to the presence of ammonium and proteins that were not degraded. For the integration of the PHA production system into the industry, the produced effluent must cope with the discharge limits in accordance with the established legislation. Further studies must be developed to include a polishing treatment of the generated effluent for nitrogen removal. In this way, autotrophic nitrogen removal processes such as anammox based ones [45] are a preferred option, since no organic matter is needed and it can be fully converted in the PHA accumulation system.

Another factor that may be further optimized is the substrate conditioning to maximize the PHA accumulation. Since the wastewater from mussel cookers processing shows a high nitrogen concentration, removing nitrogen to a fraction of the acidified wastewater previous to be fed to the accumulation reactor will be advantageous. Another option will be to mix the acidified stream with other effluent generated in the same facility, or even with the effluent of the acidification unit if the cited nitrogen removal process is implemented, in order to obtain an influent stream to the PHA-system with a higher carbon to nitrogen ratio for fostering the PHA production and increased economic attractiveness [16].

4. Conclusions

Acidified industrial wastewater from mussel cookers was demonstrated to be a suitable substrate for PHA production while its pollutant load is reduced. PHA-accumulating microorganisms were successfully selected in the enrichment SBR under high salinity and acidic conditions (pH 4 during the feast phase) in only 60 days. Microorganisms from phylum *Bacteroidetes*, and genera *Azoarcus*, *Comamonas* and *Thauera* from phylum *Proteobacteria* were identified in the enriched MMC. The mixed culture accumulated up to 25 wt% of PHA, with a HB:HV ratio of 83:17, at a salt concentration of 13.7 g NaCl/L. However, the wastewater nitrogen content promoted biomass growth and limited the PHA accumulation. Thus, the evaluation of nitrogen removal processes must be considered not only as polishing step for discharging the wastewater into the environment but before the acidification unit to maximize the PHA accumulation. The enriched culture showed a higher PHA accumulating capacity (up to 40.9 wt%) when nutrients were limited in the substrate of the accumulation assay, showing the relatively high PHA-accumulating microorganisms selection using this complex substrate. Moreover, when pH value in the accumulation reactor falls below 4, the PHA accumulation is hindered and accumulation products are consumed.

The final recovery of PHA in the global system (acidification, enrichment and accumulation units) was approximately 10% of the soluble COD contained in the raw cooked mussel processing wastewater.

Supplementary Materials: The following are available online at http://www.mdpi.com/2071-1050/12/24/10386/s1.

Author Contributions: Conceptualization, A.F.-V. and A.M.-C.; Formal analysis, A.P. and A.F.-V.; Funding acquisition, A.V.d.R. and A.M.-C.; Investigation, A.P. and A.F.-V.; Project administration, A.M.-C.; Supervision, A.M.-C.; Visualization, A.P. and A.F.-V.; Writing—original draft, A.F.-V.; Writing—review & editing, A.P., A.V.d.R. and A.M.-C. All authors have read and agreed to the published version of the manuscript.

Funding: This research was funded by the Spanish Government (AEI) through the FISHPOL (CTQ2014-55021-R) and TREASURE (CTQ2017-83225-C2-1-R) projects. The authors belong to the Galician Competitive Research Group GRC ED431C 2017/29 and to the CRETUS Strategic Partnership (ED431E 2018/01). All these programs are co-funded by the FEDER (EU).

Conflicts of Interest: The authors declare no conflict of interest.

References

1. Alkaya, E.; Demirer, G.N. Minimizing and adding value to seafood processing wastes. *Food Bioprod. Process.* **2016**, *100*, 195–202. [CrossRef]
2. Cristóvão, R.O.; Pinto, V.M.S.; Gonçalves, A.; Martins, R.J.E.; Loureiro, J.M.; Boaventura, R.A.R. Fish canning industry wastewater variability assessment using multivariate statistical methods. *Process Saf. Environ. Prot.* **2016**, *102*, 263–276. [CrossRef]
3. Garcia-Sanda, E.; Omil, F.; Lema, J.M. Clean production in fish canning industries: Recovery and reuse of selected wastes. *Clean Technol. Environ. Policy* **2003**, *5*, 289–294. [CrossRef]
4. Bello Bugallo, P.M.; Stupak, A.; Cristóbal Andrade, L.; Torres López, R. Material flow analysis in a cooked mussel processing industry. *J. Food Eng.* **2012**, *113*, 100–117. [CrossRef]
5. Tay, J.H.; Show, K.Y.; Hung, Y.T. Seafood processing wastewater treatment. In *Waste Treatment in the Food Processing Industry*; Wang, L.K., Hung, Y.T., Lo, H.H., Yapijakis, C., Eds.; Taylor & Francis Group: Boca Raton, FL, USA, 2006.
6. Méndez, R.; Omil, F.; Soto, M.; Lema, J.M. Pilot plant studies on the anaerobic treatment of different wastewaters from a fish-canning factory. *Water Sci. Technol.* **1992**, *25*, 37–44. [CrossRef]

7. Picos-Benítez, A.R.; Peralta-Hernández, J.M.; López-Hincapié, J.D.; Rodríguez-García, A. Biogas production from saline wastewater of the evisceration process of the fish processing industry. *J. Water Process Eng.* **2019**, *32*, 100933. [CrossRef]
8. Mannina, G.; Presti, D.; Montiel-Jarillo, G.; Carrera, J.; Suárez-Ojeda, M.E. Recovery of polyhydroxyalkanoates (PHAs) from wastewater: A review. *Bioresour. Technol.* **2020**, *297*, 122478. [CrossRef]
9. Fra-Vázquez, A.; Pedrouso, A.; Val del Rio, A.; Mosquera-Corral, A. Volatile fatty acid production from saline cooked mussel processing wastewater at low ph. *Sci. Total Environ.* **2020**, *732*, 139337. [CrossRef]
10. Chen, G.G.-Q. (Ed.) Industrial production of PHA. In *Plastics from Bacteria: Natural Functions and Applications;* Springer: Berlin/Heidelberg, Germany; Münster, Germany, 2010; pp. 121–132.
11. Anjum, A.; Zuber, M.; Zia, K.M.; Noreen, A.; Anjum, M.N.; Tabasum, S. Microbial production of polyhydroxyalkanoates (PHAs) and its copolymers: A review of recent advancements. *Int. J. Biol. Macromol.* **2016**, *89*, 161–174. [CrossRef]
12. Valentino, F.; Morgan-Sagastume, F.; Campanari, S.; Villano, M.; Werker, A.; Majone, M. Carbon recovery from wastewater through bioconversion into biodegradable polymers. *New Biotechnol.* **2017**, *37*, 9–23. [CrossRef]
13. Anterrieu, S.; Quadri, L.; Geurkink, B.; Dinkla, I.; Bengtsson, S.; Arcos-Hernandez, M.; Alexandersson, T.; Morgan-Sagastume, F.; Karlsson, A.; Hjort, M.; et al. Integration of biopolymer production with process water treatment at a sugar factory. *New Biotechnol.* **2014**, *31*, 308–323. [CrossRef] [PubMed]
14. Tamang, P.; Banerjee, R.; Köster, S.; Nogueira, R. Comparative study of polyhydroxyalkanoates production from acidified and anaerobically treated brewery wastewater using enriched mixed microbial culture. *J. Environ. Sci.* **2019**, *78*, 137–146. [CrossRef] [PubMed]
15. Oliveira, C.S.S.; Silva, C.E.; Carvalho, G.; Reis, M.A. Strategies for efficiently selecting PHA producing mixed microbial cultures using complex feedstocks: Feast and famine regime and uncoupled carbon and nitrogen availabilities. *New Biotechnol.* **2017**, *37*, 69–79. [CrossRef] [PubMed]
16. Rodriguez-Perez, S.; Serrano, A.; Pantión, A.A.; Alonso-Fariñas, B. Challenges of scaling-up PHA production from waste streams. A review. *J. Environ. Manag.* **2018**, *205*, 215–230. [CrossRef]
17. He, H.; Chen, Y.; Li, X.; Cheng, Y.; Yang, C.; Zeng, G. Influence of salinity on microorganisms in activated sludge processes: A review. *Int. Biodeterior. Biodegrad.* **2017**, *119*, 520–527. [CrossRef]
18. Xiao, Y.; Roberts, D.J. A review of anaerobic treatment of saline wastewater. *Environ. Technol.* **2010**, *31*, 1025–1043. [CrossRef]
19. Passanha, P.; Kedia, G.; Dinsdale, R.M.; Guwy, A.J.; Esteves, S.R. The use of NaCl addition for the improvement of polyhydroxyalkanoate production by Cupriavidus necator. *Bioresour. Technol.* **2014**, *163*, 287–294. [CrossRef]
20. Palmeiro-Sanchez, T.; Fra-Vazquez, A.; Rey-Martinez, N.; Campos, J.L.; Mosquera-Corral, A. Transient concentrations of NaCl affect the PHA accumulation in mixed microbial culture. *J. Hazard. Mater.* **2016**, *306*, 332–339. [CrossRef]
21. Krulwich, T.A.; Sachs, G.; Padan, E. Molecular aspects of bacterial pH sensing and homeostasis. *Nat. Reviews. Microbiol.* **2011**, *9*, 330–343. [CrossRef]
22. Amann, R.; Krumholz, L.; Stahl, D.A. Fluorescent-oligonucleotide probing of whole cells for determinative, phylogenetic, and environmental-studies in microbiology. *J. Bacteriol.* **1990**, *172*, 762–770. [CrossRef]
23. Daims, H.; Lucker, S.; Wagner, M. Daime, a novel image analysis program for microbial ecology and biofilm research. *Environ. Microbiol.* **2006**, *8*, 200–213. [CrossRef] [PubMed]
24. APHA-AWWA-WEF. *Standard Methods for the Examination of Water and Wastewater*, 23rd ed.; American Public Health Association/American Water Works Association/Water Environment Federation: Washington, DC, USA, 2017.
25. Bower, C.E.; Holm-Hansen, T. A salicylate–hypochlorite method for determining ammonia in seawater. *Can. J. Fish. Aquat. Sci.* **1980**, *37*, 794–798. [CrossRef]
26. Lowry, O.H.; Rosebrough, N.J.; Farr, A.L.; Randall, R.J. Protein measurement with the Folin phenol reagent. *J. Biol. Chem.* **1951**, *193*, 265–275.
27. Loewus, F.A. Improvement in anthrone method for determination of carbohydrates. *Anal. Chem.* **1952**, *24*, 219. [CrossRef]

28. Smolders, G.J.F.; van der Meij, J.; van Loosdrecht, M.C.M.; Heijnen, J.J. Stoichiometric model of the aerobic metabolism of the biological phosphorus removal process. *Biotechnol. Bioeng.* **1994**, *44*, 837–848. [CrossRef] [PubMed]
29. Beun, J.J.; van Loosdrecht, M.C.M.; Heijnen, J.J. Aerobic granulation in a sequencing batch airlift reactor. *Water Res.* **2002**, *36*, 702–712. [CrossRef]
30. Mahmoud, N.; Zeeman, G.; Gijzen, H.; Lettinga, G. Anaerobic stabilisation and conversion of biopolymers in primary sludge—Effect of temperature and sludge retention time. *Water Res.* **2004**, *38*, 983–991. [CrossRef]
31. Dionisi, D.; Majone, M.; Vallini, G.; Gregorio, S.D.; Beccari, M. Effect of the length of the cycle on biodegradable polymer production and microbial community selection in a sequencing batch reactor. *Biotechnol. Prog.* **2007**, *23*, 1064–1073. [CrossRef]
32. Carvalho, G.; Oehmen, A.; Albuquerque, M.G.E.; Reis, M.A.M. The relationship between mixed microbial culture composition and PHA production performance from fermented molasses. *New Biotechnol.* **2013**, *31*, 257–263. [CrossRef]
33. Morgan-Sagastume, F. Characterisation of open, mixed microbial cultures for polyhydroxyalkanoate (PHA) production. *Rev. Environ. Sci. Bio Technol.* **2016**, *15*, 593–625. [CrossRef]
34. Ventosa, A.; Márquez, M.; Sánchez-Porro, C.; Haba, R. Taxonomy of halophilic *Archaea* and *Bacteria* (chapter 3). In *Advances in Understanding the Biology of Halophilic Microorganisms*; Vreeland, R.H., Ed.; Springer: Dordrecht, The Netherlands, 2012; pp. 59–80.
35. Cui, Y.-W.; Zhang, H.-Y.; Lu, P.-F.; Peng, Y.-Z. Effects of carbon sources on the enrichment of halophilic polyhydroxyalkanoate-storing mixed microbial culture in an aerobic dynamic feeding process. *Sci. Rep.* **2016**, *6*, 30766. [CrossRef] [PubMed]
36. Tamis, J.; Luzkov, K.; Jiang, Y.; van Loosdrecht, M.C.; Kleerebezem, R. Enrichment of *plasticicumulans acidivorans* at pilot-scale for PHA production on industrial wastewater. *J. Biotechnol.* **2014**, *192 Pt A*, 161–169. [CrossRef]
37. Korkakaki, E.; Mulders, M.; Veeken, A.; Rozendal, R.; van Loosdrecht, M.C.M.; Kleerebezem, R. PHA production from the organic fraction of municipal solid waste (OFMSW): Overcoming the inhibitory matrix. *Water Res.* **2016**, *96*, 74–83. [CrossRef] [PubMed]
38. Chua, A.S.M.; Takabatake, H.; Satoh, H.; Mino, T. Production of polyhydroxyalkanoates (PHA) by activated sludge treating municipal wastewater: Effect of ph, sludge retention time (SRT), and acetate concentration in influent. *Water Res.* **2003**, *37*, 3602–3611. [CrossRef]
39. Villano, M.; Beccari, M.; Dionisi, D.; Lampis, S.; Miccheli, A.; Vallini, G.; Majone, M. Effect of pH on the production of bacterial polyhydroxyalkanoates by mixed cultures enriched under periodic feeding. *Process Biochem.* **2010**, *45*, 714–723. [CrossRef]
40. Oehmen, A.; Pinto, F.; Silva, V.; Albuquerque, M.G.; Reis, M.A. The impact of ph control on the volumetric productivity of mixed culture PHA production from fermented molasses. *Eng. Life Sci.* **2013**, *14*, 143–152. [CrossRef]
41. Montiel-Jarillo, G.; Carrera, J.; Suárez-Ojeda, M.E. Enrichment of a mixed microbial culture for polyhydroxyalkanoates production: Effect of pH and N and P concentrations. *Sci. Total Environ.* **2017**, *583*, 300–307. [CrossRef]
42. Oliveira, C.S.S.; Silva, M.O.D.; Silva, C.E.; Carvalho, G.; Reis, M.A.M. Assessment of protein-rich cheese whey waste stream as a nutrients source for low-cost mixed microbial PHA production. *Appl. Sci.* **2018**, *8*, 1817. [CrossRef]
43. Johnson, K.; Kleerebezem, R.; van Loosdrecht, M.C.M. Influence of ammonium on the accumulation of polyhydroxybutyrate (PHB) in aerobic open mixed cultures. *J. Biotechnol.* **2010**, *147*, 73–79. [CrossRef]
44. Tamis, J.; Mulders, M.; Dijkman, H.; Rozendal, R.; Loosdrecht, M.C.M.V.; Kleerebezem, R. Pilot-scale polyhydroxyalkanoate production from paper mill wastewater: Process characteristics and identification of bottlenecks for full-scale implementation. *J. Environ. Eng.* **2018**, *144*, 04018107. [CrossRef]
45. Val del Rio, A.; Pichel, A.; Fernandez-Gonzalez, N.; Pedrouso, A.; Fra-Vázquez, A.; Morales, N.; Mendez, R.; Campos, J.L.; Mosquera-Corral, A. Performance and microbial features of the partial nitritation-anammox process treating fish canning wastewater with variable salt concentrations. *J. Environ. Manag.* **2018**, *208*, 112–121. [CrossRef] [PubMed]

Publisher's Note: MDPI stays neutral with regard to jurisdictional claims in published maps and institutional affiliations.

sustainability

Article

Polyhydroxyalkanoates Production by Mixed Microbial Culture under High Salinity

João M. Carvalho [1,2], Bruno C. Marreiros [1,2,*] and Maria A. M. Reis [1,2]

1 UCIBIO—Applied Molecular Biosciences Unit, Department of Chemistry, NOVA School of Science and Technology, NOVA University Lisbon, 2819-516 Caparica, Portugal; jms.carvalho@fct.unl.pt (J.M.C.); amr@fct.unl.pt (M.A.M.R.)

2 Associate Laboratory i4HB—Institute for Health and Bioeconomy, NOVA School of Science and Technology, NOVA University Lisbon, 2819-516 Caparica, Portugal

* Correspondence: b.marreiros@fct.unl.pt

Abstract: The fishing industry produces vast amounts of saline organic side streams that require adequate treatment and disposal. The bioconversion of saline resources into value-added products, such as biodegradable polyhydroxyalkanoates (PHAs), has not yet been fully explored. This study investigated PHA production by mixed microbial cultures under 30 g_{NaCl}/L, the highest NaCl concentration reported for the acclimatization of a PHA-accumulating mixed microbial culture (MMC). The operational conditions used during the culture-selection stage resulted in an enriched PHA-accumulating culture dominated by the Rhodobacteraceae family (95.2%) and capable of storing PHAs up to 84.1% wt. (volatile suspended solids (VSS) basis) for the highest organic loading rate (OLR) applied (120 Cmmol/(L.d)). This culture presented a higher preference for the consumption of valeric acid (0.23 ± 0.03 $Cmol_{HVal}/(Cmol_X.h)$), and the 3HV monomer polymerization (0.33 ± 0.04 $Cmmol_{HV}/(Cmmol_X.h)$ was higher as well. As result, a P(3HB-co-3HV)) with high HV content (63% wt.) was produced in the accumulation tests conducted at higher OLRs and with 30 g_{NaCl}/L. A global volumetric PHA productivity of 0.77 gPHA/(L.h) and a specific PHA productivity of 0.21 gPHA/(gX.h) were achieved. These results suggested the significant potential of the bioconversion of saline resources into value-added products, such as PHAs.

Keywords: saline resources; halotolerant; PHA-accumulating MMC; PHAs accumulation; biopolymer; P(3HB-co-3HV)

Citation: Carvalho, J.M.; Marreiros, B.C.; Reis, M.A.M. Polyhydroxyalkanoates Production by Mixed Microbial Culture under High Salinity. *Sustainability* **2022**, *14*, 1346. https://doi.org/10.3390/su14031346

Academic Editors: José Luis Campos, Anuska Mosquera Corral, Ángeles Val del Río and Alba Pedrouso Fuentes

Received: 7 December 2021
Accepted: 18 January 2022
Published: 25 January 2022

1. Introduction

The fish and seafood segment from the agro-food industry is growing globally with a compound annual growth rate (CAGR) of 4.8% (2021–2026) towards a market value of EUR 518 billion by 2021 [1]. Although the fishery industry has been associated with high nutritional and commercial value, it has also generated a large amount of organic side streams that require adequate treatment and disposal [2,3]. Worldwide, 96.4 million tonnes of fish were generated in 2021, 90% of which came from marine environments, generating saline organic side streams that contained NaCl in the range of 3.5–46 g_{NaCl}/L, depending on the final product [4]. Companies have therefore been required to conduct expensive treatments of those saline streams before they can be safely disposed, as the conventional biological treatments for organic compound removal are less efficient due to their strong salt inhibition [5,6]. The ability to use these saline organic side streams as feedstock for biologic processes would be economically favourable for these industries by reducing treatment costs while also reducing the environmental burden caused by the commonly used chemicals during treatment [7]. Moreover, the conversion of these resources into value-added products, such as polyhydroxyalkanoates (PHAs), would also decrease the costs of treatment and contribute towards a circular economy.

PHAs are biodegradable polymers produced by certain microorganisms as intracellular carbon and energy reserves in nutrient-limited environments, and the production of this polymer is of growing interest in today's society as they are feasible substitutes of conventional plastics, due to similar physicochemical properties and a wide range of potential applications [8]. The uses of PHAs are therefore increasing, with a CAGR of 14.2%, and it is estimated that this sector will reach a global market value of EUR 107 million by 2025 [9]. Nonetheless, the production costs of PHAs are still limiting wider commercialization and industrialization [8] of this bioplastic as the production costs are estimated to be four times higher (EUR 4–5/kg) than conventional plastics.

The use of mixed microbial cultures (MMCs) is a strategy to reduce costs of production as: (i) aseptic conditions are not required, (ii) it allows the use of inexpensive/cheap raw materials such as saline waste streams [10], and (iii) the MMCs are more amenable to deal with complex matrices due to the diversity of microorganisms and pathways.

A typical PHA production process by MMCs comprises three stages [8]: (i) acidogenic fermentation of the organic resources for the production of volatile fatty acids (VFAs), which are precursors for PHA biosynthesis; (ii) culture selection, where the inoculum is enriched in PHA-accumulating MMC by applying selective pressure (e.g., feast and famine (F/f) regime); and (iii) PHA production, where the selected PHA-accumulating MMC from the second stage are fed with VFA-rich streams produced in the first stage to promote PHA accumulation of up to the culture's maximum capacity.

The selection stage is considered the core of the process [11], as the obtained PHA-accumulating MMC will determine the efficiency of the production process, namely in terms of PHA storage capacity and PHA productivity [8]. This stage focuses not only in obtaining an MMC enriched with PHA-accumulating organisms, capable of producing high amounts of PHAs, but also in the ability to sustain a high growth of the enriched MMC for the PHA production stage, in order to increase the volumetric productivity of the global PHA production process [8]. However, a low volumetric productivity is normally obtained due to low biomass concentrations reached in the selection reactor [10]. This is related to the fundamentals of the generally adopted F/f selection strategy, where the culture is subjected to periods of carbon substrate excess (feast phase) followed by periods of carbon starvation (famine phase), so that the organisms that do not store PHAs are not able to grow during the famine period and are eliminated from the reactor [10]. Nutrient supplementation is an important factor for the success of the selection reactor, as some feedstocks, although rich in carbon compounds, are frequently poor in nutrients or have reduced nutrient availability, namely in nitrogen that is required to support culture growth. The uncoupled feedings of carbon and nitrogen, where the latter is added after the exogenous carbon is exhausted, have shown to lead to a better selection performance with lower F/f ratios, higher PHA concentrations, and reduced culture instability [10,12].

The ability to use saline resources for PHA production could also promote the economic viability of PHA production. Moreover, the use of saline organic streams could open the possibility of using seawater ($\approx$3.5% NaCl w/v) as a washing and diluting agent in the PHA production process. Previous works, using cheese whey as feedstock for the production of biodegradable food packaging with PHAs, have shown that the production of PHAs has been associated with excess freshwater usage [13]. This represents not only a significant consumption of freshwater, as the rapid decrease in freshwater availability is a major concern worldwide, but also brings surplus operation costs to the process. It has also been suggested that PHA production at high salinity concentrations can facilitate the recovery of produced PHAs from halotolerant microorganisms, as the cells can be lysed in distilled water, reducing purification costs by avoiding the utilization of commonly used organic solvents for PHA extraction [14]. Therefore, the proposed measures, besides being eco-friendly, can be economically advantageous for industries located at costal sites.

Overall, the use of MMCs for the production of PHAs at high saline concentrations has potential benefits; however, PHA production using MMCs in the presence of saline conditions has not yet been fully established.

Xiao et al. provided a comprehensive review of different strategies for the anaerobic treatment of saline wastewaters [15]. Among these studies, the bacterial consortia collected from saline environments (e.g., marine sediments) have been highlighted as promising in anaerobic processes for the treatment of saline wastewaters. Moreover, the successful conversion of saline substrates into VFAs at NaCl concentrations up to 70 g/L was recently indicated. He et al. tested the acidogenic fermentation of a mixture of food waste that was artificially salted in a NaCl concentration up to 70 g/L. The highest VFA production was obtained at 10 g_{NaCl}/L (0.54 $g_{VFA}/g_{Feedstock}$) but was still high at the highest salt concentration applied (0.44 $g_{VFA}/g_{Feedstock}$) [16]. In another study, a mixture of food waste, brine, and wastewater derived from a biodiesel production facility was used to produce VFAs. The acidogenic fermentation of this salted feedstock (12–18 g_{NaCl}/L) was feasible with an acidification degree up to 46% [17]. Fra-Vasquez et al. evaluated the conversion of wastewater from a cooked mussel processing factory (22 g_{NaCl}/L and pH 4.6) into VFAs. In this study, an extensive carbohydrate degradation of 96% and a maximum acidification degree of 43% were obtained [18].

With respect to the selection of PHA-accumulating mixed cultures in saline conditions, only a few studies can be found with salinities up to 20 g_{NaCl}/L. Palmeiro-Sánchez et al. tested a PHA-accumulating MMC enriched using fermented tuna-processing wastewater in accumulation assays in the absence of NaCl and in the presence of 21.6 g_{NaCl}/L. Not only the PHA content decreased abruptly from 51.3% in the non-saline environment to 8.4% in the presence of NaCl, but also the HB–HV ratio changed, from 62:38 to 72:28 (% wt.) [19]. Conversely, in the work by Wen et al., the positive influence of NaCl on PHA storage in a culture-enriched reactor was reported [20]. Wen et al. studied the effects of NaCl at 0, 5.0, 10.0, and 15.0 g_{NaCl}/L on the enrichment of PHA-accumulating MMC using food waste. A maximum PHA content of 14.3% wt. was obtained under the condition at 15 g_{NaCl}/L, indicating the positive influence of NaCl; however, the maximum specific PHA production, biomass growth, and substrate consumption rates were found at 5 g_{NaCl}/L. In the PHA accumulation assays at the different NaCl concentrations with respective enriched cultures, the maximum PHA content of 50.5% wt. was also obtained at 5 g_{NaCl}/L and accumulations with higher NaCl led to lower PHA contents, as a PHA content of 42.6% wt. was obtained at 15.0 g_{NaCl}/L. Overall, Wen Q. et al. suggested that osmotic stress may trigger the synthesis of PHA; however, at higher NaCl contents, bacteria consume PHAs in response to high osmotic stress, leading to a lower maximum PHA content [20]. Alba Pedrouso A. et al. [21] were also able to obtain an MMC enriched in PHA-accumulating microorganisms at 5 g_{NaCl}/L using real feedstock obtained from the acidogenic fermentation of cooked mussel processing wastewater. This culture was able to reach a PHA content of 25% wt. using the same feedstock and with uncontrolled pH; however, when mimicking the feedstock using a synthetic mixture, the culture was able to reach 40% wt. Using the same feedstock, Argiz et al. [22] were able to select a more efficient PHA-accumulating MMC by removing the undesired substances (proteins and carbohydrates) with the addition of a settling stage and subsequent supernatant discharge after the end of the feast period of the cycle of an SBR. This culture was able to reach a higher PHA content of 60% wt. using a synthetic mixture as feedstock. In another work with the same feedstock but with high salinity levels, Roibás-Rozas et al. [23] were able to slowly adjust (throughout 100 days) the NaCl concentration in the selection reactor up to 20 g_{NaCl}/L, starting from a previously PHA-accumulating MMC selected at 5 g_{NaCl}/L. This culture was then also used for accumulation assays where the NaCl concentration was gradually increased throughout the assays until it reached 28 g_{NaCl}/L, reaching a PHA content of 41% wt.

The main aim of the present study was to assess the feasibility of selecting an efficient PHA-accumulating MMC under conditions of high salinity (30 g_{NaCl}/L). To date, this was the highest salinity value reported in the literature for the acclimatization of a PHA-accumulating MMC. This work was conducted at lab-scale, in a three-stage process for PHA production by MMC, where two reactors were operated for culture selection and PHA accumulation.

2. Materials and Methods

2.1. Culture Selection

A sequence batch reactor (SBR) with a working volume of 2 L was used for culture selection. The SBR was inoculated with sediments collected from a saline area of Rio Tejo (Samouco, Portugal) that were passed through a filtration sieve (350 μm).

The SBR was operated under aerobic conditions with a hydraulic retention time (HRT) of 16 h, a solids retention time (SRT) of 3 days and 8 h cycles. Each cycle period consisted of 5 steps: (i) influent feeding/filling (5 min); (ii) aerated phase (434 min); (iii) purge (5 min); (iv) settling (30 min); and (v) supernatant withdrawal (6 min).

The reactor was operated in a temperature-controlled room (19–21 °C). Mixing was kept at 100 rpm using a one-blade impeller with a six-blade Rushton turbine, and pH was controlled at pH 8.4 $\pm$ 0.5 through automatic dosing of 0.5 M HCl. Air was supplied through fine bubble diffusers and the aeration rates (L/L) were adjusted throughout the operation so that the dissolved oxygen (DO) was not limited in the reactor. DO concentration and pH were monitored online.

A salted (30 g_{NaCL}/L) synthetic VFA mixture composed of acetic (HAc), propionic (HPro), butyric (HBut), and valeric (HVal) acids (25%Cmol each) was used as carbon source. Together with the synthetic mixture, a salted (30 g_{NaCl}/L) mineral solution was fed to the SBR in order to have the following concentrations of the following components (mg/L): ATU (10); EDTA-2Na (50); $MgSO_4$ (50); $CaCl_2$ (50); $FeCl_3 \cdot 6H_2O$ (1.5); H_3BO_4 (0.15); $CuSO_4 \cdot 5H_2O$ (0.03); KI (0.03); $MnCl_2 \cdot 4H_2O$ (0.12); $Na_2MoO_4 \cdot 2H_2O$ (0.06); $ZnSO_4 \cdot 7H_2O$ (0.12); and $CoCl_2 \cdot 6H_2O$ (0.15). The feeding was composed of 100 mL carbon solution plus 900 mL dilution mineral solution.

The SBR followed an aerobic dynamic feeding, namely F/f regime with uncoupling of the carbon and nitrogen availabilities in order to obtain selection pressure for PHA-accumulating organisms. A C/N/P (Cmol basis) ratio of 100:7.5:1 was initially applied; later, this ratio was changed to 100:10:1 and then to an optimal ratio value of 100:5:1. An ammonia solution was fed to the reactor 1 h after carbon feeding. Phosphorous was given through addition to the mineral solution. The OLR was initially 60 Cmmol/(L.d) and was increased to 120 Cmmol/(L.d) as a stable culture enriched in PHA-accumulating organisms was obtained. The study had a total duration of 145 days.

2.2. PHAs Accumulation

MMC PHA-accumulation performance of the selected culture at two different OLRs (ACC-60 and ACC-120 for 60 and 120 Cmmol/(L.d), respectively), was assessed in a fed-batch reactor. A total of 1 L of biomass was collected from the corresponding SBR at the end of famine phase and used as inoculum. The accumulation reactor had a working volume of 1 L, and the accumulation procedure consisted in a pulse-wise feeding strategy of salted synthetic mixture (30 g_{NaCl}/L) under nitrogen-limiting conditions and using a food to microorganism (F/M) ratio of 1.5 times the one found in the SBR. Whenever the maximum PHA capacity of the selected culture was attained, biological activity was stopped by quenching to pH 2–3 using sulfuric acid. Assays were carried out with the same controlled conditions of pH, temperature, aeration, and stirring that were used for culture selection.

2.3. Analytical Procedures

Total solids (TS), volatile solids (VS), total suspended solids (TSS), and volatile suspended solids (VSS) were determined according to standard methods [24].

VFAs were quantified in filtered samples (0.20 μm) using a high-performance liquid chromatography (HPLC) in a VWR Hitachi Chromaster chromatographer equipped with a Pump 5160, an auto sampler 5260, a Column Oven 5310, a Diode Array Detector 5430, a RI Detector 5450, a Biorad 125-0129 30 $\times$ 4.6 mm pre-column, and an Aminex HPX-87H 300 $\times$ 7.8 mm column. The following conditions were used: column temperature 60 °C, 0.01 M H_2SO_4 eluent, flow rate 0.6 mL/min, and injection volume 99 μL. The

VFA concentrations were calculated from standard calibration curves (4–1000 mg/L of each compound).

Ammonia concentrations were determined in filtered samples (0.20 μm) using a segmented continuous flow analyser (Skalar SNA++). PHAs were extracted and hydrolyzed. The monomers of PHAs were esterified into 3-hydroxyacyl methyl esters to be quantified by gas chromatography (GC-FID, Brucker) using a method described in Oliveira et al. [11].

Lyophilized biomass was weighted and incubated with 1 mL chloroform and 1 mL acidic methanol (20% H_2SO_4) (for methanolysis) through digestion at 100 °C for 3.5 h. After the digestion step, the organic phase (methylated monomers dissolved in chloroform) was extracted and injected (2 μL) into a gas chromatograph equipped with a flame ionization detector (Bruker 430-GC) and a BR-SWax column (60 m, 0.53 mm internal diameter, 1 mm film thickness, Bruker, USA), using helium as carrier gas at 1.0 mL/min. The temperature regime started at 40 °C and increased stepwise to 100 °C at a rate of 20 °C/min, to 175 °C at a rate of 3 °C/min, and to 220 °C at a rate of 20 °C/min (cleaning step of the column after each injection). Injector and detector temperatures were 280 °C and 230 °C, respectively. Determination of 3-hydroxybutyrate (3HB) and 3-hydroxyvalerate (3HV) concentrations was made through the use of two calibration curves, one for 3HB and other for 3HV, using standards (0.1–10 g/L) of a commercial P(3HB-co-3HV) (88%: 12%, Sigma) and corrected using heptadecane as internal standard (concentration of ≈1 g/L). Intracellular PHA granules were identified using Nile blue staining as described in Oliveira et al. [11] and observed with epifluorescence microscope Olympus BX51 equipped with an Olympus XM10 camera (Cell-F software). Microbial community assessment samples were collected for microbial community analysis at the beginning of the study (inoculum), from cycles throughout operation of the reactors and from the accumulation assays.

Biomass samples from sediments collected from a saline area of Rio Tejo (Samouco, Portugal) (inoculum) and from the culture selected at OLR of 60 and 120 Cmmol/(L.d) were collected and were phylogenetically characterized on high-throughput sequencing of the 16S V1-3 rRNA gene. DNA extraction, gene sequencing and bioinformatics processing was carried out by DNASense (Aalborg, Denmark) as described in [25].

2.4. Calculations

The F/f ratio was calculated as the ratio between the period lengths of feast and famine phases of the SBR cycle. The PHA content in VS (g_{PHA}/g_{VS}) was calculated by multiplying the PHA content in TS (g_{PHA}/g_{TS}, given by GC analysis) by the VS/TS ratio obtained for the lyophilized pellets used for GC analysis. The PHA content in VSS (g_{PHA}/g_{VSS}) was calculated by multiplying the PHA content in VS by the VSS/VS ratio obtained in the reactor (VSS/VS ≈ 1). The PHA content in the biomass was determined in terms of percentage of VSS on mass basis (% wt., g_{PHA}/g_{VSS}). VSS were considered to be constituted of active biomass (X) and PHA. For determining cell growth, the generic chemical formula for MMC ($CH_{1.8}O_{0.5}N_{0.2}S_{0.02}P_{0.02}$) [26], with a molecular weight (MW) of 25.30 g/Cmol, was used. The ΔPHAs was determined as the maximum PHA content (PHA_{max},% wt.) minus the PHA content at the beginning (PHA_0,% wt.) of the experiment. Stoichiometric and kinetic performance parameters were determined for the SBR and PHA accumulation assays in pseudo-steady-state conditions with the reactors operated at both OLRs.

The specific substrate consumption rates (-qVFA, $Cmol_{VFA}/(Cmol_X.h)$); specific PHA storage (qPHA, $Cmol_{PHA}/(Cmol_X.h)$) and consumption (-qPHA, $Cmol_{PHA}/(Cmol_X.h)$) rates; and specific growth rates (qX, $Cmol_X/(Cmol_X.h)$) were determined from the linear regression of the experimental data of the substrates, PHA, and X-specific concentrations (i.e., instant concentration divided by the biomass concentration), respectively, and plotted over time. Specific storage yield ($Y_{PHA/S}$, $Cmol_{PHA}/Cmol_{VFA}$) was calculated as the ratio between qPHAs and the -qVFAs.

Growth yields on stored PHAs ($Y_{X/PHA}$, $Cmol_x/Cmol_{PHA}$) were calculated as the ratios between the specific growth rate during the famine phases (qX_{famine}) and -qPHA.

In the accumulation assays, the specific rates and yields were calculated, as described before, for each pulse. In the accumulation tests, the first three pulses' average values of each parameter were considered. Volumetric PHA productivity ($g_{PHA}/(L.h)$) was calculated as the ratio of cumulative produced PHAs (ΔPHA) in 1 L of working volume per unit of time (hour). Specific PHA productivity ($g_{PHA}/(g_X.h)$) was calculated as the ratio of cumulative produced PHAs per X at the beginning of the assay (X_0) per unit of time (hour).

Standard errors associated with the determined parameters were estimated using standard errors propagation formulae.

3. Results and Discussion

3.1. Culture Selection: PHA-Accumulating MMC

The culture selection stage for a PHA-accumulating mixed culture under conditions of high salinity (30 g_{NaCl}/L) could indicate the potential value of saline resources as well as the benefit of using seawater as a washing and diluting agent in the PHA production process.

In order to select a halotolerant PHA-accumulating MMC, sediments collected from a saline area (Samouco, Portugal) were used to inoculate the SBR with halotolerant organisms and then submitted to an SRT for three days and to an aerobic dynamic feeding, namely the conventional F/f regime and the decoupled carbon and nitrogen feeding, throughout the operation. An artificially salted (30 g_{NacL}/L) equimolar (Cmol basis) synthetic mixture was used as saline-fermented feedstock. The F/f ratio was continuously monitored (Figure 1) since it is a useful performance indicator of the culture selection process as an F/f below 0.2 h/h has been thought to boost the selection of an efficient PHA-accumulating culture [8].

The culture selection was started with an OLR of 60 Cmmol/(L.d) and a C/N/P ratio of 100 Cmol:7.5 Nmol:1 Pmol. An F/f ratio below 0.2 h/h was obtained after 14 days of operation, suggesting that the selective pressure was correctly being applied. However, the selected culture was unable of complete nitrogen consumption at the initial (100 Cmol: 7.5 Nmol) or higher (100 Cmol: 10 Nmol) C/N applied. The C/N ratio was then optimized to 100 Cmol: 5 Nmol (Figure 1), and a stable PHA-accumulating MMC was attained (F/f < 0.2 h/h). With regards to suspended solids (VSS and TSS), a strong increase in the TSS over VSS was observed for the first period of acclimatization (Figure 1), resulting in a VSS/TSS ratio of 29 $\pm$ 5% wt. for the culture selected at an OLR of 60 Cmmol/(L.d), which suggested the intracellular accumulation of inorganic compounds. Several halophile microorganisms have been reported to accumulate a variety of small molecules, both inorganic (e.g., Na^+, K^+, Cl^-) and organic (e.g., ectoine, glycine betaine, 3-hydroxybutyric acid), in the cytoplasm to counteract the external osmotic pressure [27].

The selected culture enriched at OLR 60 Cmmol/(L.d) showed a cycle with a typical profile for the aerobic dynamic feeding strategy (F/f regime and uncoupled carbon and nitrogen availabilities) (Figure 2A). As carbon, in the form of VFAs, was added to the reactor, cells began consuming the VFAs and storing carbon in the form of PHAs. As the VFAs were completely consumed (end of the feast), PHA production halted, and a small decrease in PHA concentration in the reactor occurred due to cell metabolic activity. After 1 h of feeding, nitrogen, in the form of ammonia, was added to the reactor, allowing the cell growth of PHAs. This caused a decrease in PHA content as it was consumed for growth and other metabolic activity. As nitrogen in the media was depleted, the PHA consumption rates decreased throughout the remaining cycle. The VSS concentration increased with the increase in PHA content during the feast period, but as the VFAs were depleted and the PHAs were consumed for cell maintenance and growth, the VSS concentration decreased.

In order to further increase volumetric PHA productivity, the OLR was doubled to 120 Cmmol/(L.d) whilst maintaining selective pressure for PHA-accumulating organisms (Figure 1). Subsequently, the concentration of active biomass (X) increased $\approx$1.7 times from 1.92 $\pm$ 0.04 to 3.26 $\pm$ 0.34 g/L (Figure 1) while also increasing the PHA storage capacity from 35 $\pm$ 2% wt. to 49 $\pm$ 3% wt., indicating a good culture adaptation to higher organic load and even leading to better culture performance. The obtained PHA contents at the end of the feast period were already quite high and comparable with other reported accumulation assays [23], which suggested the possibility of skipping the third stage of the traditional MMC PHA-accumulating process in order to lower operating costs. This culture showed a cycle profile very similar to that obtained in an OLR of 60 Cmmol/(L.d) (Figure 2B), whereas the VSS/TSS ratio shifted to 52 ± 0% wt.

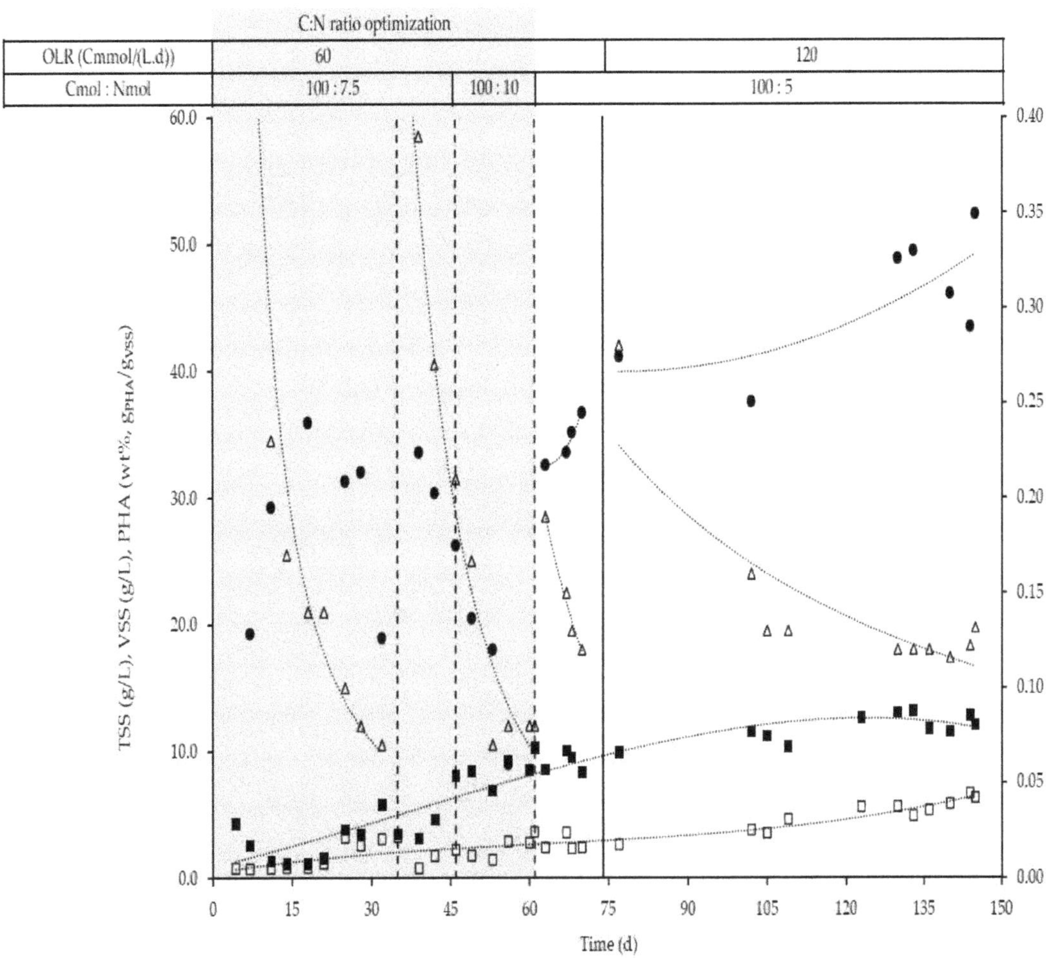

Figure 1. Profile of the mixed culture adaptation throughout the operation. A feast and famine (F/f) below 0.2 h/h is considered to boost the selection of an efficient PHA-accumulating culture. Gray-filled areas represent different C/N ratios used and the white area is reactor operation at optimal C/N. The vertical solid line represents the organic loading rate increase. TSS (■); VSS (□); F/f(Δ); and PHAmax (•).

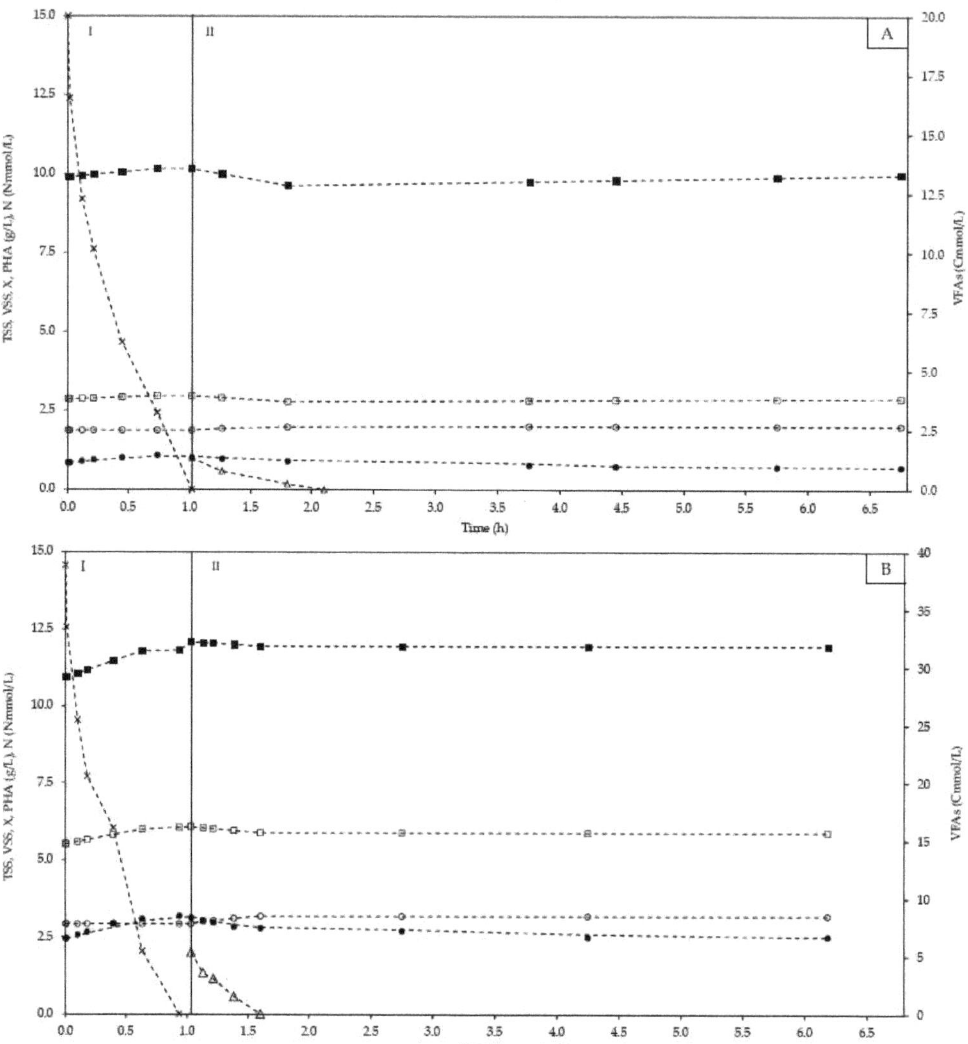

Figure 2. Representative SBR cycle conducted at organic loading rate (OLR) 60 Cmmol/(L.d) (**A**) and 120 Cmmol/(L.d) (**B**). The vertical solid line limit feast (I) and famine (II) periods. TSS (■); VSS (□); NH_4^+(Δ); PHAs (•); X (o); and VFAs (X).

The culture was then subjected to another OLR increase to 180 Cmmol/(L.d); however, the system became unstable as the biomass settling characteristics suffered a change, increasing the sludge volume index up to a point that the used reactor could not support, and the operation was stopped.

With regards to PHA polymer, a composition of an average HB/HV ratio of 48%:52% wt. (46%:54%Cmol) was obtained for an OLR of 60 Cmmol/(L.d), whereas a polymer with higher 3HV content, 35%:65% wt. (33%:67%Cmol), was produced at a higher OLR. Considering that HV monomers can be produced by a single molecule of valeric acid or by a combination of acetic and propionic acids in a ratio of 1:1 molecules [28], approximately 46% ± 2% (≈6.1 $Cmol_{HV}$/13.3 $Cmol_{HVprecursors}$) and 82% ± 2% (≈21.9 $Cmol_{HV}$/26.7 $Cmol_{HVprecursors}$) of the HV precursors were used towards HV synthesis for OLR 60 and 120 Cmmol/(L.d), respectively. The recruitment of 3HB as a compatible solute to counteract the high external osmotic pressure and to be a potent compatible solute with chaperoning activity [27,29] could be the explanation/promoter for the higher HV polymerization and the obtained HB:HV ratio.

Overall, the general behaviour of the different OLRs was similar; however, the specific consumption and production rates found for the tested conditions had noticeable differences (Table 1). The consumption rates of acetic and propionic acids were similar, 0.08 ± 0.00 vs. 0.10 ± 0.01 $Cmol_{HAc}/(Cmol_X.h)$ and 0.10 ± 0.02 vs. 0.11 ± 0.03 $Cmol_{HPro}/(Cmol_X.h)$, whereas consumption rates of butyric and valeric acids were higher under conditions with higher OLR, increasing to 78% from 0.09 ± 0.01 to 0.16 ± 0.03 $Cmol_{HBut}/(Cmol_X.h)$ and to 77% from 0.13 ± 0.02 to 0.23 ± 0.03 $Cmol_{HVal}/(Cmol_X.h)$ (OLR 60 vs. OLR 120). The higher consumption-specific rates of these VFAs led to a slight difference in the overall VFA consumption rates, which increased from 0.40 ± 0.03 $Cmol_{VFA}/(Cmol_X.h)$ at an OLR of 60 Cmmol/(L.d) to 0.60 ± 0.04 $Cmol_{VFA}/Cmol_X$ at an OLR of 120 Cmmol/(L.d). With regards to PHA production, there was a considerable difference in the specific production rate of the 3HV monomer, being three times higher (0.11 ± 0.01 and 0.33 ± 0.04 $Cmol_{HV}/(Cmol_X.h)$), whereas the 3HB production was slightly higher, 0.09 ± 0.00 and 0.15 ± 0.02 $Cmol_{HB}/(Cmol_X.h)$, which could explain the sizeable differences in the composition of the copolymer produced by the culture. The determined PHA production yield was also considerably higher for the higher OLR (0.60 ± 0.01 vs. 0.75 ± 0.04 $Cmol_{PHA}/Cmol_{VFA}$).

Table 1. Operating conditions applied and performance parameters determined for the culture selection stage considering the average of two monitored batches carried out at organic loading rate (OLR) 60 and 120 Cmmol/(L.d).

Parameter (Unit)	Average ± Standard Deviation	
OLR ($Cmmol_{VFA}/(L.d)$)	60	120
FP profile (HAc/HPro/HBut/HVal,% Cmol basis)	25:25:25:25	25:25:25:25
Feast/famine (h/h)	14.0 ± 0.00	0.13 ± 0.01
X @cycle start ($g_X/L/Cmol_X/L$)	1.92 ± 0.04/75.8 ± 1.70	3.26 ± 0.34/129 ± 13.6
PHA_{max} (% wt., VSS basis)	35.1 ± 1.56	49.2 ± 3.13
ΔPHAs ((% wt., VSS basis)	5.70 ± 0.93	9.6 ± 1.76
HB/HV ratio (% wt. basis/Cmol basis)	48:52/46:54	35:65/33:67
-qVFA ($Cmmol_{VFA}/(C\text{-}mmol_X.h)$)	0.40 ± 0.03	0.60 ± 0.04
-qHAc ($Cmmol_{HAc}/(Cmmol_X.h)$)	0.08 ± 0.00	0.10 ± 0.01
-qHPro ($Cmmol_{HPro}/(Cmmol_X.h)$)	0.10 ± 0.02	0.11 ± 0.03
-qHBut ($Cmmol_{HBut}/(Cmmol_X.h)$)	0.09 ± 0.01	0.16 ± 0.03
-qHVal ($Cmmol_{HVal}/(Cmmol_X.h)$)	0.13 ± 0.02	0.23 ± 0.03
qPHAs ($Cmmol_{PHA}/(Cmmol_X.h)$)	0.20 ± 0.01	0.46 ± 0.01
qHB ($Cmmol_{HB}/(Cmmol_X.h)$)	0.09 ± 0.00	0.12 ± 0.01
qHV ($Cmmol_{HV}/(Cmmol_X.h)$)	0.11 ± 0.01	0.33 ± 0.04
qX_{famine} ($Cmmol_X/(Cmmol_X.h)$)	0.06 ± 0.00	0.15 ± 0.02
-qPHAs ($Cmmol_{PHA}/(Cmmol_X.h)$)	0.10 ± 0.03	0.36 ± 0.01
-qHB ($Cmmol_{HB}/(CmmolX.h)$)	0.08 ± 0.02	0.13 ± 0.02
-qHV ($Cmmol_{HV}/(Cmmol_X.h)$)	0.09 ± 0.03	0.23 ± 0.03
$Y_{PHA/VFA}$ ($Cmmol_{PHA}/Cmmol_{VFA}$)	0.60 ± 0.01	0.75 ± 0.04
$Y_{X/PHAs}$ ($Cmmol_X/Cmmol_{PHA}$)	0.70 ± 0.18	0.44 ± 0.03

During the famine phase, the specific consumption of PHAs (-qPHA) and the biomass growth (q_{Famine}) were higher in the higher OLR, being 0.36 ± 0.01 $Cmmol_{PHA}/(Cmmol_X.h)$ and 0.15 ± 0.02 $Cmmol_X/(Cmmol_X.h)$ vs. 0.10 ± 0.03 $Cmmo_{PHA}/(Cmmol_X.h)$ and 0.06 ± 0.00 $Cmmol_X/(Cmmol_X.h)$, respectively, for the culture with an OLR of 60 Cmmol/(L.d). However, the obtained yields for the biomass growth were 0.70 ± 0.18 $Cmol_X/Cmol_{PHAs}$ and 0.44 ± 0.03 $Cmol_X/Cmol_{PHAs}$ for the higher OLR condition. In this respect, the recruitment of 3HB, as a player in the osmotic balance, could also explain the differences found in growth yields.

Comparing the obtained results with the ones obtained by Oliveira et al., using real non-saline feedstock and an uncoupled feast and famine strategy, the culture in the present study had higher qPHAs (0.33 ± 0.04 vs. 0.24 ± 0.0 $Cmol_X/Cmol_{PHA}$) even though it was in presence of 30 g_{NaCl}/L [10]. However, the real feedstock used also had in the composition a small amount of ammonia and proteins that may have contributed to the slightly lower qPHAs values. The obtained values in the present study were also higher, as compared to a study by Alba Pedrouso et al., where the culture that had acclimatized at 5 g_{NaCl}/L reached a qPHA of 0.24 $Cmmol_{PHA}/(Cmmol_X.h)$ [21]. It was also considerably higher, as compared to the slowly adapted culture to 20 g_{NaCl}/L in the work of Roibás-Rozas et al., where the qPHA was only 0.05 $Cmmol_{PHA}/(Cmmol_X.h)$ [23]. The use of an inoculum obtained from a saline environment, a strategy that other studies have not followed, may have played a major role in the swiftness and favourable outcome of the enrichment reactor as the microorganisms were already adapted to their natural saline environments and were, therefore, able to outperform cultures that had not been naturally adapted.

The differences found between the PHA storage capacity, the specific rates of the two applied OLRs, as well as the shift of the HB/HV ratio of the produced polymer suggested a slight culture change in the OLR of 120 Cmmol/(L.d). In this respect, culture samples were taken for the identification of the selected microorganisms just before the change of the OLR and at the end of the operation. The 16S rRNA gene sequencing revealed the presence of a diverse, mixed microbial community in the sediments collected from the saline area of Rio Tejo and that the applied selective pressure for PHA-accumulating organisms had led to a shift in the mixed microbial profile (Supplementary Material, Figure S1). The culture selected at an OLR of 60 Cmmol/(L.d) was enriched on a microorganism from the Alphaproteobacteria phylum, dominated by a species from Rhodobacteraceae (74.5%) as well as a few species from the Phyllobacteriaceae (13.3%) family, whereas at the highest OLR operation (120 Cmmol/(L.d)), the Rhodobacteraceae family dominated even more, representing 95.2% of the selected culture. Rhodobacteraceae are aquatic bacteria that have a vast global distribution [30] and frequently thrive in marine environments, which was in accordance with the origin of the used inoculum and the saline applied conditions. Rhodobacteraceae family members are known to have PhaC class I PHAs synthase, producing short-chain length PHAs based on VFA precursors availability [31,32]. It has also been reported that the synthesis of both P3HB and P(3HB-co-3HV) polymers were from unrelated carbon sources [33].

Overall, an efficient and stable halotolerant PHA-accumulating culture was successfully selected under high salinity conditions (30 g_{NaCl}/L), and to the best of our knowledge, this may be the first report of culture selection for PHA-accumulating MMCs at such a high salt concentration.

3.2. PHAs Accumulation Assays

After the cycles were characterized, half of the biomass of the selection reactor at the end of the famine phase was harvested for an accumulation assay in a fed-batch reactor using a pulse-wise feeding strategy (Figure 3). Pulses of the VFA mixture without ammonia and with a food to microorganism ratio (Cmol/gVSS) of 1.5 times the one found in the selection reactor were given to the culture as the DO concentration started to increase, so that the culture was consistently consuming VFAs and accumulating PHAs. For both accumulation assays, the biomass concentrations were considered constant as no nitrogen was fed to the system, and the specific rates of the three initial pulses were aligned with the presented averages. The obtained results are presented in Table 2.

The accumulation reactor that operated at an OLR of 60 Cmmol/(L.d) (ACC-60, Figure 3A) was fed with 4 pulses, reaching a maximum PHA content of 55.3% wt. with a polymer composition of 49%: 51% (% wt./wt., HB/HV) was similar to the one found in the preceding monitored cycles.

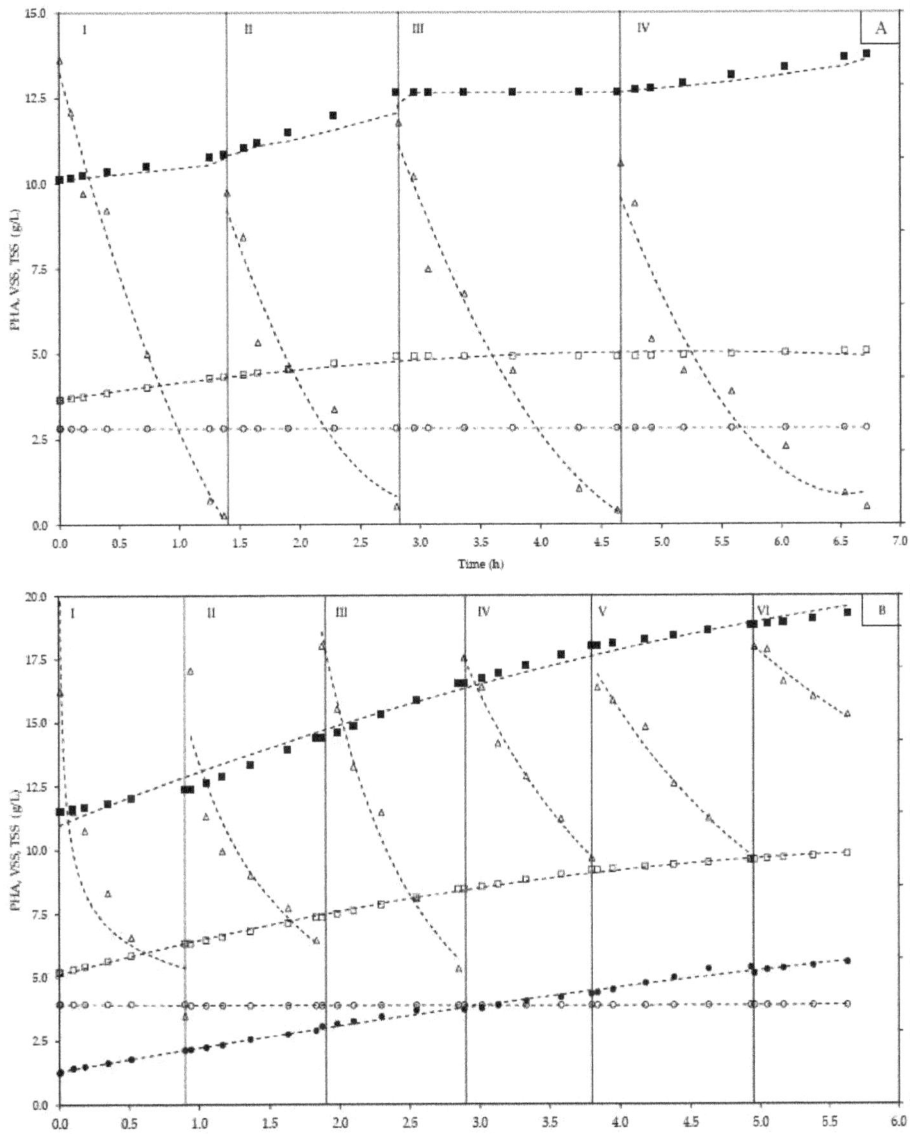

Figure 3. Profile of fed-batch accumulations conducted with the cultures selected at OLRs of 60 Cmmol/(L.d) (**A**) and 120 Cmmol/(L.d) (**B**). Pulses limited by the vertical solid lines and numbered between I and IV for ACC-60 and I and VI for ACC-120. TSS (■); VSS (□); PHAs (•); X (o); and VFAs (Δ).

The accumulation with the culture selected at an OLR 120 Cmmol/(L.d) (ACC-120) had a slightly different behavior and provided carbon pulses closer to the main PHA precursors depletion (valeric and butyric acids) and, therefore, completed a 6-pulse accumulation assay in less time than the 4 pulses of ACC-60. The higher consumption rates of valeric and butyric acids obtained for the culture selected at an OLR 120 Cmmol/(L.d) (Table 1) allowed us to identify their real-time depletion based on the DO profile (Supplementary Material,

Figure S2). However, due to the feeding strategy applied, acetic and propionic acids were accumulated throughout the assay at approximately 95% of the total remnant VFA concentration (53 Cmmol/L) in the effluent when the assay was stopped. The accumulation at an OLR of 120 Cmmol/(L.d) (ACC-60, Figure 3B) was able to reach a PHA content of 84.1% wt.

Table 2. Operating conditions applied and performance parameters determined for the PHA-accumulation stage considering one monitored batch carried out with the culture selected at OLRs of 60 and 120 Cmmol/(L.d). Only the first four pulses were considered for the average values, as during the fifth pulse, the exogenous carbon was mainly channelled for growth.

Parameter (Unit)	Average ± Standard Deviation	
OLR ($Cmmol_{VFA}$/(L.d))	60	120
FP profile (HAc/HPro/HBut/HVal,% Cmol basis)	25:25:25:25	25:25:25:25
X @inoculum (g_X/L/$Cmol_X$/L)	2.73/108	3.72/147
PHAs @inoculum (% wt., VSS basis)	22.3	35.4
PHA_{max} (% wt., VSS basis)	55.3	84.1
HB/HV ratio (% wt. basis/Cmol basis)	49:51/47:53	37:63/35:65
-qVFA[b] ($Cmmol_{VFA}$/(C-$mmol_X$.h))	0.41 ± 0.02	0.45 ± 0.04
qPHA[b] ($Cmmol_{PHA}$/($Cmmol_X$.h))	0.24 ± 0.01	0.30 ± 0.01
$Y_{PHA/VFA}$ ($Cmmol_{PHA}$/$Cmmol_{VFA}$)	0.60 ± 0.05	0.67 ± 0.05
Volumetric PHA productivity (g_{PHA}/(L.h))	0.30 [a] (0.32 [b])	0.77 [a] (0.84 [b])
Specific PHA productivity(g_{PHA}/(g_X.h))	0.11 [a] (0.12 [b])	0.21 [a] (0.28 [b])

[a] considering 4 pulses for ACC-60 and 6 pulses for ACC-B. [b] considering the first 3 pulses of the accumulation test.

The HB/HV polymer composition of ACC-120 was also similar to the one found in the cycle monitorizations that preceded the accumulation at 37%:63% (% wt./wt., HB/HV) and 35%:65% (Cmmol basis; HB/HV). As seen in the comparison of the monitorizations at the different OLRs, the qPHA was higher in the ACC-120 with a value of 0.22 ± 0.03 $Cmol_{HV}$/($Cmol_X$.h) than the 0.16 ± 0.01 $Cmol_{HV}$/($Cmol_X$.h) obtained for ACC-60. The PHAs storage yield was also higher in ACC-120 with a value of 0.67 ± 0.05 $Cmol_{PHA}$/$Cmol_{VFA}$ while 0.60 ± 0.05 $Cmol_{PHA}$/$Cmol_{VFA}$ was the yield found for ACC-60. Overall, a higher volumetric (0.77 g_{PHA}/(L.h)) and specific productivity (0.21 g_{PHA}/(g_X·h)) were obtained in ACC-120, as compared to ACC-60 (0.3 g_{PHA}/(L.h)) and 0.11 mg_{PHA}/(g_X.h), respectively). The obtained values, namely in ACC-120, were considerably higher than the presented results of Roibas-Rozas et al., where an adapted MMC to a lower NaCl concentration (close to 20 g_{NaCl}/L) was able to reach a PHA concentration of 41.5% wt. with a specific productivity of 68.1 mg_{PHA}/(g_X.h) [23]. These results were, however, obtained by using a real feed, rich in nitrogen, and the salt concentration was increased throughout the assay, starting from 5 g_{NaCl}/L, and the VFA mixture used was 43:7:42:8 (Cmol basis) acetic/propionic/butyric/valeric acids [23]. The obtained values in the present study were also significantly higher, as compared to those obtained by Oliveira et al. of 33% wt. using the acclimatized culture with an uncoupled F/f strategy and real feedstock (as previously described). Nevertheless the qPHAs and specific productivity were slightly higher than those obtained in the present study with 0.41 $Cmol_{PHA}$/($Cmol_X$.h) and 0.25 g_{PHA}/(g_X.h), respectively [10].

The high HV content in the copolymer (Table 2) obtained at the highest OLR was unusually high and surpassed the lowest melting temperature range obtained for P(3HB-co-3HV) accordingly to the work by Chan et al. [34]. In that study the melting temperature decreased to 60–70 °C for copolymers with an HV content up to 35–44% Cmol and increased to a range of 80–100 °C for higher HV contents. The morphological and physical properties of the obtained polymer enriched in HV will be determined in future studies, which will give insights about potential applications of the polymer.

4. Conclusions

The bioconversion of VFAs into PHAs by MMC under high salinity was shown to be feasible in our study. An efficient halotolerant PHA-accumulating MMC dominated by Rhodobacteraceae (93%) was selected at a salinity of 30 g_{NaCl}/L by submitting the culture to an aerobic dynamic feeding, namely an F/f regime with uncoupled carbon and nitrogen availabilities throughout the operation. This culture showed good adaptation to high salinity, likely due to the activation of the metabolic pathways to counteract the osmotic pressure, namely through the assimilation of inorganic compounds and the recruitment of 3HB as osmolytes.

The culture demonstrated a high PHA storage capacity, reaching 84.1% wt. of P(3HB-co-3HV) with an HV content of 63% wt. The produced sort-chain-length copolymer was unusual due to its high HV content. This efficient culture also presented a notable volumetric productivity of 0.77 g_{PHA}/(L.h) and specific productivity of 0.21 g_{PHA}/(gX.h).

To best of our knowledge, this is the first study of a successful acclimatization of an efficient PHA-accumulating MMC with near-sea-water salinity. It suggested the potential for using saline side streams as feedstock for PHA production as well as for using sea water in the process, which could lower production costs and the environmental burden as well as promote the integration of this biopolymer in the market.

Supplementary Materials: The following supporting information can be downloaded at: https://www.mdpi.com/article/10.3390/su14031346/s1, Figure S1: Heatmap of the most abundant bacterial microbial communities present in the inoculum and the selected cultures for OLR of 60 and 120 Cmmol/(L.d).; FigureS2: Representative oxygen profile consumption for a cycle at OLR of 60 Cmmol/(L.d) (dash line) and at OLR of 120 Cmmol/(L.d) (full line).

Author Contributions: Conceptualization, B.C.M. and M.A.M.R.; investigation, J.M.C. and B.C.M.; visualization, J.M.C. and B.C.M.; writing—original draft preparation, J.M.C.; writing—review and editing, B.C.M. and M.A.M.R.; supervision, B.C.M. and M.A.M.R.; project administration, B.C.M. and M.A.M.R.; funding acquisition, M.A.M.R. All authors have read and agreed to the published version of the manuscript.

Funding: This research was funded by FCT—Fundação para a Ciência e a Tecnologia, I.P. (SaltiPHA project: PTDC/BTA-BTA/30902/2017, Research Unit on Applied Molecular Biosciences—UCIBIO: UIDP/04378/2020 and UIDB/04378/2020, Associate Laboratory Institute for Health and Bioeconomy—i4HB: LA/P/0140/2020).

Institutional Review Board Statement: Not applicable.

Informed Consent Statement: Not applicable.

Data Availability Statement: Not applicable.

Acknowledgments: The authors also acknowledge Elisabete Freitas, Elsa Mora, and Mónica Centeio for their assistance with the chemical analysis.

Conflicts of Interest: The authors declare no conflict of interest.

References

1. Fish & Seafood—Worldwide | Statista Market Forecast. Available online: https://www.statista.com/outlook/cmo/food/fish-seafood/worldwide (accessed on 29 November 2021).
2. Ching, Y.C.; Redzwan, G. Biological treatment of fish processing saline wastewater for reuse as liquid fertilizer. *Sustainability* **2017**, *9*, 1062. [CrossRef]
3. Cleaner Production Assessment in Fish Processing. Available online: https://digitallibrary.un.org/record/441680 (accessed on 21 May 2021).
4. Anh, H.T.H.; Shahsavari, E.; Bott, N.J.; Ball, A.S. Options for Improved Treatment of Saline Wastewater from Fish and Shellfish Processing. *Front. Environ. Sci.* **2021**, *9*, 236. [CrossRef]
5. Lefebvre, O.; Moletta, R. Treatment of organic pollution in industrial saline wastewater: A literature review. *Water Res.* **2006**, *40*, 3671–3682. [CrossRef] [PubMed]
6. Dan, N.P.; Visvanathan, C.; Basu, B. Comparative evaluation of yeast and bacterial treatment of high salinity wastewater based on biokinetic coefficients. *Bioresour. Technol.* **2003**, *87*, 51–56. [CrossRef]

7. Srivastava, A.; Parida, V.K.; Majumder, A.; Gupta, B.; Gupta, A.K. Treatment of saline wastewater using physicochemical, biological, and hybrid processes: Insights into inhibition mechanisms, treatment efficiencies and performance enhancement. *J. Environ. Chem. Eng.* **2021**, *9*, 105775. [CrossRef]
8. Kourmentza, C.; Plácido, J.; Venetsaneas, N.; Burniol-Figols, A.; Varrone, C.; Gavala, H.N.; Reis, M.A.M. Recent advances and challenges towards sustainable polyhydroxyalkanoate (PHA) production. *Bioengineering* **2017**, *4*, 55. [CrossRef]
9. Polyhydroxyalkanoate (PHA) Market Global Forecast to 2025 | MarketsandMarkets. Available online: https://www.marketsandmarkets.com/Market-Reports/pha-market-395.html?gclid=CjwKCAjw9MuCBhBUEiwAbDZ-7tH0vuR1VsKfgUvzT6jPuDry18A_H8vWVihQJFN_iN7OotkL2PVH7xoCDykQAvD_BwE (accessed on 29 November 2021).
10. Oliveira, C.S.S.; Silva, C.E.; Carvalho, G.; Reis, M.A. Strategies for efficiently selecting PHA producing mixed microbial cultures using complex feedstocks: Feast and famine regime and uncoupled carbon and nitrogen availabilities. *New Biotechnol.* **2017**, *37*, 69–79. [CrossRef]
11. Oliveira, C.S.S.; Silva, M.O.D.; Silva, C.E.; Carvalho, G.; Reis, M.A.M. Assessment of protein-rich cheese whey waste stream as a nutrients source for low-cost mixed microbial PHA production. *Appl. Sci.* **2018**, *8*, 1817. [CrossRef]
12. Silva, F.; Campanari, S.; Matteo, S.; Valentino, F.; Majone, M.; Villano, M. Impact of nitrogen feeding regulation on polyhydroxyalkanoates production by mixed microbial cultures. *New Biotechnol.* **2017**, *37*, 90–98. [CrossRef]
13. Ecoefficient Biodegradable Composite Advanced Packaging | ECOBIOCAP Project | FP7 | CORDIS | European Commission. Available online: https://cordis.europa.eu/project/id/265669 (accessed on 20 November 2020).
14. Koller, M.; Niebelschütz, H.; Braunegg, G. Strategies for recovery and purification of poly[(R)-3-hydroxyalkanoates] (PHA) biopolyesters from surrounding biomass. *Eng. Life Sci.* **2013**, *13*, 549–562. [CrossRef]
15. Xiao, Y.; Roberts, D.J. A review of anaerobic treatment of saline wastewater. *Environ. Technol.* **2010**, *31*, 1025–1043. [CrossRef] [PubMed]
16. He, X.; Yin, J.; Liu, J.; Chen, T.; Shen, D. Characteristics of acidogenic fermentation for volatile fatty acid production from food waste at high concentrations of NaCl. *Bioresour. Technol.* **2019**, *271*, 244–250. [CrossRef] [PubMed]
17. Duarte, M.S.; Oliveira, J.V.; Pereira, C.; Carvalho, M.; Mesquita, D.P.; Alves, M.M. Volatile Fatty Acids (VFA) Production from Wastewaters with High Salinity—Influence of pH, Salinity and Reactor Configuration. *Fermentation* **2021**, *7*, 303. [CrossRef]
18. Fra-Vázquez, A.; Pedrouso, A.; Val del Rio, A.; Mosquera-Corral, A. Volatile fatty acid production from saline cooked mussel processing wastewater at low pH. *Sci. Total Environ.* **2020**, *732*, 139337. [CrossRef]
19. Palmeiro-Sánchez, T.; Oliveira, C.S.S.; Gouveia, A.R.; Noronha, J.P.; Ramos, A.M.; Mosquera-Corral, A.; Reis, M.A.M. NaCl presence and purification affect the properties of mixed culture PHAs. *Eur. Polym. J.* **2016**, *85*, 256–265. [CrossRef]
20. Wen, Q.; Ji, Y.; Hao, Y.; Huang, L.; Chen, Z.; Sposob, M. Effect of sodium chloride on polyhydroxyalkanoate production from food waste fermentation leachate under different organic loading rate. *Bioresour. Technol.* **2018**, *267*, 133–140. [CrossRef]
21. Pedrouso, A.; Fra-Vazquez, A.; Del Rio, A.V.; Mosquera-Corral, A. Recovery of Polyhydroxyalkanoates from Cooked Mussel Processing Wastewater at High Salinity and Acidic Conditions. *Sustainability* **2020**, *12*, 10386. [CrossRef]
22. Argiz, L.; Fra-Vázquez, A.; del Río, Á.V.; Mosquera-Corral, A. Optimization of an enriched mixed culture to increase PHA accumulation using industrial saline complex wastewater as a substrate. *Chemosphere* **2020**, *247*, 125873. [CrossRef]
23. Roibás-Rozas, A.; Val del Rio, A.; Hospido, A.; Mosquera-Corral, A. Strategies for the valorisation of a protein-rich saline waste stream into polyhydroxyalkanoates (PHA). *Bioresour. Technol.* **2021**, *334*, 124964. [CrossRef]
24. APHA/AWWA. *Standard Methods for the Examination of Water and Wastewater*, 20th ed.; APHA American Public Health Association: Washington, DC, USA, 1998.
25. Wang, X.; Oehmen, A.; Freitas, E.B.; Carvalho, G.; Reis, M.A.M. The link of feast-phase dissolved oxygen (DO) with substrate competition and microbial selection in PHA production. *Water Res.* **2017**, *112*, 269–278. [CrossRef]
26. Heinzle, E.; Biwer, A.P.; Cooney, C.L. *Development of Sustainable Bioprocesses*; John Wiley & Sons, Ltd: Chichester, UK, 2006.
27. Obruca, S.; Sedlacek, P.; Koller, M.; Kucera, D.; Pernicova, I. Involvement of polyhydroxyalkanoates in stress resistance of microbial cells: Biotechnological consequences and applications. *Biotechnol. Adv.* **2018**, *36*, 856–870. [CrossRef] [PubMed]
28. Prados, E.; Maicas, S. Bacterial Production of Hydroxyalkanoates (PHA). *Univers. J. Microbiol. Res.* **2016**, *4*, 23–30. [CrossRef]
29. Soto, G.; Setten, L.; Lisi, C.; Maurelis, C.; Mozzicafreddo, M.; Cuccioloni, M.; Angeletti, M.; Ayub, N.D. Hydroxybutyrate prevents protein aggregation in the halotolerant bacterium Pseudomonas sp. CT13 under abiotic stress. *Extremophiles* **2012**, *16*, 455–462. [CrossRef] [PubMed]
30. Pujalte, M.J.; Lucena, T.; Ruvira, M.A.; Arahal, D.R.; Macián, M.C. The family *Rhodobacteraceae*. In *The Prokaryotes*; Springer: Berlin/Heidelberg, Germany, 2014; pp. 439–512. [CrossRef]
31. Mezzolla, V.; D'Urso, O.F.; Poltronieri, P. Role of PhaC type I and type II enzymes during PHA biosynthesis. *Polymers* **2018**, *10*, 910. [CrossRef]
32. Perez-Zabaleta, M.; Atasoy, M.; Khatami, K.; Eriksson, E.; Cetecioglu, Z. Bio-based conversion of volatile fatty acids from waste streams to polyhydroxyalkanoates using mixed microbial cultures. *Bioresour. Technol.* **2021**, *323*, 124604. [CrossRef]

33. Van-Thuoc, D.; Huu-Phong, T.; Minh-Khuong, D.; Hatti-Kaul, R. Poly(3-Hydroxybutyrate-co-3-Hydroxyvalerate) Production by a Moderate Halophile *Yangia* sp. ND199 Using Glycerol as a Carbon Source. *Appl. Biochem. Biotechnol.* **2015**, *175*, 3120–3132. [CrossRef]
34. Chan, C.M.; Vandi, L.J.; Pratt, S.; Halley, P.; Ma, Y.; Chen, G.Q.; Richardson, D.; Werker, A.; Laycock, B. Understanding the effect of copolymer content on the processability and mechanical properties of polyhydroxyalkanoate (PHA)/wood composites. *Compos. Part A Appl. Sci. Manuf.* **2019**, *124*, 105437. [CrossRef]

sustainability

MDPI

Article

Municipal Wastewater Reuse: Is it a Competitive Alternative to Seawater Desalination?

Dafne Crutchik * and José Luis Campos

Faculty of Engineering and Sciences, Universidad Adolfo Ibáñez, Av. Diagonal Las Torres 2640, Santiago 7941169, Chile; jluis.campos@uai.cl

* Correspondence: dafne.crutchik@uai.cl; Tel.: +56-(2)-2331-1934

Abstract: Water scarcity is becoming a global challenge to attempts to narrow the water demand–supply gap. To overcome this problem, it is sensible to consider alternative technologies that can exploit non-conventional water resources. The choice of such technologies should be, however, carefully analyzed, because any choice might be unfeasible from an economic point of view. In this work, a methodology to select the most appropriate non-conventional water resource, out of municipal wastewater and seawater, was proposed. Specifically, we attempted to determine which alternative provides cheaper water supply and production costs for domestic uses, depending on the wastewater treatment system used and the water plant capacity. The production of water under three scenarios was analyzed: (i) a city that has a conventional wastewater treatment plant (WWTP); (ii) a city that uses primary treatment and submarine outfalls to treat municipal wastewater; (iii) seawater desalination. The proposed methodology was tested in Chilean cities that are located in areas where water is a scarce resource. The results showed that the reuse of municipal wastewater represents a cost-competitive alternative to seawater desalination, mainly when municipal wastewater is treated in a conventional WWTP and when water flow demand is higher than 1500 m^3/d. In contrast, seawater desalination becomes more profitable than wastewater reuse when the treatment of municipal wastewater is based on the use of submarine outfalls. This study provides a useful economic tool for promoting municipal wastewater reuse as a non-conventional water source for supplying water to cities that suffer from water scarcity in Chile and in similar areas of the world.

Keywords: economic analysis; non-conventional water resources; resource recovery; water; water scarcity

Citation: Crutchik, D.; Campos, J.L. Municipal Wastewater Reuse: Is it a Competitive Alternative to Seawater Desalination? *Sustainability* **2021**, *13*, 6815. https://doi.org/10.3390/su13126815

Academic Editor: Ashwani Kumar Tiwari

Received: 28 April 2021
Accepted: 8 June 2021
Published: 16 June 2021

1. Introduction

Water scarcity has been recognized as a serious global issue of this century. At present, around 25% of the world's population live in areas that experience water stress, and it is estimated that more than half of the global population will experience severe water scarcity in the coming years [1]. As the world's population continues to grow and the impacts of climate change intensify, water shortages are expected to increase in the near future, particularly in arid and semi-arid regions [1,2]. In the context of water scarcity, it is imperative to use water more efficiently. However, this is not enough, and it is necessary to explore alternative and sustainable technologies to exploit non-conventional water resources (e.g., seawater, wastewater, rainwater, fog, among others) [3–5].

Among these alternative water supply technologies, membrane-based separation processes, particularly reverse osmosis, provide an opportunity to deal with water scarcity in many water-stressed regions around the world [6–9]. At present, seawater reverse osmosis dominates the desalination market; around 66 million m^3/d were produced worldwide, which accounts for 69% of the total desalinated water [4]. Nevertheless, regardless of the success of this desalination process, it may be limited by the consumption of energy related to the filtration process and later water transportation costs [10,11]. In this context, it is recognized that reverse osmosis is an energy-intensive process, especially when desalinating seawater [12–14], but continued improvements, such as energy recovery or coupling membrane systems with renewable energy sources, have allowed a reduction in economic costs, thus making it more economically viable [10,15,16]. The typical production costs of seawater desalination are between 0.45 and 2.5 USD/m^3 (excluding the costs of water transportation) for desalination plant capacities between 60,000 and 1000 m^3/d, respectively, and it is expected that these costs will reduce by around 20% over the next 5 years and by more than half in the next 20 years [16–18].

Furthermore, the produced water must be transported from the production facility to the water use site (e.g., water purification plant), and it may be expensive to transport water across long-distances [14]. In fact, when water demand sites are located far away to the coast and/or at high altitudes, exploiting other conventional or non-conventional water resources may be more economical than obtaining water from the sea (Table 1) [9,19]. Therefore, water transportation costs can significantly contribute to total water production costs, affecting the economic viability of the seawater desalination process [11,14,20]. Thus, the total costs to produce water are one of the most critical factors that influence the implementation of these kind of projects [21,22], and these costs depend on several factors, such as plant production capacity, the quality of feedwater, technology, the location of water plants, energy costs, plant lifetime, among others [9,16,23]. In fact, most previous studies have reported a wide range of water production costs because they have been developed for a particular water demand site with a specific plant capacity, water production technology, and feedwater, and consequently these costs were restricted to the particular conditions in which they were determined, which can make it difficult to compare among them.

Table 1. Transportation cost of desalinated water for different cities [9].

City, Country	Distance (km)	Elevation (m)	Cost (USD/m^3)
México city, México	225	2500	2.44
Sana, Yemen	135	2500	2.38
Beijing, China	135	100	1.13
Crateus, Brazil	240	350	1.33
Phoenix, USA	280	320	1.34
Delhi, India	1050	500	1.90
Zaragoza, Spain	163	500	1.36

Taking into account that about 66% of the global urban population lives in urban centers bordering the ocean [14,16,24], the potential implementation of seawater desalination plants is therefore especially significant. However, it may not be a viable solution for water-stressed regions that are located a long distance from the coast or at a high altitude [9,24]. The reuse of municipal wastewater could be a viable alternative to address water scarcity for these cases [6,14,25]. In this context, there are already European countries that reuse treated wastewater for non-potable and potable uses [2]. The conventional treatment of municipal wastewater is usually based on primary treatment followed by secondary treatment, which usually involves a biological process to remove organic matter from wastewater, in order to meet the standards needed for its discharge. In order to reuse treated municipal wastewater, a tertiary treatment (e.g., membrane-based separation processes) is needed to remove the remaining pollutants from secondary treated effluent, such as inorganic and organic compounds, pathogens, or nutrients, in order to meet water standards [26]. The reuse of municipal wastewater allows for an increase in the water supply flow rate, but its additional costs, consisting of both the extra treatment needed to reach the water quality requirement and the transportation of the produced water to the reuse site, should be considered.

Based on this background, this work proposed a methodology to select the most appropriate non-conventional water resource between municipal wastewater and seawater. Specifically, we consider which alternative provides a cheaper water supply for domestic uses, particularly in water-stressed regions. The proposed methodology was tested in Chilean cities that are located in areas where water is a scarce resource. Chile was chosen as the subject of this research because it is one of the countries most vulnerable to the impacts of climate change, due mainly to its geographical location. At present, water scarcity is one of the most significant effects of climate change in several regions of Chile, particularly in the northern and central regions, and it is expected that their water demand will increase significantly over the next decades [27]. Several desalination plants have been implemented to supply water for domestic and/or industrial uses in the north of the country in recent years [28,29], where water shortages are more severe. However, as previously mentioned, the economic feasibility of these desalination projects is limited by the costs associated with water transportation, particularly for cities located in inland areas, where the industrial sector (e.g., mining) is only able to pay these water costs, causing important socio-economic conflicts related to water use [30]. Therefore, it is necessary to explore more economical water supply options for domestic use—for instance, the reuse of municipal wastewater.

This paper contributes to the current literature by proposing a novel methodology that helps to select, from an economic point of view, the most appropriate non-conventional water resource, between municipal wastewater and seawater, to supply water, when assessing the implementation of water production projects.

2. Materials and Methods

2.1. Proposed Scenarios

Three scenarios to produce water from municipal wastewater and seawater are proposed in this work:

(a) Scenario 1. This scenario is based on a city whose wastewater is treated by a WWTP with both primary and secondary treatments. In this case, a post-treatment of the WWTP effluent would be implemented by means of a hybrid ultrafiltration (UF) and reverse osmosis (RO) system. First, the secondary treated effluent is passed through the UF membrane to remove particles and colloids and then through RO membrane to remove the remaining pollutants. It was considered that UF and RO membranes are periodically cleaned using reagents to maintain a suitable pressure drop. The recovery ratio was set to 70% in relation to the inlet wastewater volume [31].

(b) Scenario 2. This scenario is based on a coastal city whose wastewater is treated by a primary treatment and its effluent is sent to sea by means of a submarine outfall. For this reason, in this proposed scenario, the construction of an activated sludge unit followed by a hybrid UF-RO system was considered.

(c) Scenario 3. This scenario involves the complete implementation of a seawater desalination plant based on the reverse osmosis process. The recovery ratio of the reverse osmosis system was set to 50% in relation to the initial volume of the feedwater (seawater) [14].

2.2. Methodology for Economic Assessment

The proposed methodology to determine the best option to supply water to the water-demanding city was schematically outlined throughout the following steps. Figure 1 illustrates the methodology proposed in this work.

1. The required water flow rate was calculated. The annual flow of the water demand (m^3/year) of the studied city was estimated based on its population and water consumption (m^3/cap/year).
2. The total water production costs (USD/m^3) for the proposed scenarios were estimated. First, total production costs were calculated based on the total capital costs (CAPEX) and the operating and maintenance costs (OPEX) (Equation (1)).

$$\text{Total production cos ts} = \frac{(\text{CAPEX} + \text{OPEX})\ (\text{USD/year})}{\text{feedwater flow}\ (\text{m}^3/\text{year})} \quad (1)$$

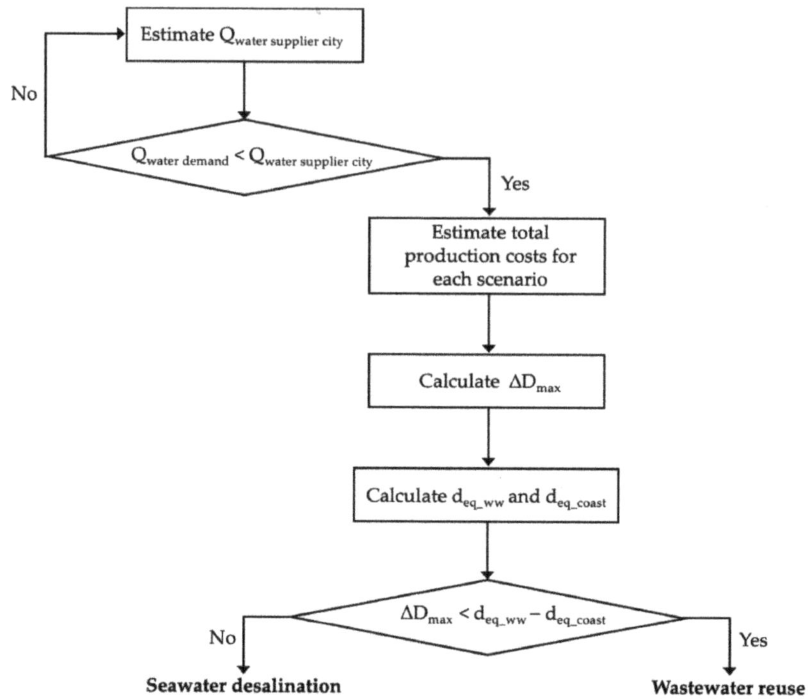

Figure 1. Procedure to select the most appropriate water resource to supply water for domestic uses (Q is water flow rate (m^3/d), ΔD_{max} is the maximum distance that produced water can be transported (km), d_{eq_ww} and d_{eq_coast} are the equivalent distance ((km) from the water source to water demand site for wastewater reuse scenarios and seawater desalination, respectively).

The total capital costs and the operating and maintenance costs for each scenario are detailed below:

(a) Scenario 1. The total capital costs included the cost related to the purchase of equipment (UF and RO units), the required equipment for piping, instrumentation/electricity, engineering costs and civil works (Table 2). The operating and maintenance costs included energy consumption, reagents consumption, membrane replacement, and maintenance and labor costs (Table 2). The capital costs and operating and maintenance costs related to the UF-RO system were calculated using cost functions that were developed based on the data reported by Plumlee et al. (2014) [32].

(b) Scenario 2. The total capital costs included costs related to the construction or purchase of the equipment (activated sludge system, UF and RO units) and the required equipment for piping, instrumentation/electricity, engineering costs, and civil works (Table 2). The costs associated with the operation of the activated sludge system were energy and reagents consumption, labor, waste management and maintenance. The total capital costs and operating and maintenance costs for the UF-RO system included the items described for scenario 1 (Table 2). The total capital costs and operating and maintenance costs of the activated sludge were calculated based on the data reported by Guo et al. (2014) [33] and Molinos-Senante et al. (2010) [34], respectively.

(c) Scenario 3. The total capital costs for the desalination plant included 5 cost items, which were construction and infrastructure costs (main equipment, piping, instrumentation/electricity, among others), land acquisition costs, engineering costs, and development and management costs (Table 2). The operating and maintenance costs included energy consumption, membrane replacement, maintenance, reagents consumption and labor costs (Table 2). Based on the salinity concentration and average temperature of the coast of Chile, around 35,000 ppm and 17 °C, respectively [35], it was assumed that the energy consumption of seawater desalination was 3.5 kWh/m^3. The capital costs and operating and maintenance costs were estimated using cost functions that were developed based on the results reported by Molinos-Senante and González (2019) [35]. The cost functions and the economic parameters for the proposed scenarios are given in Table 2. The price of electricity was set at 0.109 USD/kWh [36]. All costs used in this work were normalized to the USD of 2019.

Table 2. Summary of the cost functions for total capital costs (USD) and operating and maintenance costs (USD/year) for water production (y is cost, x is capacity (m^3/d)).

Item	Cost Function	Reference
Capital costs		
1. Municipal wastewater		
Reverse osmosis system	$y = 272.54 \cdot x + 4.9835 \cdot 10^6$	
Ultrafiltration system	$y = 136.38 \cdot x + 2.4859 \cdot 10^6$	
Yard piping	$y = 40.97 \cdot x + 7.3826 \cdot 10^5$	[32]
Sitework land scaping	$y = 20.38 \cdot x + 3.7647 \cdot 10^5$	
Site electrical and controls	$y = 81.86 \cdot x + 1.4916 \cdot 10^6$	
Activated sludge system	$\log(y) = 0.256 \cdot (\log(x))^{1.556} + 4.545$	[33]
2. Seawater		
Construction and infrastructure	$y = 8.996 \cdot 10^5 \cdot x + 6.20 \cdot 10^6$	
Land acquisition	$y = 17.995 \cdot x + 1.2363 \cdot 10^5$	[35]
Engineering	$y = 31.53 \cdot x + 2.1608 \cdot 10^5$	
Development and management	$y = 4.5263 \cdot x + 3.0165 \cdot 10^4$	
Operating and maintenance costs		
1. Municipal wastewater		
Labor	0.02 USD/m^3	
Reagents UF	$y = 3.1224 \cdot 10^{-2} + x \cdot 2.2448 \cdot 10^{-5}$	
Membrane replacement UF	$y = 4.6073 \cdot 10^{-3} + x \cdot 8.9988 \cdot 10^{-6}$	
Energy consumption UF	$y = -5.4386 \cdot 10^{-3} + x \cdot 4.0363 \cdot 10^{-6}$	[32]
Reagents RO	$y = 2.2126 \cdot 10^{-2} + x \cdot 2.2727 \cdot 10^{-5}$	
Membrane replacement RO	$y = 1.1905 \cdot 10^{-2} + x \cdot 8.8019 \cdot 10^{-6}$	
Energy consumption RO	$y = -3.0484 \cdot 10^{-2} + x \cdot 4.0087 \cdot 10^{-5}$	
Activated sludge system		
Energy consumption	0.033 USD/m^3	
Reagents	0.025 USD/m^3	
Labor	0.060 USD/m^3	[34]
Maintenance	0.038 USD/m^3	
Waste management	0.029 USD/m^3	
2. Seawater		
Energy consumption	$y = 1.461 \cdot 10^{-3} \cdot x + 4.946 \cdot 10^6$	
Membrane replacement	$y = 8 \cdot 10^{-2} \cdot x - 1.57 \cdot 10^{-1}$	
Reagents	$y = 4 \cdot 10^{-2} \cdot x - 7.85 \cdot 10^{-2}$	[35]
Labor	$y = 1.496 \cdot 10^{-2} \cdot x + 1.44 \cdot 10^5$	
Maintenance	$y = 8.086 \cdot 10^{-5} \cdot x + 7.883 \cdot 10^3$	

Then, the minimum cost of produced water that makes the net present value (NPV) zero (Equation (2)) for the proposed scenarios was calculated:

$$NPV = \sum_{t=1}^{T} \frac{(B_t - C_t) \cdot (1+i)^t}{(1+r)^t} - \text{Total capital cos t} \quad (2)$$

where: B_t are the benefits due to the water sale, C_t is the sum of the operating and maintenance costs, i is the inflation rate (3%), r is the interest rate (5%), and T is the payback time (20 years).

3. The best scenario to produce water was determined. A pairwise comparison in terms of the total production costs for the proposed scenarios was developed in order to determine the most profitable scenario. The total production costs for the proposed scenarios were compared and used to estimate the maximum distance that produced water can be transported from the water plant production to the water demand city if the cheaper scenario were selected. This distance (ΔD_{max}) was expressed as a function of the total production costs (USD/m^3), the transportation costs for horizontal distance (a, 0.05 USD/m^3/km/year) and the lifetime of the facility (t, 20 years) (Equation (3)). The transportation costs, a, was determined as a function of the piping and pumping costs for the horizontal distance, using the data reported by Caldera et al. (2018) [11] and ESCWA (2009) [37]. It should be noted that scenarios 1 and 2 have not been compared among them because it is only possible to implement one of these scenarios for a particular city, and their selection depends on the actual wastewater treatment system.

$$\Delta D_{max} = \left(\frac{\text{total production cos ts}_{\text{scenario 3}} - \text{total production cos ts}_{\text{scenario 1 or 2}}}{a} \right) \quad (3)$$

4. The selection of possible water sources for the studied city was carried out. Once the suitable scenario was determined, the selection of the potential water sources was developed based on the flow rate of water that would be supplied and the distance between the water source and water demand site. It should be noted that water transportation distance is comprised of the horizontal and vertical distances, respectively, and they have a different impact on the water production cost. Vertical distance has a larger impact on water transportation costs than horizontal distance [19], and thereby an equivalent distance from the water source to water demand site for wastewater reuse scenarios (d_{eq_ww}) and seawater desalination (d_{eq_coast}) was determined (Equation (4)):

$$d_{eq} = X \cdot \left(1 + \frac{b}{a} \cdot tg\left(\frac{Y}{X} \right) \right) \quad (4)$$

where: X and Y are the horizontal and vertical distances (km) from the water source to the water demand site, and a and b are the transportation costs for horizontal and vertical distances, 0.05 and 1.82 USD/m^3/km/year, respectively. The values of a and b were determined as a function of the piping and pumping costs for the vertical and horizontal distances, respectively, and were calculated based on the data reported by Caldera et al. (2018) [11] and ESCWA (2009) [37]. The horizontal and vertical distances between the water source and water demand site were determined using Google Earth Pro.

5. Finally, the best option to supply water to the water-demanding city was selected based on the maximum distance that produced water can be transported from the water plant production to the water-demanding city (ΔD_{max}), and the equivalent distance that water should be transported was obtained for the potential water sources. Therefore, if $\Delta D_{max} > (d_{eq\_ww} - d_{eq\_coast})$, the reuse of wastewater is more favorable than seawater desalination. In contrast, if $\Delta D_{max} < (d_{eq\_ww} - d_{eq\_coast})$, the desalination of seawater is more economical than wastewater reuse.

2.3. Case Study

Chile is a developing country that experiences great climatic variation throughout the country, and consequently, it has a highly unequal water distribution, with the northern regions being mostly arid and semi-arid areas and the south regions being temperate, ranging from Mediterranean to marine west coast areas [38]. The northern and central regions of Chile are economically and socially important due to the development of the main economic activities, mining and agriculture activities. However, water scarcity has been an important issue for these regions over the last decade. A sample of 8 Chilean cities was used to validate the methodology proposed in this work. These cities are located in the north and central regions of the country (Figure 2), where water scarcity is more accentuated [39].

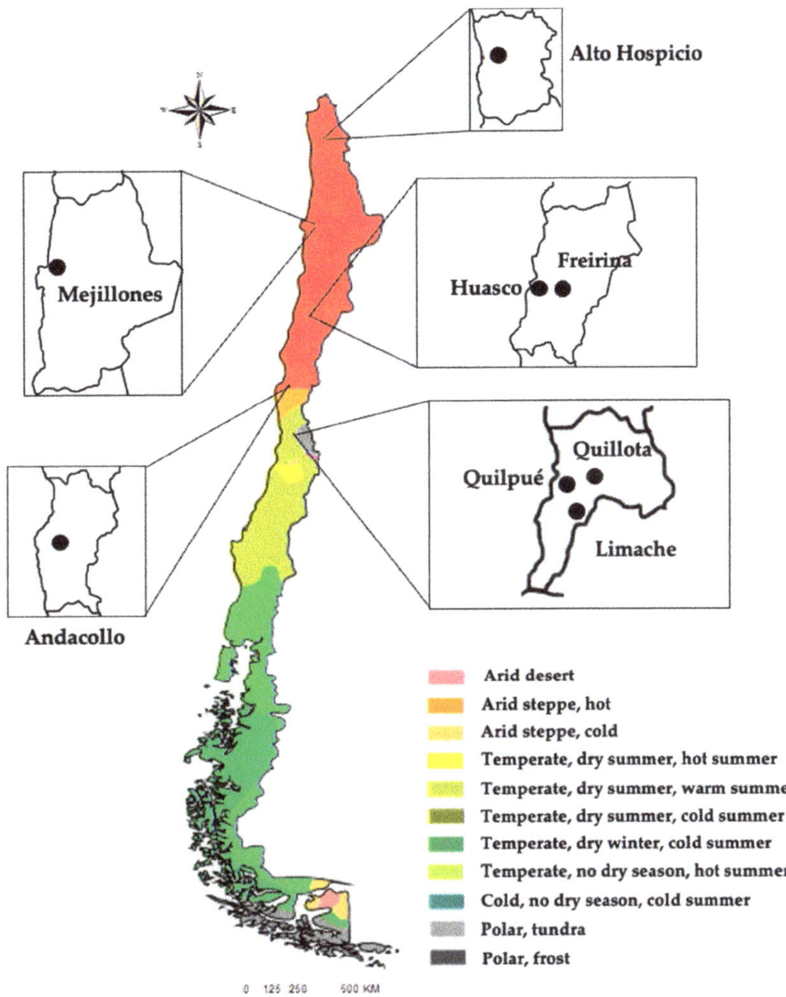

Figure 2. Localization of studied cities of Chile.

Two types of municipal wastewater treatment configurations can be differentiated in the wastewater treatment system of Chile: wastewater treatment plants that use primary settling followed by a biological treatment (hereafter referred to as conventional WWTP); and those that use submarine outfalls preceded by preliminary treatment for sanitary discharge (hereafter referred to as submarine outfalls WWTP). As previously mentioned, it was assumed that the recovery ratio to produce water from municipal wastewater was set at 70%, and thereby the reuse of municipal wastewater in the same city was not studied in this work because the produced water would not be sufficient to meet its water requirements. Therefore, the production and transportation of produced water to a nearby city which suffers from water stress was the strategy studied for these Chilean cities.

3. Results and Discussion

3.1. Cost Associated with Water Production

Based on the production plant capacity, the capital costs and operating and maintenance (O&M) costs to produce water flow rates between 500 and 100,000 m^3/d were calculated for each proposed scenario (Table 3). As expected, as water production capacity decreased, capital and operating and maintenance costs increased, regardless of the proposed scenario. Overall, the production of water from municipal wastewater by the use of the UF-RO system (scenario 1) was found to be less capital-intensive than seawater desalination, for water production capacities higher than 5000 m^3/d (Table 3). However, for the studied range of water production capacities, the capital costs of desalinated seawater were far lower than the cost to produce water from municipal wastewater by the implementation of an activated sludge unit followed by a hybrid UF-RO system (scenario 2) (Table 3).

Table 3. Total capital costs (USD/m^3/year), operating and maintenance costs (USD/m^3), and total production costs (USD/m^3) estimated for the proposed scenarios.

Water Production Capacity (m^3/d)	Scenario 1			Scenario 2			Scenario 3		
	Capital Costs	O&M Costs	Total Costs	Capital Costs	O&M Costs	Total Costs	Capital Costs	O&M Costs	Total Costs
100,000	0.071	0.264	0.334	0.135	0.450	0.585	0.135	0.452	0.587
90,000	0.072	0.264	0.335	0.136	0.450	0.586	0.136	0.452	0.587
80,000	0.073	0.264	0.337	0.138	0.450	0.588	0.136	0.452	0.588
70,000	0.074	0.264	0.338	0.139	0.450	0.589	0.137	0.452	0.589
60,000	0.076	0.264	0.340	0.141	0.450	0.592	0.138	0.452	0.590
50,000	0.079	0.264	0.343	0.145	0.450	0.595	0.140	0.452	0.591
45,000	0.080	0.264	0.345	0.147	0.450	0.597	0.141	0.452	0.592
40,000	0.083	0.265	0.347	0.149	0.451	0.600	0.142	0.452	0.594
35,000	0.086	0.265	0.350	0.153	0.451	0.603	0.143	0.452	0.595
30,000	0.089	0.265	0.354	0.157	0.451	0.608	0.146	0.452	0.598
25,000	0.095	0.265	0.360	0.163	0.451	0.615	0.149	0.452	0.601
20,000	0.103	0.266	0.369	0.173	0.452	0.624	0.153	0.452	0.605
15,000	0.116	0.267	0.383	0.188	0.453	0.641	0.161	0.453	0.613
10,000	0.143	0.269	0.411	0.218	0.455	0.672	0.176	0.453	0.629
7500	0.169	0.270	0.440	0.247	0.456	0.704	0.191	0.454	0.644
5000	0.223	0.274	0.497	0.305	0.460	0.765	0.221	0.455	0.676
4000	0.263	0.277	0.540	0.348	0.463	0.811	0.243	0.456	0.699
3000	0.330	0.281	0.611	0.420	0.467	0.887	0.281	0.457	0.738
2000	0.463	0.290	0.753	0.561	0.476	1.037	0.356	0.460	0.816
1500	0.597	0.299	0.896	0.673	0.485	1.158	0.431	0.463	0.894
1000	0.864	0.346	1.209	0.978	0.532	1.509	0.581	0.469	1.050
500	1.664	0.457	2.121	1.801	0.643	2.444	1.030	0.487	1.518

Regarding the reuse of municipal wastewater, the capital costs were mainly given by the UF and RO equipment for scenario 1, regardless of the plant capacity. These items represented around 25% and 49% of the total capital costs, respectively. The other items of capital costs (i.e., yard piping, sitework landscaping, site electrical and controls) represented around 26% of the overall capital costs for this scenario. If they are standardized considering the water production capacity, values of 0.07 and 1.66 USD/m^3 are obtained for plant capacities of 100,000 and 500 m^3/d, respectively (Table 3). Meanwhile, for scenario 2, in which an activated sludge system followed by a hybrid UF-RO system to treat wastewater was studied, capital costs were distributed mainly among the activated sludge system (39%), the UF (20%), and the RO (41%) units. For this scenario, the standardized total capital costs were between 0.14 and 1.80 USD/m^3, ranging in size from 100,000 to

500 m^3/d, respectively (Table 3). Finally, for the implementation of seawater desalination for producing water, the most relevant items in the total capital costs were the construction and infrastructure costs. For this scenario, the unitary total capital costs ranged between 0.14 and 1.03 USD/m^3 for water plant capacities of 100,000 and 500 m^3/d, respectively (Table 3). The capital costs for the proposed scenarios are shown in Tables S1–S3 in the Supplementary Material.

Regarding operating and maintenance costs, they increased as the water production capacity decreased, regardless of the proposed scenario (Table 3). For scenario 1, the operating and maintenance costs ranged from 0.26 to 0.46 USD/m^3 for water plant capacities of 100,000 and 500 m^3/d, respectively. The consumption of energy and reagents were the most expensive items in the operating and maintenance costs, comprising around 34% and 42% of the total operating and maintenance costs, respectively, for this scenario. Meanwhile, the operating and maintenance costs estimated for scenario 2 were around 1.5 times higher than those of scenario 1. They were 0.45 and 0.64 USD/m^3 for water production capacities of 100,000 and 500 m^3/d, respectively (Table 3). The costs related to the operation of activated sludge contributed around 40% of the overall operating and maintenance costs, while energy and reagents consumption represented around 25% and 20% of these costs, respectively.

Moreover, the operating and maintenance costs for scenario 3 (seawater desalination) were similar to those obtained for the implementation of an activated sludge and UF-RO system to produce water from municipal wastewater. These costs were 0.45 and 0.49 USD/m^3 for desalination plant capacities of 100,000 and 500 m^3/d, respectively (Table 3), and the consumption of energy accounted for approximately 68% of the operating and maintenance costs. The costs associated with the membrane replacement and reagents consumption were other important items in the operating and maintenance costs, representing around 18% and 9% of these costs, respectively. Recently, literature has reported that operating and maintenance costs for seawater desalination plants are between 1.21 and 1.12 USD/m^3 for plant production capacities ranging from 1000 to 10,000 m^3/d, and they are in the range of 1.04 and 0.30 USD/m^3 for desalination plant capacities between 15,000 and 100,000 m^3/d [14,35].

Thus, these results indicated that the operating costs to produce water from municipal wastewater were lower than those estimated for seawater desalination, if the wastewater treatment system is based on primary settling followed by a biological treatment (scenario 1). The operating and maintenance costs for the proposed scenarios are shown in Tables S4–S6 in the Supplementary Material.

3.2. Economic Analysis

In order to compare the results obtained for the proposed scenarios, the total production costs and the minimum price of the produced water were determined. Overall, the results indicated that these decreased with an increasing plant production capacity, regardless of the used feedwater (Table 3 and Figure 3). In this regard, the total production costs of the desalinated water were between 0.59 and 1.52 USD/m^3 for desalination plant capacities of 100,000 and 500 m^3/d, respectively. Meanwhile, for the use of municipal wastewater as feedwater, the total water production costs were between 0.33 and 2.12 USD/m^3 for scenario 1, and they ranged from 0.59 to 2.44 USD/m^3 for scenario 2 for production plant capacities of 100,000 and 500 m^3/d, respectively (Table 3). Different wastewater treatment costs can be found in the literature ranging from 0.40 and 1.26 USD/m^3, which can be associated with different combinations of tertiary treatments, such as coagulation-flocculation, filtration, ultrafiltration, disinfection, reverse osmosis, among others [40–44]. Among these studies, Fundación Chile (2016) [44] evaluated the reuse of municipal wastewater from submarine outfalls WWTPs for agricultural activities. These authors reported operating costs ranging from 0.79 to 1.02 USD/m^3, including water transportation costs. However, these authors did not indicate the technologies that were being considered to replace the current wastewater treatment system. This wide range of potential used technologies prevents the comparison of these obtained values with the previously reported ones. In any case, the total costs obtained in this work are within this range.

Furthermore, the price at which produced water would have to be sold in order to recover the internal costs of the project for the conventional WWTP ranged between 0.53 and 4.27 USD/m^3 for production plant capacities of 100,000 and 500 m^3/d, respectively. Meanwhile, this was found to range between 0.91 and 4.77 USD/m^3 for the production of water from a submarine outfalls WWTP with plant capacities of 100,000 and 500 m^3/d, respectively. The minimum price of the produced water was also calculated for the scenario where seawater desalination was evaluated, obtaining values ranging from 1.30 to 4.03 USD/m^3 to produce water, for plant capacities of 100,000 and 500 m^3/d, respectively (Figure 3). Therefore, results would indicate that WWTP retrofitting would be more favorable than the implementation of seawater desalination when the city has conventional WWTP to treat its wastewater rather than a submarine outfalls WWTP.

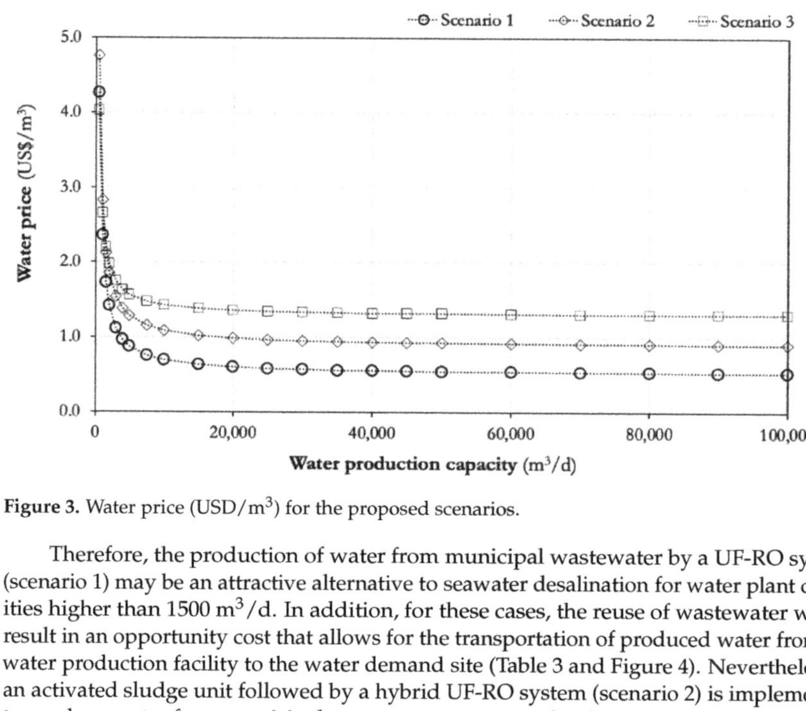

Figure 3. Water price (USD/m^3) for the proposed scenarios.

Therefore, the production of water from municipal wastewater by a UF-RO system (scenario 1) may be an attractive alternative to seawater desalination for water plant capacities higher than 1500 m^3/d. In addition, for these cases, the reuse of wastewater would result in an opportunity cost that allows for the transportation of produced water from the water production facility to the water demand site (Table 3 and Figure 4). Nevertheless, if an activated sludge unit followed by a hybrid UF-RO system (scenario 2) is implemented to produce water from municipal wastewater, seawater desalination becomes more profitable than the reuse of municipal wastewater for production plant capacities lower than 70,000 m^3/d. In this case, the selection of seawater desalination instead of the reuse of wastewater allows for the production of water and its transport to the water demand site (Figure 4). It should be noted that the maximum distance that produced water can be transported will depend on the water flow to be supplied (Figure 4).

3.3. *Water Production from Non-Conventional Water Resources: Case Studies from Chile*

Here, we illustrate the use of the proposed methodology in eight different cities of Chile that are placed in regions that are prone to suffer from water stress (Figure 2), and they, collectively, represent a sample of water security problems that can be found in arid, semi-arid or Mediterranean areas of fast-growing economies. These cities are located near to the coast or in inland areas, and their water demand is between 1033 and 33,659 m^3/d (Table 4 and Figure 2). The production of water, either by the reuse of municipal wastewater or by the desalination of seawater to provide water for these Chilean cities was studied.

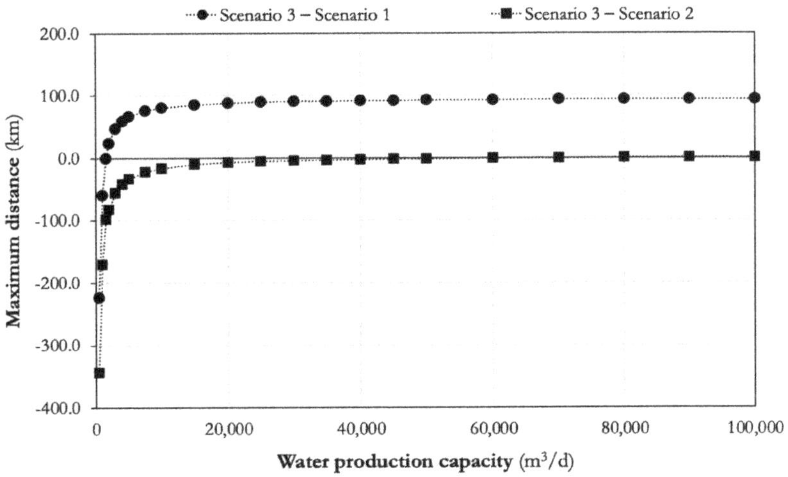

Figure 4. Estimated maximum distance (km) for the pairwise comparison of the proposed scenarios.

Table 4. Estimated values obtained to supply water from potential water supplier cities to the studied cities.

City	Water Demand (m^3/d)	WWTP	Potential Water Supplier City	Produced Water (m^3/d)	d_{eq_ww} (km)	d_{eq_coast} (km)	ΔD_{max} (km)
Alto Hospicio	23,824	conventional	Iquique Pozo Almonte	28,719 1533	22.4 61.4	3.4	86.1
Mejillones	2915	submarine outfalls	Antofagasta Tocopilla	53,788 3683	65.6 132.0	0.7	−46.3
Freirina	1033	conventional	Huasco Vallenar	1352 6990	20.2 43.8	20.6	−54.0
Huasco	2003	submarine outfalls	Vallenar Freirina	6990 697	62.8 20.2	0.8	−65.9
Andacollo	2252	conventional	La Serena Ovalle Tongoy	30,403 13,297 835	78.1 68.2 92.0	69.0	32.4
Limache	8813	conventional	Villa Alemana Quillota Olmué	19,037 11,898 1865	23.6 21.2 10.9	30.3	86.9
Quillota	17,624	conventional	Viña del Mar La Calera Concón	50,772 7393 5986	49.8 19.5 33.1	25.1	89.0
Quilpué	33,659	submarine outfalls	Viña del Mar Villa Alemana Limache	50,772 21,937 5950	34.6 6.0 28.4	11.9	−2.7

As shown in Tables 4 and 5, results indicated that the production of water from conventional WWTP is an attractive alternative to seawater desalination for water production capacities ranging from 2200 to around 24,000 m^3/d, regardless of the distance between the coast and the city. For these cities, the reuse of municipal wastewater allows for the transport of produced water for a maximum distance between 32.4 and 89.0 km, respectively (Table 4 and Figure 5). For instance, for the city of Alto Hospicio, cities located within an equivalent distance of 86.1 km from this city could be candidates for the position of their water supplier. Therefore, the reuse of the municipal wastewater generated in the city of Iquique could be a potential water supplier city because it is located at a distance of around 22.4 km from Alto Hospicio. However, the reuse of municipal wastewater would not be a good option for cities with a water flow demand lower than 2200 m^3/d. In this regard, the city of Freirina has a water flow demand of around 1000 m^3/d, so seawater desalination would be the appropriate scenario for the supply of water to this city (Table 4). As previously mentioned, the desalination of seawater becomes more profitable than municipal wastewater reuse at low water flows (production plant capacity < 1500 m^3/d) (Tables 4 and 5). Moreover, seawater desalination would be more profitable to produce water than wastewater reuse when the city uses submarine outfalls WWTP to treat municipal wastewater, regardless of the water flow demand (Tables 4 and 5). As shown in Table 4, the maximum distance that desalinated water can be transported is between 46.3 and 65.9 km for cities with water flows ranging between 2000 and 3000 m^3/d. It should be noted that these cities are located near to the coast (distance from the coast < 1 km), and this transport distance decreases with an increasing water plant capacity as well as an increasing distance from the coast (Table 4 and Figure 5).

Table 5. Summary of the selected cities to supply water to the water-demanding cities.

City	Water Resource	Water Price (USD/m³)
Alto Hospicio	Wastewater reuse (from Iquique)	0.63
Mejillones	Seawater desalination	2.23
Freirina	Seawater desalination	2.62
Huasco	Seawater desalination	2.66
Andacollo	Wastewater reuse (from Ovalle)	1.31
Limache	Wastewater reuse (from Quillota)	0.69
Quillota	Wastewater reuse (from Viña del Mar)	0.66
Quilpué	Seawater desalination	1.37

The minimum price of the produced water would range from 0.63 to 1.31 USD/m^3 for cities supplied by the water produced from municipal wastewater, and it would be between 1.37 and 2.66 USD/m^3 for the cities where seawater desalination is used to supply the demanded water (Table 5). It should be noted that the estimated price of water supplied by municipal wastewater is lower than the current market price in Chile, which is between 1.09 and 3.20 USD/m^3 [45]. Meanwhile, the price to produce water from seawater desalination is within the current range of Chilean prices. Thus, the results indicated that the reuse of municipal wastewater would be more economical to produce and supply water compared to the desalination of seawater for water flow demands higher than around 2000 m^3/d, particularly for cities that use conventional WWTP to treat municipal wastewater.

Therefore, these results fit quite well with the expected results. Thus, it can be inferred that the proposed methodology should be useful in deciding on the best non-conventional water resource, between seawater and municipal wastewater, to supply water to cities that suffer from water stress. Additionally, it is important to note that the scenarios proposed in this work should be evaluated according to the local characteristics of the regions where these water production systems will be installed.

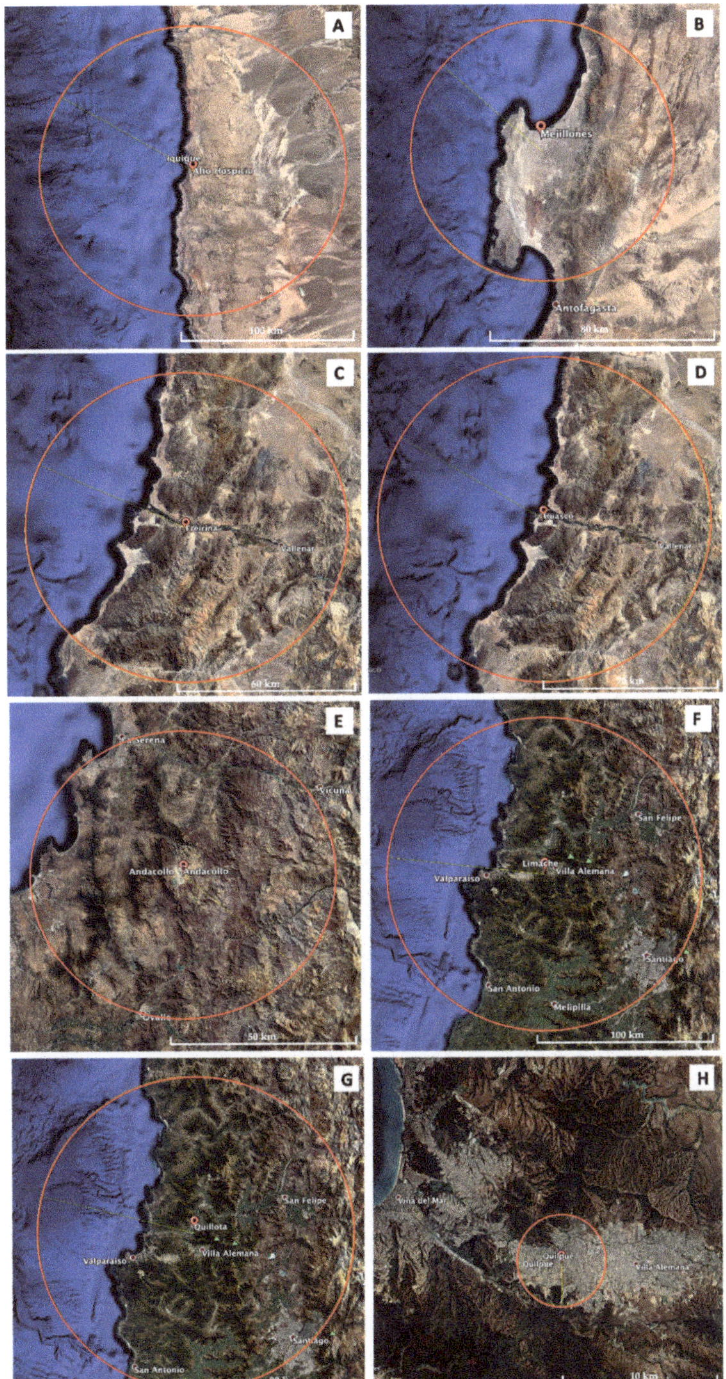

Figure 5. Maximum equivalent distance to transport produced water for studied cities: (**A**) Alto Hospicio; (**B**) Mejillones; (**C**) Freirina; (**D**) Huasco; (**E**) Andacollo; (**F**) Limache; (**G**) Quillota; (**H**) Quilpué.

4. Conclusions

Even when seawater desalination has considerable water supply potential, particularly for countries like Chile due to its abundance and availability, the reuse of municipal wastewater represents an economical alternative to seawater desalination for the production of water. In this regard, the retrofitting of conventional WWTPs in order to reuse municipal wastewater can be more profitable than seawater desalination for water plant capacities higher than 1500 m^3/d. In contrast, when the treatment of municipal wastewater is based on the use of submarine outfalls, seawater desalination becomes more profitable than wastewater reuse for the production of water flow rates lower than 70,000 m^3/d.

From a policy perspective, the proposed methodology should be of great interest to water authorities interested in promoting municipal wastewater reuse as a non-conventional water source for the supplying of water to cities that are located in water scarce areas.

Moreover, it should be noted that the reuse of municipal wastewater in the same city has not been considered in the proposed methodology. This methodology is based on the reuse of municipal wastewater from nearby cities that generate wastewater flows higher than the water required for the water-demanding site. Hence, future research on this issue should focus on including the reuse of municipal wastewater from more than one city (e.g., from nearby cities and from the same city) in order to meet the water requirements of the water-demanding site. Additionally, other local non-conventional water resources, such as rainwater and fog, could be included as feedwater to produce water in this methodology.

Supplementary Materials: The following are available online at https://www.mdpi.com/article/10.3390/su13126815/s1, Table S1: Capital costs to produce water from municipal wastewater by a UF-RO system (scenario 1), Table S2: Capital costs to produce water from municipal wastewater by an activated sludge unite followed by a UF-RO system (scenario 2), Table S3: Capital costs to produce water by the desalination of seawater (scenario 3), Table S4: Operating and maintenance costs to produce water from municipal wastewater by a UF-RO system (scenario 1). Table S5: Operating and maintenance costs to produce water from municipal wastewater by an activated sludge unite followed by a UF-RO system (scenario 2), Table S6: Operating and maintenance costs to produce water by the desalination of seawater (scenario 3).

Author Contributions: Conceptualization, J.L.C. and D.C.; methodology, D.C. and J.L.C.; writing-original draft preparation and performed the calculations, D.C.; data analysis, D.C. and J.L.C.; writing-review and editing, D.C. and J.L.C. All authors contributed to the writing and proofreading of the paper. All authors have read and agreed to the published version of the manuscript.

Funding: This research was funded by Universidad Adolfo Ibáñez and the Chilean Government through the Projects ANID/FONDECYT/1200850 and CRHIAM Centre grant number ANID/FONDAP/15130015.

Institutional Review Board Statement: Not applicable.

Informed Consent Statement: Not applicable.

Data Availability Statement: Not applicable.

Conflicts of Interest: The authors declare no conflict of interest.

References

1. UN (United Nations Water). World Water Development Report 2019. Available online: https://unwater.org/publications/world-water-development-report-2019/ (accessed on 4 February 2021).
2. Voulvoulis, N. Water reuse from a circular economy perspective and potential risks from an unregulated approach. *Curr. Opin. Environ. Sci. Health* **2018**, *2*, 32–45. [CrossRef]
3. Hussain, M.I.; Muscolo, A.; Farooq, M.; Ahmad, W. Sustainable use and management of non-conventional water resources for rehabilitation of marginal lands in arid and semiarid environments. *Agric. Water Manag.* **2019**, *221*, 462–476. [CrossRef]
4. Jones, E.; Qadir, M.; van Vliet, M.T.H.; Smakhtin, V.; Kang, S. The state of desalination and brine production: A global outlook. *Sci. Total Environ.* **2019**, *657*, 1343–1356. [CrossRef] [PubMed]
5. Qadir, M.; Jiménez, G.C.; Franum, R.L.; Dodson, L.L.; Smakhtin, V. Fog Water Collection: Challenges beyond Technology. *Water* **2018**, *10*, 372. [CrossRef]

6. Yusuf, A.; Sodiq, A.; Giwa, A.; Eke, J.; Pikuda, O.; De Luca, G.; Di Salvo, J.; Chakraborty, S. A review of emerging trends in membrane science and technology for sustainable water treatment. *J. Clean. Prod.* **2020**, *266*, 121867. [CrossRef]
7. Aliyu, U.M.; Rathilal, S.R.; Isa, Y.M. Membrane desalination technologies in water treatment: A review. *Water Pract. Technol.* **2018**, *13*, 738–752. [CrossRef]
8. Amy, G.; Ghaffour, N.; Li, Z.; Francis, L.; Valladares Linares, R.; Missimer, T.; Lattemann, S. Membrane-based seawater desalination: Present and future prospects. *Desalination* **2017**, *401*, 16–21. [CrossRef]
9. Gude, V.G. Desalination and sustainability-an appraisal and current perspective. *Water Resour.* **2016**, *89*, 87–106. [CrossRef]
10. Ahmed, F.E.; Hashaikeh, R.; Hilal, N. Hybrid technologies: The future of energy efficient desalination—A review. *Desalination* **2020**, *495*, 114659. [CrossRef]
11. Caldera, U.; Bogdanov, D.; Breyer, C. Desalination costs using renewable energy Technologies. In *Renewable Energy Powered Desalination Handbook*; Elsevier: Amsterdam, The Netherlands, 2018; pp. 287–329.
12. Elsaid, K.; Kamil, M.; Sayed, E.T.; Abdelkareem, M.A.; Wilberforce, T.; Olabi, A. Environmental impact of desalination technologies: A review. *Sci. Total Environ.* **2020**, *748*, 141528. [CrossRef]
13. Okamoto, Y.; Lienhard, J.H. How RO membrane permeability and other performance factors affect process cost and energy use: A review. *Desalination* **2019**, *470*, 114064. [CrossRef]
14. World Bank. *The Role of Desalination in an Increasingly Water-Scarce World*; World Bank: Washington, DC, USA, 2019.
15. Mollahosseini, A.; Abdelrasoul, A.; Sheibany, S.; Amini, M.; Salestan, S.K. Renewable energy-driven desalination opportunities—A case study. *J. Environ. Manag.* **2019**, *239*, 187–197. [CrossRef]
16. Ghaffour, N.; Missimer, T.M.; Amy, G.L. Technical review and evaluation of the economics of water desalination: Current and future challenges for better water supply sustainability. *Desalination* **2013**, *309*, 197–207. [CrossRef]
17. Garcia, X.; Pargament, D. Reusing wastewater to cope with water scarcity: Economic, social and environmental considerations for decision-making. *Resour. Conserv. Recy.* **2015**, *101*, 154–166. [CrossRef]
18. Al-Karaghouli, A.; Kazmerski, L.L. Energy consumption and water production cost of conventional and renewable-energy-powered desalination processes. *Renew. Sustain. Energy Rev.* **2013**, *24*, 343–356. [CrossRef]
19. Zhou, Y.; Tol, S.J. Evaluating the costs of desalination and water transport. *Water Resour. Res.* **2005**, *41*, W03003, 1–10. [CrossRef]
20. Abdulbaki, D.; Al-Hindi, M.; Yassine, A.; Najm, M.A. An optimization model for the allocation of water resources. *J. Clean. Prod.* **2017**, *164*, 994–1006. [CrossRef]
21. Goh, P.S.; Matsuura, G.T.; Ismail, A.F.; Ng, B.C. The water-energy nexus: Solutions towards energy-efficient desalination. *Energy Technol.* **2017**, *5*, 1136–1155. [CrossRef]
22. Pinto, F.S.; Marques, R.C. Desalination projects economic feasibility: A standardization of cost determinants. *Renew. Sustain. Energy Rev.* **2017**, *78*, 904–915. [CrossRef]
23. Papapetrou, M.; Cipollina, A.; La Commare, U.; Micale, G.; Zaragoza, G.; Kosmadakis, G. Assessment of methodologies and data used to calculate desalination costs. *Desalination* **2017**, *419*, 8–19. [CrossRef]
24. Livi-Bacci, M. *A Concise History of World Population*; John Wiley & Sons: Hoboken, NJ, USA, 2012.
25. Ventura, D.; Consoli, S.; Barbagallo, S.; Licciardello, F.; Cirelli, G.L. How to overcome barriers for wastewater agricultural reuse in Sicily (Italy)? *Water* **2019**, *11*, 335. [CrossRef]
26. EPA (U.S. Environmental Protection Agency). *Guidelines for Water Reuse*; U.S. Environmental Protection Agency: Washington, DC, USA, 2012.
27. Vicuña, S.; Vargas, X.; Boisier, J.P.; Mendoza, P.A.; Gómez, T.; Vásquez, N.; Cepeda, J. Impacts of climate change on water resources in Chile. In *Water Resources of Chile*; World Water Resources; Fernández, B., Gironás, J., Eds.; Springer Nature: Cham, Switzerland, 2021. [CrossRef]
28. Alvez, A.; Aitken, D.; Rivera, D.; Vergara, M.; McIntyre, N.; Concha, F. At the crossroads: Can desalination be a suitable public policy solution to address water scarcity in Chile's mining zones? *J. Environ. Manag.* **2020**, *258*, 110039. [CrossRef]
29. Herrera-León, S.; Cruz, C.; Kraslawski, A.; Cisternas, L.A. Current situation and major challenges of desalination in Chile. *Desalination Water Treat.* **2019**, *171*, 93–104. [CrossRef]
30. CNID (Consejo Nacional de Innovación para el Desarrollo). Evaluation of the Socio-Economic Conflicts of Large Projects with Focus on Water and Energy for 1998–2015. Available online: http://www.cnid.cl/wp-content/uploads/2017/04/Informe-final-CNID-Evaluacio%CC%81n-de-Conflictos-Socioambientales-1.pdf (accessed on 24 February 2021).
31. Yang, J.; Monnot, M.; Ercolei, L.; Moulin, P. Membrane-Based Processes Used in Municipal Wastewater Treatment for Water Reuse: State-Of-The-Art and Performance Analysis. *Membranes* **2020**, *10*, 131. [CrossRef] [PubMed]
32. Plumlee, M.; Stanford, B.D.; Debroux, J.F.; Hopkins, D.C.; Snyder, S.A. Cost of advanced treatment in water reclamation. *Ozone Sci. Eng.* **2014**, *36*, 485–495. [CrossRef]
33. Guo, E.; Englehardt, J.; Wu, T. Review of cost versus scale: Water and wastewater treatment and reuse processes. *Water Sci. Technol.* **2014**, *69*, 223–234. [CrossRef]
34. Molinos-Senante, M.; Hernández-Sancho, F.; Sala-Garrido, R. Economic feasibility study for wastewater treatment: A cost–benefit analysis. *Sci. Total Environ.* **2010**, *408*, 4396–4402. [CrossRef] [PubMed]
35. Molinos-Senante, M.; González, D. Evaluation of the economics of desalination by integrating greenhouse gas emission costs: An empirical application for Chile. Renew. *Energy* **2019**, *133*, 1327–1337. [CrossRef]

36. ENEL. Tarifas Vigentes. Available online: https://www.enel.cl/es/clientes/informacion-util/tarifas-y-reglamentos/tarifas.html (accessed on 4 February 2021).
37. ESCWA (Economic and Social Commission for Wester Asia). *ESCWA Water Development Report 3: Role of Desalination in Addressing Water Scarcity*; United Nations Publications: New York, NY, USA, 2009.
38. Sarricolea, P.; Herrera-Ossandon, M.; Meseguer-Ruiz, O. Climatic regionalisation of continental Chile. *J. Maps* **2017**, *13*, 66–73. [CrossRef]
39. Aitken, D.; Rivera, D.; Godoy-Faúndez, A.; Holzaptel, E. Water Scarcity and the Impact of the Mining and Agricultural Sectors in Chile. *Sustainability* **2016**, *8*, 128. [CrossRef]
40. Murashko, K.; Nikku, M.; Sermyagina, E.; Vauterin, J.J.; Hyppänen, E.; Pyrhónen, J. Techno-economic analysis of a decentralized wastewater treatment plant operating in closed-loop. A Finnish case study. *J. Water Process. Eng.* **2018**, *25*, 278–294. [CrossRef]
41. Melgarejo, J.; Prats, D.; Molina, A.; Trapote, A. A case study of urban wastewater reclamation in Spain: Comparison of water quality produced by using alternative processes and related costs. *J. Water Reuse Desalination* **2016**, *6*, 72–81. [CrossRef]
42. Plappally, A.K.; Lienhard, J.H. Costs for water supply, treatment, end-use and reclamation. *Desalination Water Treat.* **2012**, *51*, 1–33. [CrossRef]
43. Valladares Linares, R.; Li, Z.; Yangali-Quintanilla, V.; Ghaffour, N.; Amy, G.; Leiknes, T.; Vrouwenvelder, J.S. Life cycle cost of a hybrid forward osmosis e low pressure reverse osmosis system for seawater desalination and wastewater recovery. *Water Res.* **2016**, *88*, 225–234. [CrossRef] [PubMed]
44. Fundación Chile. *Aguas Residuales como Nueva Fuente de Agua. Diagnóstico del Potencial Reúso de Aguas Residuales en la Región de Valparaíso*; Fundación Chile: Santiago-Valparaíso, Chile, 2016.
45. SISS. Superintendencia de Servicios Sanitarios. Informe de Gestión del Sector Sanitario. 2019. Available online: https://www.siss.gob.cl/586/articles-17955_recurso_1.pdf (accessed on 4 February 2021).

sustainability

MDPI

Review

Treatment Technologies for Cooling Water Blowdown: A Critical Review

Mariam Soliman [1], Fadwa Eljack [1,*], Monzure-Khoda Kazi [1], Fares Almomani [1], Elalim Ahmed [2] and Ziad El Jack [2]

[1] Department of Chemical Engineering, College of Engineering, Qatar University, Doha P.O. Box 2713, Qatar; mariam.soliman@qu.edu.qa (M.S.); kazi0001@qu.edu.qa (M.-K.K.); falmomani@qu.edu.qa (F.A.)
[2] Ministry of Municipality and Environment (MME), Infrastructure Planning Department (IPD), Doha P.O. Box 2727, Qatar; EAAHMED@mme.gov.qa (E.A.); zeljack@hotmail.com (Z.E.J.)
* Correspondence: Fadwa.Eljack@qu.edu.qa

Abstract: Cooling water blowdown (CWBD) generated from different industries and district cooling facilities contains high concentrations of various chemicals (e.g., scale and corrosion inhibitors) and pollutants. These contaminants in CWBD streams deem them unsuitable for discharge into surface water and some wastewater treatment plants. The pollutants present in CWBD, their sources, and the corresponding impacts on the ecosystem are discussed. The international and regional (Gulf states) policies and regulations related to contaminated water discharge standards into water bodies are examined. This paper presents a comprehensive review of the existing and emerging water treatment technologies for the treatment of CWBD. The study presents a comparison between the membrane (membrane distillation (MD), reverse osmosis (RO), nanofiltration (NF), and vibratory shear enhanced membrane process (VSEP)) and nonmembrane-based (electrocoagulation (EC), ballasted sand flocculation (BSF), and electrodialysis (ED)) technologies on the basis of performance, cost, and limitations, along with other factors. Results from the literature revealed that EC and VSEP technologies generate high treatment performance (EC~99.54% reduction in terms of silica ions) compared to other processes (membrane UF with reduction of 65% of colloidal silica). However, the high energy demand of these processes (EC~0.18–3.05 kWh/m^3 and VSEP~2.1 kWh/m^3) limit their large-scale applications unless connected with renewable sources of energy.

Keywords: cooling water blowdown; treatment technologies; contaminants; membrane technologies; emerging technology

Citation: Soliman, M.; Eljack, F.; Kazi, M.-K.; Almomani, F.; Ahmed, E.; El Jack, Z. Treatment Technologies for Cooling Water Blowdown: A Critical Review. *Sustainability* **2022**, *14*, 376. https://doi.org/10.3390/su14010376

Academic Editors: José Luis Campos, Anuska Mosquera Corral, Ángeles Val del Río and Alba Pedrouso Fuentes

Received: 30 November 2021
Accepted: 27 December 2021
Published: 30 December 2021

1. Introduction

Scarcity of freshwater is one of the eminent dangers that can be found in many countries around the world. It can be considered a global problem as a result of the rapid growth of population; industrialization; and, more importantly, the pollution caused to fresh water from many sources. The water supply could be increased beyond the natural hydrological cycle by implementing energy-efficient and sustainable technologies for recycling and reusing the wastewater instead of releasing it into water bodies, which may cause further problems and diseases. When all systems that consume water optimize their usage and apply developed technologies, it will ensure that water will be conserved and secured, with considerable savings in freshwater consumption being realized. Although water is consumed by many sectors and processes in different industries, cooling tower systems require and consume a significant amount of water in industries, power plants, universities, and government buildings [1]. Cooling towers are units that provide an energy-efficient and cost-effective operation for devices in need of cooling [2,3]. During the process of cooling, water is continuously recirculated, while some water evaporates; this leads to an increase in the concentration of salt and contaminants to high levels. As the number of recirculation cycles increases, the solubility of various solids is reduced; consequently, solids will form a shale shape on the warm surface of the condenser pipes. The formed scales in the cooling tower unit cause a reduction in the heat transfer efficiency as they insulate the metal surface of the tower [4]. With further recirculation of the concentrated water, permanent damage can occur to the cooling system [1]. Therefore, this highly concentrated water stream is discharged out of the system as a cooling water blowdown water (CWBD). The discharge may contain iron oxides, calcium phosphate, calcium carbonate, magnesium silicate, silica, and many other contaminants and pollutants [4]. CWBD discharge helps enhance heat transfer efficiency since the concentration of the silica and hardness ions in the circulated water is kept under the level where scales can be formed [5]. A make-up stream of fresh water is used to compensate for the amount of water lost in evaporation and CWBD discharge.

Water shortages and the rising prices of freshwater have encouraged many industries to reduce their dependence on freshwater and focus on using treated water as an alternative [4]. For example, in GCC countries, lack of freshwater resources forces these countries to use the treated sewage effluent (TSE) as CWBD. On the other hand, various treatment technologies have been used or proposed to treat any type of wastewater and sustain the available water resources. Different technologies have been used to treat the CWBD. Reverse osmosis (RO), electrodialysis (ED), nanofiltration (NF), electrocoagulation (EC), vibratory shear enhanced membrane process (VSEP), ballasted sand flocculation (BSF), and membrane distillation (MD) were usually proposed as suitable technologies. Saha et al. [6] evaluated the effect of operational parameters and the type of electrodes on the removal of organic pollutants from CWBD when electrochemical oxidation is applied as a pre-treatment technology. In a different paper, Saha et al. [7] studied the treatment of CWBD by combining constructed wetlands with the electrochemical oxidation process. The study showed that the integrated system of VFCW-EO had a better ability to remove organic chemicals, such as TOC, COD, and the corrosion inhibitor benzotriazole, than the EO and VFCW systems.

Wagner et al. [8] reviewed the possibility of using electrochemical oxidation for the removal and conditioning of chemicals out of CWBD, and results showed an excellent treatment performance represented by removal rates of 85% and 51% for COD and TOC respectively, with the BDD-anode. Additionally, Li et al. [9] recently conducted a toxicity assessment, technical performance, and economic evaluation for the treatment of CWBD water by implementing adsorption-electrocatalytic oxidation. Results showed that PANI/TiO_2, which is polyaniline-modifiedTiO_2, was a promising adsorbent for phosphorus and organics removal from CWBD.

The aforementioned literature reviews showed that the conducted studies, reviews, and published work on the treatment of CWBD water are limited. There is a clear gap in knowledge related to the performance of different treatment technologies in tackling the CWBD problem. The impact of water quality parameters and the discharge volume on the process's performance and the energy demand required further investigation. Reviewing various treatment technologies will provide an idea about the possible alternatives to treat the effluent streams from cooling towers, which will make decision-makers aware of the potential processes for treating purposes. Additionally, evaluating critical criteria, such as the technologies' cost, efficiency, and effluent quality, is highly important in selecting the most sustainable, green, and economical option. Regulations and standards are another important evaluation factor as they determine the maximum level of contaminates concentration permitted at the endpoint or the discharging area. Therefore, this work presents a comprehensive review and comparison between the existing and emerging water treatment technologies for the treatment of CWBD. Different treatment technologies of CWBD were reviewed and evaluated on the basis of process scale, maintenance, chemical additive requirements, energy consumption, quality of permeate and sludge, and more importantly removal ability of contaminants and cost analysis. The impacts of contaminants presented in CWBD on the environment and human health, as well as the operations of cooling tower systems, are presented and discussed. In addition, the regional and international policies and regulations related to contaminated water discharge standards into water bodies are explored and discussed. The paper highlights regulations and standards of wastewater discharged to sewage treatment plants, the marine environment, and irrigation purposes.

2. Overview of CWBD Pollutants and Impacts

As a part of the cooling processes in cooling towers, a portion of the concentrated water is discharged out of the system as CWBD to control the concentration level of different ions and compounds in the cooling tower. In this section, pollutants found in CWBD are reviewed, with a focus on the ones of concern and their impacts.

2.1. CWBD Contaminants

Studying the available pollutants in CWBD will help in reusing it or selecting an end of pipe solution to reduce these pollutants before discharging the effluent into water bodies. The type and the quantity of contaminants available in that effluent vary from one system to another and depend on many factors. For instance, the source of the inlet and make-up water to the cooling tower has a great impact on the presence of various contaminants over others. The types of chemicals used for treatment purposes, as inhibitors or even as an anti-corrosion inside the towers, can significantly affect the composition of chemicals found in water. Stratton et al. [10] investigated the water quality parameters of the blowdown and make-up water from 11 cooling towers. The results showed that the chemical composition of the water varied greatly between the cooling towers for many reasons, such as the make-up water chemistry and the chemicals used for treatment.

Table 1 presents the most common contaminants in CWBD stream from different recent references. Ahmed et al. [11] showed an excellent representation of the CWBD characteristics regarding the available contaminates and their corresponding concentrations for different streams and references. Common contaminants are found in all or at least most effluents, including calcium ions, magnesium ions, and chloride (see Table 1). Sulphate, phosphate, iron and sodium ions, and TDS, among others, can be also found in the CWBD. It is noticeable that the presence of some contaminants differs between the effluent streams. One primary reason could be that the studies did not conduct complete characterization analysis for all available contaminants in CWBD, and that can be due to the scope of the focus of their studies. For example, the target of Abdel-shafy and his colleagues in their paper [4] was to treat CWBD from calcium, magnesium, and silica ions using magnesium electrodes in electrocoagulation (EC) technology; hence, their analysis focus was only on these ions. Hong and other authors in a thesis project [1] focused on reducing the total dissolved solids (TDS) level in CWBD to be equivalent to tap water. Differences in levels of contaminants are because of the amount and type of chemicals used in the cooling tower as well as the cyclic concentration. Moreover, increasing the evaporation rate in cooling towers will increase the concentration of ions [12,13]. The differences in rates of evaporation depending on the design and the efficiency of the tower. The quality of air passed in the cooling tower also impacts the characterization of CWBD water because air with a high amount of dust will increase the total amount of suspended solids and turbidity of the CWBD stream compared to filtered air [11].

2.2. Contaminants of Concerns in CWBD and Their Impacts

The effluent from the cooling tower contains a wide range and various types of contaminants that can significantly affect the environment and human life. Total dissolved solids (TDS) at high concentrations are considered as one of the major contaminants. Dickerson and Vinyard [14] reported that the elevated concentrations of TDS caused the extinctions of two nonindigenous species of fish in Walker Lake, Nevada. In a standard desalination plant, 50,000 to 70,000 ppm are considered as a common range for the effluent TDS in CWBD [1,15]. In addition, elevated levels of TDS in water streams can cause scaling and corrosion to the pipelines [1], consequently affecting transport efficiency and increasing maintenance costs.

Weber-Scannell et al. [16] discussed the effects of TDS on aquatic organisms. Discharging CWBD water with a high amount of phosphate ($PO4^{3-}$) into water bodies increases the growth of algae, leading to oxygen depletion in the water [17], and eventually mortality of aquatic creatures such as fish, flora, and fauna [18].

Groundwater is a freshwater resource that can be highly impacted by contaminants found in CWBD. Ions of sodium (Na^+), magnesium (Mg^{2+}), sulfate ($SO_4{}^{2-}$), potassium (K^+), chloride (Cl^-), calcium (Ca^{2+}), and bicarbonate ($HCO_3{}^-$) are the main inorganic ions present in natural waters. However, increasing their concentrations can cause health as well as environmental issues. For example, physical inconveniences such as diarrhea and skin irritation can be caused when the groundwater is highly concentrated with sulphate [19]. Increasing the level of magnesium and calcium as a result of injecting wastewater streams such as CWBD can cause water hardness. For some developing countries, groundwater is their main or only source of drinking water, and such contaminants threaten their water security. Such hardness ions do not only affect groundwater, but they can cause scaling and other mechanical problems to the cooling towers when recycled without treatment [20]. Other contaminants such as arsenic in groundwater with elevated levels can cause cancer [21], loss of limbs, or even lead to death in critical cases [19,22]. The drinking of fluoride-concentrated groundwater can have an adverse impact on the growth of children and at extreme concentrations, it can lead to death because of its toxicity [19]. Trace of heavy metals such as zinc, chromium, lead, nickel, silver, aluminum, copper, cadmium, and cobalt also exist naturally in groundwater, and their concentrations can be increased with human activities [23,24], consequently affecting the ecosystem as well as making groundwater unfit for human consumption. Zinc is considered a poisoning metal that causes skin irritations, anemia, and other infections [18,25]. The presence of lead in animal and human bodies impacts the synthesis process of hemoglobin that can lead to anemia and more severe problems [18]. Cadmium is toxic for organisms that live in the aquatic environment and can cause problems to the kidney and liver. In contrast, chromium can cause cancer for humans as well as skin irritation [18].

Table 1. Common contaminants available in CWBD water.

Parameter	Unit	[4]	[11,26] (J. Löwenberg, 2015)	[7]	[1] ****
pH		8.2	7.9	6.8 ± 0.2	
Calcium (Ca^{2+})	mg/L	392 **	1204 **	338 ± 7.6	125.9
Magnesium (Mg^{2+})	mg/L	280 **	259 **	58 ± 2.4	12.5
Silica (SiO_2)	mg/L	27 ***	0.9		
Chloride (Cl^-)	mg/L	162	500	458 ± 10	205.32
Zinc (Zn^{2+})	mg/L	1.2			
Phosphate ($PO_4{}^{3-}$)	mg/L	6.61 *	5.9		
Iron (Fe^{2+})	mg/L	0.1			0.6343
Sulphate ($SO_4{}^{2-}$)	mg/L	711		1043 ± 52	469.05
Aluminium (Al^{3+})	mg/L				0.0596
Barium (Ba^{2+})	mg/L		0.145		0.1142
Potassium (K^+)	mg/L			75 ± 1.3	8.1
Sodium (Na^+)	mg/L			334 ± 2.9	262.8
Copper (Cu)	mg/L				0.224
Strontium (Sr^{2+})	mg/L		1.500		1.0853
Bromide (Br^-)	mg/L				43.35
Fluorine (F)	mg/L				8.1
Total suspended solids (TSS)	mg/L		12		
Total dissolved solids (TDS)	mg/L	1297			1329
Total organic carbon (TOC)	mg/L			41 ± 1.3	
Chemical oxygen demand (COD)	mg/L			107 ± 6.4	
Nitrate (NO^{3-})	mg/L		86.7	57 ± 1.8	
Turbidity	NTU	7.93	7.3		
CWBD water source		Effluent of a urea fertilizer plant	From a cooling tower (CT) next to the Dow premises in Terneuzen (The Netherlands)	From (CT) of Dow Benelux BV (Terneuzen, The Netherlands)	From CSULB cooling towers

* Total phosphate as PO_4. ** As $CaCO_3$. *** Silicates as SiO_2. **** Unit changed from μg/L to mg/L.

Different types of anti-corrosion chemicals such as chromates, nitrites, molybdates, and tungstates [27], as well as biocides such as glutaraldehyde and isothiazolin [28] are added to cooling towers to inhibit the growth of algae, fungi, and bacteria available in the cooling water. Such chemicals and compounds are highly toxic for all living things and the environment when dumped into water bodies [1].

Using wastewater and CWBD as an example for irrigation applications is commonly applied in many countries to reduce the water scarcity problem [29]. However, the effect of available contaminates in such streams should be considered since they can contaminate soil and crops with lasting impacts on the whole biological chain [30]. Ingestion of such food can result in accumulated levels of contaminants that lead to many of the above-mentioned diseases [29,31].

Aside from the direct usage of wastewater, some facilities discharge it to sewers to be treated in sewage treatment plants. The design of the treatment system differs on the basis of the type of sewage. The availability of grit chambers, screens, sedimentation tanks, and other units will greatly affect the treatment ability of the sewage treatment plants in terms of many aspects [32]. The efficiency, maintenance costs, and the penetration of contaminants with the effluent water could be some of the impacts of industrial wastewater on the sewage treatment plants that can adversely affect the environment. The sewage treatment systems are designed on the basis of technologies that can handle certain types and concentrations of contaminants, and variations can lead to the production of water with low quality or cause a damage to the system. According to the study performed by Pophali et al., influent with high TDS can interfere with the oxygen transfer essential for biological metabolism, hence affecting the efficiency of the activated sludge process.

Considering the environmental, health, and operational, undesirable impacts of CWBD contaminants on different water systems, there is a need to consider water treatment technologies as means of managing CWBD. The following sections of the paper present reviews of existing technologies and evaluate them on the basis of their technical, environmental, and economic performance.

3. Overview on CWBD Water Treatment Technologies

The technologies that will be reviewed in this section are extensively implemented for the treatment of wastewater and CWBD as a type of wastewater. There are many pre-treatment processes used for CWBD water that can screen solids and remove other contaminants to reduce the load on the major treatment technologies, which are used to remove dissolved contaminants, suspended solids, etc. However, the focus here will be on the treatment processes, which were found to be applicable and suitable for CWBD treatment according to conducted studies and the literature; the technologies are presented in Figure 1. Non-membrane-based technologies include EC and BSF, while membrane-based technologies are MD, ED, RO, NF, and an emerging technology called VSEP membrane process. All these technologies are reviewed in this section, then evaluated on the basis of their ability for removal of contaminants, cost, and other factors in Section 5.

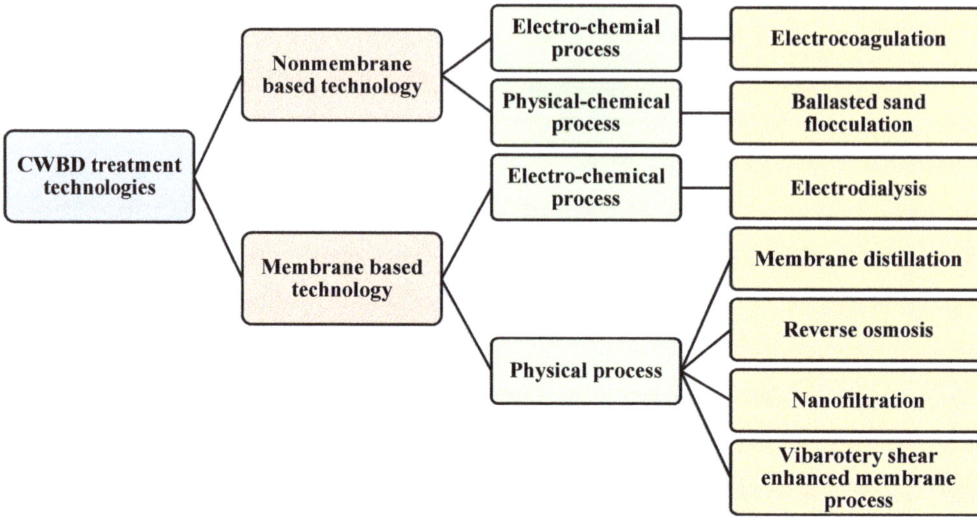

Figure 1. A summary of CWBD treatment technologies.

3.1. Membrane-Based Technologies

Membrane-based technologies are used to separate contaminants out of various streams such as wastewater. These processes do not require the addition of chemicals, have relatively low energy, and can be used and operated easily [33]. The principle of membrane processes is mainly based on the semi-permeable membranes that act as a filter that allow water to flow through and catch other contaminants. Substances can penetrate through the membrane under certain conditions, such as high pressure and the presence of electric potential [33]. Although the working principle is common between the membrane technologies, several differences make these technologies unique; major differences are highlighted in the following sections for each process.

3.1.1. Electrodialysis Process (ED)

The electrodialysis process for treating cooling water blowdown is an electrochemical [1] and membrane-based process. Ions in this process are transported and separated selectively by electrical field across several ion-exchange membranes [34]. At the end of the process, the concentration of ions increases in the concentrate compartment and decreases in the dilute compartment [1]. Further information about ED in the food, nutraceutical, beverage industries, as well as other industrial and municipal wastewater industries and designs, are covered within References [34–36]. The authors of [1] reported the usage of the ED process for the treatment and removal of TDS and other contaminants from CWBD efficiently. Major outcomes showed the ability of ED in a two-chambered cell to reduce TDS by 91.3%, 84.6%, 83.7%, and 93.4%, for trials 1, 2, 3, and 4, respectively. Additionally, ED had the ability to reduce the sulfate content in CWBD by 96% for each sample; reduce chloride for all samples by 91.9%; and an average removal of sodium and calcium 93.8% and 95.7%, respectively.

3.1.2. Membrane or Thermal Membrane Distillation (MD or TMD)

Membrane distillation is another membrane-based separation and physical technology used for the treatment of solutions that contain mainly water, such as CWBD, as studied by [37,38]. In MD, there is a direct contact between the aqueous solution and the microporous membrane, which is hydrophobic, at least from one side of that membrane [39]. MD is a thermally driven membrane [37], wherein the temperature difference between the two sides of the membrane induces a partial pressure gradient that leads to mass transfer of molecules through the pores of the membrane [37–39]. More well represented and reviewed details about this process can be found in the following papers: [37–45]. Generally, MD can treat many wastewater streams such as CWBD and water effluent from the process. For CWBD treatment, MD can utilize the waste heat from cooling towers in the process to create the temperature and pressure gradient as the driving force for the separation [37,38].

3.1.3. Reverse Osmosis (RO)

Reverse osmosis (RO) is one of the commonly used membrane-based technologies to treat wastewater stream effluent from different industrial processes as well as other sectors. This technology can be implemented to treat CWBD and usually requires a pre-treatment process to limit and reduce the membrane fouling as reviewed and studied in the literature [11,26,46–49]. Scaling phenomena occur when the dissolved salts become concentrated due to the extraction of clean water, which can lead to fouling of RO membrane. Examples of scaling constituents present in CWBD are calcium and magnesium hardness, fluoride, bicarbonates, barium, sulfates, strontium, silicate, and phosphates. As previously mentioned, pretreatment is one of the scaling control strategies in the RO process. In addition, conventional techniques such as the addition of scale inhibitors (e.g., polyacrylamide (PAM), polyacrylic acid (PAA), and polymaleic anhydride), or by adjusting the operational parameters such as time. Recently, novel techniques have been developed to control the scaling in RO process such as nanofiltration and feed flow reversal. RO is a membrane desalination process driven by pressure, where a semipermeable membrane is used to allow only water to penetrate, leaving behind the dissolved ions and salt. Several pre-treatment processes have been reviewed by Ahmed et al. [11]; they include constructed wetlands, coagulation settling and filtration, microfiltration, powdered activated carbon (PAC) adsorption, and ultrafiltration (UF). The result out of the review showed that physio-chemical processes are practical for CWBD pre-treatment. Among the prefiltration processes, ultrafiltration is the most popular option; however, MF can be a better alternative to UF. Löwenberg et al. [26] conducted experiments and investigated the suitability of using PAC adsorption, coagulation, and UF as pre-treatment processes before RO for the treatment CWBD. The main output of this study showed that coupling PAC with UF is the best combination as a pre-treatment process to enhance the performance of the RO process. In another study, Hossein et al. [47] investigated the suitability of coagulation-filtration and UF as pre-treatment processes before nanofiltration (NF) and RO to treat CWBD. Results showed that both pre-treatment processes are efficient; however, UF may face fouling, and hence it requires a pre-treatment to overcome this issue.

3.1.4. Nanofiltration (NF)

Nanofiltration is a membrane-based technology used for the treatment of waters, such as desalination of brackish water and seawater. Moreover, it can treat wastewater streams from different applications such as textile, industrial, and pharmaceutical [50]. With a pore size between 1 to 10 nm, small ions and organic substrates can be selectively removed by NF with low consumption of energy [50–52]. Olariu et al. [20] mentioned that the pore size of NF membrane could be between 1 to 10 nm, which is capable of removing large organics, as well as monovalent and divalent ions. Further details about the process, the working mechanism of NF technology, and other information are well presented in the following papers: [50–53]. NF can be implemented in industries where CWBD water requires treatment. Olariu et al. [20] used the NF process as a treatment process of CWBD water in a pilot plant with other pre-treatment steps. Results showed that around 97% of salts were rejected. In addition, Hossien et al. [47] experimented as a part of their study on the effect of using RO or NF as a post-treatment process. The results showed that both were applicable and produced high-quality water for reuse.

3.1.5. Vibratory Sheared Enhanced Membrane Process (VSEP)

VSEP is one of the emerging membrane-based separation technologies used for the treatment of wastewater streams discharged from different applications. VSEP is similar to conventional membrane technologies; however, the membrane is mechanically vibrated for better removal efficiency of contaminants as well as reducing fouling problems associated with membrane processes due to the high sheared stress applied on the surface of the membrane [54,55]. VSEP can be used in various applications, such as for the treatment of brine as discussed by Balasubramanian et al. [56] and the treatment of landfill leachates as experimented by Zouboulis et al. [57]. The use of VSEP is for the treatment of CWBD water streams for water recovery and reuse, as well as for reducing the volume of CWBD streams.

3.2. Non-Membrane Based-Technologies

3.2.1. Electrocoagulation Process (EC)

Electrocoagulation is a non-membrane-based and electrochemical separation process used to remove different pollutants by applying chemical and physical mechanisms [58–60]. The supplied electricity to the system can destabilize emulsified, dissolved, or suspended pollutants and contaminants in an aqueous medium [58]. This process can be used in many industrial applications to treat the effluents out of the processes, such as manufacturing [61] and petrochemical industries [62]. It is used to remove contaminants such as hardness ions [63], nickel [64], iron [65], chromium [66], fluoride [67], and phosphate [68]. More descriptions of the mechanisms used in the process and the reaction are available in the following references: [58,69–72]. Recently, El-khateeb et al. [4] studied the possibility of using magnesium rod-electrodes in the EC process to treat the effluent stream blowdown from cooling towers. Results showed that the system was able to remove hardness ions and silica with efficiencies of 51.80 and 93.70%, respectively. Other studies have shown supporting evidence that the use of the EC process to treat CWBD is effective [4,5,73].

3.2.2. Ballasted Sand Flocculation Process (BSF)

BSF is one of the physical-chemical non-membrane-based processes used for the treatment of water and wastewater streams effluent from various sectors. It is capable of removing many contaminants, such as total suspended solids TSS, and a wide range of heavy metals, as well as reducing COD and BOD indicators [74]. BSF can be used for the treatment of urban run-off water [75], stormwater run-off [76], CWBD [77], and others. Three main processes followed in BSF technology include injection of micro-sand, coagulant, and polymer to the system, followed by a maturation process, and finally settling the mixture and separation [74]. Further details and experimental work about BSF can be found in the following references: [74–76,78,79].

4. Evaluation of CWBD Water Treatment Technologies

Implementing one of the previously reviewed technologies for the treatment of CWBD water is highly applicable. However, considering the most suitable, green, sustainable, and highest performance technology is the main objective targeted by industries. Before implementation, screening and evaluating suitable technologies is necessary and requires a clear definition of the performance criteria. In this section, key criteria are used to compare and assess the treatment systems for CWBD; Table 2 shows a summary of the findings. The criteria considered are the scale of the process; maintenance requirements; chemical additive requirements for the system; energy consumption; permeate (effluent) quality; sludge characteristics; and, most importantly, the ability to remove CWBD contaminants and cost. All of these criteria are discussed and evaluated in this section.

Most of the seven technologies presented earlier were implemented in different scales, including laboratory, pilot, and commercial or industrial. However, ED and RO processes are considered one of the most established and well-known processes and are widely used [80–83]. NF, VSEP, BSF, and EC can be considered as emerging technologies, and this is due to limitations in their performance and cost, as is highlighted below.

Required maintenance for a system is often a factor that the industry considers. This criterion is impacted by the material used in the system, the number of moving parts, availability of membranes and associated fouling problems, and many others. EC technology requires maintenance mainly related to the periodic replacements of the electrodes used in the system [58]. It is considered a low maintenance system as compared to membrane-based technologies such as RO, ED, and NF. As some require high operating pressure and have issues with fouling. ED process has a longer membrane lifetime, and therefore maintenance will be lower than RO process [84]. Applying a pre-treatment process ahead to these technologies can reduce the fouling problems, consequently reducing the maintenance requirements. MD and VSEP processes require low maintenance; the latter is designed with a vibrating membrane to minimize fouling. The needed maintenance by VSEP is mainly associated with the few moving parts in the system [85].

The use of chemical additives in the process is an inherent part of some water treatment systems such as ED, RO, MD, and NF. Moreover, often, chemical treatment is required for the regeneration and cleaning of membrane-based processes. In the studied water treatment systems, BSF technology requires a high quantity of chemical dosages. The optimum values are 5–150 mg/L of alum, 40–190 mg/L of $FeCl_3$ (ferric chloride), 0.3–1 mg/L of polymer, and 3–12 mg/L of sand.

Energy requirements and consumption is critical aspect that affects the operational cost of the process directly impacts on the environment in terms of emission. Electricity is the main source of power used in these processes; however, in MD, most of the consumed power in the process is in terms of heat, with a small amount of electricity for running pumps [37]. Both NF(0.3–1 kWh/m^3) [86] and ED processes (depending on the level of TDS) require less energy compared to the RO process (1.5–6 kWhe/m^3) [87,88], mainly due to lower pressure requirement [89,90]. However, increased energy requirements are observed for ED for influent streams with higher salt content [91]. VSEP requires higher energy than RO process mainly due to the need for intense shear requirements on the membrane; under the same conditions, at TDS of 500 mg/L, motor/pump efficiency of 85%, and feed water recovery of 75%, the energy required by RO is 0.7 kWh/m^3. In contrast, VSEP requires 2.1 kWh/m^3. In the BSF process, hydro-cyclone is used for the separation and recirculation of micro-sand back to the process, which requires high-pressure input, and consequently higher energy consumption [92]. Finally, for EC, the operation of the process depends mainly on a continuous source of electricity. However, many studies were conducted to reduce the consumption of power by using more effective electrodes or changing their configurations for lower energy, as indicated in Table 2. The alternative available for EC is to consider using of a renewable energy source such as a solar system to reduce environmental impacts from energy consumption [93].

The quality and characterization of the permeate stream (treated) and the sludge or concentrated stream (rejected with contaminants) are key factors in selecting a suitable wastewater treatment technology. EC process produces high-quality effluent with low content of TDS and has neither color nor odor. EC system removes hardness and silica ions with different types of electrodes such as Zn, Fe, and Al electrodes with a removal efficiency of 38.63% and 95.62%, 36.99% and 98.93%, and 55.36% and 99.54% for the total hardness and silica ions, respectively. Although the water quality effluent from RO process is high, ED process has a higher recovery rate in comparison. For MD process, low recovery or flux of water is produced as compared to RO process, but the salt concentration in permeate is approximately zero. NF process has stable flowrate and clean permeate as in Table 2; in fact, many industries are using the technology and some are replacing their RO process [50]. BSF is comparable to conventional processes, e.g., RO and ED, and in some cases outperforms others in terms of permeate quality [74]. VSEP is the only technology that produces a permeate stream free of solids in comparison to the studied alternatives.

Table 2. Evaluation of the treatment technologies according to the selected criteria.

Criteria	Electrocoagulation (EC)	Electrodialysis (ED)	Membrane Distillation (MD)	Ballasted Sand Flocculation (BSF)	Reverse Osmosis (RO)	Nanofiltration (NF)	Vibratory Sheared Enhanced Membrane Process (VSEP)
Process scale	All levels [70]	Large-scale [80]	Pilot scale, large scale are implemented mostly in desalination process [42]	Pilot and large-scale [74]	Large-scale [48]	Implemented in large scale for coking wastewater treatment [97]; Pilot scale implementation for rubber wastewater	Large-scale VSEP as a recovery system for CWBD water in several facilities such as gas production, coal gas-fired power plant, and biomass plants [98,99]
Maintenance	Low [100]	Maintenance and the cost of maintenance associated only with the membrane [1]. Low maintenance [101]	Low [102]	High maintenance Required for example for the hydro-cyclone [103,104]	High maintenance requirement due to fouling [26]; Low maintenance if pre-treatment process is used [105]	Low maintenance if pre-treatment process is used [105]	Low maintenance because system has few moving parts [85]
Chemicals additives to the process	None [94]	None [34,106]	None [42]	Require high amount of chemicals, e.g., Alum and $FeCl_3$ compared to traditional processes [74]	None [107]; Requires chemicals for membrane cleaning to prevent fouling [108]	None [109]	None [85]
Utility and energy requirements	0.18–3.05 kWh/m^3 with magnesium electrodes for CWBD treatment [4]; Requires electricity [5]; Al electrode and monopolar-parallel connection, the energy consumption is less compared to bipolar-series [73]	From 1.1 to 2.9 kWh/m^3 for the treatment of almond industry wastewater; Requires a source of electricity; Less energy intensive compared to RO and thermal processes [1]; A lot of electricity is consumed in case of high level of salt in the influent stream [91];	Required energy is mainly in form of heat, and with a small amount of electricity for pumps [37]; The process requires high energy [39,110]	High energy for hydro-cyclone [92]	Relative: energy demand increases as a result of the fouling issue [26]	Less energy compared to RO processs [90]	With same conditions, the energy consumption of VSEP process is three times higher than RO process [111]
The quality of permeate (effluent water)	Low content of TDS, odorless, and colorless effluent [73]	Higher water recovery rate than effluent from RO process [110]; Requires post-treatment to remove the remaining sludge [73]	Low water recovery [110]; Low permeate flux compared to RO process [43]; Low (salt concentration near zero) [38]	Comparable quality to conventional treatment technologies [74]	High quality [112]	Produce clean, high-quality water, and the permeate has a stable flowrate [51,113]	High quality [34,114]; Solid-free permeate [85]

Table 2. *Cont.*

Criteria	Electrocoagulation (EC)	Electrodialysis (ED)	Membrane Distillation (MD)	Ballasted Sand Flocculation (BSF)	Reverse Osmosis (RO)	Nanofiltration (NF)	Vibratory Sheared Enhanced Membrane Process (VSEP)
The sludge quality (rejection)	Low sludge discharge, stable, and non-toxic [71,94]; Sludge contains mainly metallic oxides/hydroxides [73]	High sludge discharge [73]	Non-volatile compounds, macromolecules, and inorganic ions are all highly rejected from the water stream to the sludge with (99–100%) and the separation theoretically can reach 100% [115–117]	Flocs easily eliminated by settling [74]; Sludge layer is distinct, clear, and supernatant [78]	A concentrated stream with high salinity [118]	Low discharge sludge [109]; Retentate concentrations lower than RO for low value salts in the influent stream [109]; Rejection efficacy altered when fouling occurs [50]	Concentrated waste stream with 30 to 35% total solids (higher than the feed) [95,96]

In terms of sludge or the concentrated stream produced from these technologies, EC has non-toxic and low amounts of sludge with no brine formation compared to the membrane, ion-exchange, and conventional technologies [94]. On the other hand, ED produces a high amount of sludge and RO has a highly saline concentrated stream. The discharged volume and the retentate concentrations of NF process are lower as compared to RO process. For BSF process, the generated flocs can be easily eliminated. VSEP concentrated stream contains a high amount of salts (30–35%) compared to the process influent [95,96]. Finally, the MD process rejects various types of contaminants with a very high percentage close to 100% to form a concentrated stream of non-volatile compounds, macromolecules, and inorganic ions.

Contaminant removal effectiveness of the technology is a top criterion in selecting the CWBD treatment system. The literature has presented the removal and treatment capability of EC, ED, and MD processes for several contaminates in CWBD stream. It is important to clarify that the conducted experiments by EC and ED processes analyzed the removal or reducing levels of targeted contaminants such as magnesium, calcium, and silica for EC and reducing the level of TDS for ED. Hence, the results presented in Table 3 reflect the published data on the ability of the technology to remove contaminants and do not reflect the technology's ability to remove all the other contaminants in Table 1. As shown in Table 3, all three methods (ED, EC, MD) can remove major common contaminants in CWBD streams such as calcium, magnesium, chloride, and sulphate. TDS, which is considered one of the crucial contaminants of concern, can be removed by both EC and ED processes. According to the literature, only EC is reported to remove silica, zinc, and phosphate, and only MD is reported to remove copper and manganese. Additionally, ED can remove bromides and fluorine ions from CWBD water. Moreover, sodium and potassium can be removed using ED and MD, but Fe can be removed only by EC and MD. The removal efficiency differs for each technology, and it depends on the concentration of the feed and the operating conditions as reported by the authors of that literature.

Table 3. The contaminants that can be removed by CWBD treatment technologies.

Contaminants	**CWBD Treatment Technologies**		
	EC [4]	**ED [1]**	**MD [37]**
Total dissolved Solids (TDS)	√	√	
Calcium (Ca^{2+})	√	√	√
Magnesium (Mg^{2+})	√	√	√
Silica (SiO_2)	√		
Chloride (Cl^-)	√	√	√
Zinc (Zn^{2+})	√		
Phosphate ($PO_4{}^{3-}$)	√		
Iron (Fe^{2+})	√		√
Sulphate ($SO_4{}^{2-}$)	√	√	√
Aluminium (Al^{3+})			
Barium (Ba^{2+})			
Potassium (K^+)		√	√
Sodium (Na^+)		√	√
Copper (Cu)			√
Strontium (Sr^{2+})			
Bromide (Br^-)		√	
Fluorine (F)		√	
Manganese (Mn^{2+})			√
Nitrate (NO^{3-})			

The applicability for the treatment of CWBD streams using BSF, RO, NF, and VSEP technologies has been reported in the literature, even though there is a lack of published experimental data related to the specific contaminants that can be removed by these systems [11,26,47,99]. Reference [77] discussed the fact that the BSF process can be used for the CWBD effluent treatment, and Ahmed et al. [11] studied the potential for CWBD water recovery by reverse osmosis, discussing the fouling parameters as well as implemented various pre-treatment processes ahead of RO. Löwenberg et al. [26] conducted experimental work to evaluate various pre-treatment processes before applying RO as the final treatment step for CWBD. For NF and RO, Hossein et al. [47] also studied reusing CWBD after recovering it by NF or RO processes and investigated various pre-treatment steps to control fouling problems. As an emerging technology, VSEP technology website [99] reported that this process could reduce the volume of CWBD stream for either disposal or recycling it back to the cooling towers on the basis of pilot plant data.

In general, the reviewed technologies can remove a wide range of contaminants from many water and wastewater streams. For example, RO can remove almost all contaminants of concern such as TDS, ammonia (as N), iron, lead, nitrate (as N), sulphate, chloride, phosphate, calcium, magnesium, and others [119,120]. BSF can remove TP, TSS, iron, lead, zinc, etc. [74]. NF process can remove iron, manganese, calcium, magnesium, fluoride, sulphate, and more [121]. For VSEP technology, TDS, chloride, sulfate, calcium, magnesium, fluoride, nitrate, and others can be removed [111].

Conducting a cost analysis is another criterion to evaluate the technologies since the economic aspect should be justified and viable. Table 4 summarizes the operational and capital costs of the treatment technologies on the basis of color-coding. EC and MD technologies have low capital and operating costs, and consequently, lower total costs. The operating cost of EC process is mainly attributed to the fact that the power consumption in the form of electricity is high since it is one of the major requirements for the system to run and remove contaminants. For MD technology, it contains various designs and configurations for that the cost differs. For example, Air Gap membrane distillation (AGMD) is considered an effective method compared to others as it has how operational and maintenance costs. For the capital cost, MD generally has a lower capital cost than reverse osmosis, and such lower costs make the total cost low. Moving to RO and VSEP technologies requires high capital and operating costs, hence relatively high overall total cost. The high operational cost of RO process can be attributed to the fouling problems that shorten the membrane's lifetime and requires replacing it; moreover, the RO process operates at high pressure, which consumes high energy, leading to a higher cost of operation. For the emerging technology VSEP, the high operational cost is because of the high energy requirements to generate shear and vibrating the membrane. Subramani et al. [111] reported out of the conducted study that the consumption of energy by VSEP process is three times (2.1 kWh/m^3) higher than RO process (0.7 kWh/m^3), reaching more than 10 years, all of which contribute to lowering the operational cost [84]. However, ED has a high capital cost, which makes the total cost at a moderate level. The BSF process has a higher operational cost than conventional or traditional processes, and this can be said to be mainly due to the high dosages of chemicals needed for the process [74]. On the other hand, for the capital cost, BSF requires smaller sedimentation units because it has high settling rates; additionally, the BSF requires less land size, and hence lower capital cost, leading to a moderate total cost. For NF, which is another membrane-based technology, the process operates at low pressure and requires less energy consumption; hence, it has low operational costs. However, for the capital cost, the implementation of NF technology at large scales causes the capital cost to be high, which limits NF applications for treatment purposes in industries and makes the total cost relatively moderate. In the end, the total cost depends on many factors such as the size of the plant, the concentration of the influent stream, and the maintenance and energy requirements. Reducing the total cost can be done by enhancing the treatment technology by using efficient material with low cost and by utilizing renewable sources of energy instead of a direct source of power.

Table 4. Cost analysis of CWBD treatment technologies.

	EC	ED	MD	BSF	RO	NF	VSEP
Operational cost	Green	Green	Green	Orange/red	Orange/red	Green	Orange/red
Capital cost	Green	Orange/red	Green	Green	Orange/red	Orange/red	Orange/red
Total cost	Green	Yellow	Green	Yellow	Orange/red	Yellow	Orange/red
References	[5,122,123]	[122,124]	[44,125]	[74,126]	[48,122]	[127,128]	[129]
Color coding	Meaning						
Green	Low						
Yellow	Moderate						
Orange/red	High						

5. Regulations of Wastewater Effluent from Industries

CWBD is considered a wastewater stream discharged from cooling towers into the marine environment or sewage treatment plants, or reused in applications such as irrigation. Each of these discharge points has certain regulations and permissible limits for the water contaminants; the constraints are there to avoid negative and long-term consequences on the environment and society. Table 5 shows standards regulated by the United States Environmental Protection Agency (EPA) and some GCC agencies. Each agency's standards vary depending on the endpoint or application, and there is a noticeable difference in limits between agencies for the application. For example, the concentration of the pollutant by EPA for irrigation purposes differ only slightly from the one regulated by Qatar for boron, cadmium, cobalt, manganese, and zinc. However, the concentrations of aluminum, fluoride, iron, and lead are considerably different when both standards are compared. One major reason behind these differences is attributed to the type of soil; in Qatar, as in many other GCC countries such as the UAE, the soil is sandy, and this plays a role in trace metal adsorption and translocation in the soil–plant environment. The accumulation of heavy metals such as Zn, Cd, Cu, Cr, Fe, and Pb in the soil will be minimal because of the high infiltration of the sandy soil, deep percolation losses, and high evaporation rate [130].

Discharging the treated wastewater into marine environments as compared to irrigation has more stringent levels for some contaminants such as fluoride, TDS, and sulphate in addition to the pollutant indicators such as COD. This can be because of the negative direct impacts of these contaminants on living creatures such as fishes, flora, and fauna [18]. For other contaminants such as nickel, zinc, chromium, cobalt, and manganese, the standards for irrigation are more stringent than for the marine environment. This can be attributed to the fact the levels are already small, and the concentrations of these pollutants will be further diluted in the marine compared to irrigation. Comparing the standards of both the UAE and Qatar, one can notice that for some contaminants such as boron, cobalt, nickel, and zinc, Qatar permits slightly higher concentrations than the UAE and lower concentrations for other contaminants such as aluminum, fluoride, and iron. For Oman, almost all regulated contaminant levels are the same as the UAE, except for cyanide, phosphate, zinc, and cadmium. In general, GCC countries have comparable limits on contaminants concentrations. Using diffusers with single or multi ports, and the depth of these diffusers determines the level of dilution for the discharged pollutants [131]. It is also known that discharging using diffusers, in general, dilutes the contaminants more and faster compared to single-point discharges [131]. The noted similarities of regulations between GCC countries for discharging to marine life can be mainly attributed to the fact that countries have limited freshwater resources, and protecting the marine environment is of higher priority.

Table 5. Standards and regulations of treated wastewater streams for discharging or reusing.

	Standards of Water and Wastewater for Irrigation		Standards of Discharging into Marine			Standards of Discharging Liquid Waste to Public Foul Sewage Networks	
Parameter	EPA Irrigation ppm [132]	Qatar Irrigation ppm [133]	Qatar Discharges into Water Environment or Marine ppm [133]	UAE Discharges to Marine at Point of Discharge ppm [134]	Oman Discharges to Marine ppm [135]	Qatar Discharged Liquid Waste for Treatment by Public Sewage Work ppm [133]	Qatar Industrial Effluent Discharged to Sewers ppm [133]
Aluminium	5.0	15	3	20		30	-
Arsenic	0.10	0.1	-	0.05	0.05	5	5
Beryllium	0.10	-	-	0.05		5	-
Boron	0.75	1.5	1.5	1		-	-
Cadmium	0.01	0.05	0.05	0.05	0.50	2	10
Chromium	0.1	0.01	0.2	0.2		5	2
Cobalt	0.05	0.2	2	0.2		-	-
Copper	0.2	0.2	0.5	0.5	0.50	5	4
Fluoride	1	15	1	10		-	-
Iron	5	1	1	2	2.0	25	-
Lead	5	0.1	0.1	0.1	0.10	5	5
Lithium	2.5	-	-	-		-	-
Manganese	0.2	0.05	0.2	0.2		-	-
Molybdenum	0.01	-	-	-		-	-
Nickel	0.2	0.2	0.5	0.1	0.10	5	-
Selenium	0.02	-	0.02	0.02	0.02	-	-
Vanadium	0.1	-	-	-		-	-
Zinc	2	0.5	2	0.5	0.10	10	4
TDS	-	2000	1500	-		4000	-
Sulphate	-	400	0.1	-		1000	1000
Phosphate as P (PO_4^{-3})	-	30	2	2	0.10	-	-
Ammonia as N	-	-	3	2		-	-
Chlorine residual	-	0.1	0.05	-		-	-
Cyanide (total)	-	-	0.1	0.05	0.10	1	1
Nitrates	-	-	-	30		-	-
COD	-	150	100	100		3000	3000
BOD5	-	10	50	30	30	1000	-
Total organic carbon (TOC)	-	-	-	75		1000	-

Discharging the effluent wastewater from industries into sewage treatment plants is one of the management practices performed by many countries as it can treat various wastewater streams. However, caution should be taken regarding the concentration of the contaminants influent to sewage systems because some are designed for certain types of pollutants and at a specific limited range of concentration, as it may cause various harmful consequences as previously mentioned in Section 2. In Table 5, it is noted that the standard concentrations of contaminants influent to sewage treatment plants or sewer systems have higher tolerance compared to irrigation and marine discharging, especially in terms of TDS as well as the pollutant indicators such as BOD and COD. This is expected as treatment plants are designed to handle high concentration of contaminants that cannot be discharged to natural environments.

Generally, treatment and management of CWBD is necessary, and the suitable options in terms of treatment depend on desired effluent quality. This quality is dictated by the endpoint of discharge or treated water application as regulated by standards of the country, as has been presented and discussed in this work.

6. Conclusions

CWBD is a very concentrated wastewater stream with various types of contaminants that can affect the entire ecosystem if it is not handled and treated properly. Various processes can be implemented to treat CWBD stream. Some of these technologies are already established, such as RO, ED, MD, and BSF, while others are emerging, such as EC, VSEP, and NF. In terms of choosing the best technology, being green and environmentally friendly are amongst the key considerations, besides the cost and the treatment performance. On the basis of the performed evaluation, membrane-based technologies result in high-quality treated water, but some of them, such as NF and RO, are prone to fouling problems, resulting in higher maintenance requirements. Additionally, EC and VSEP also produce high-quality permeate; however, they are energy-intensive processes. BSF is the only process that requires a large quantity of chemicals as a part of the system. For the most cost-effective technologies, EC and MD should be considered; ED and NF can also be considered if a pre-treatment step is available. EC has a high treatment performance ($\approx$99.54% in terms of silica ions) compared to UF membrane method (reduction of 65% of colloidal silica). For the treatment of contaminants, ED, MD, and EC processes treat a wide range of contaminants in CWBD. However, for a fair comparison, further studies should be conducted on the ability of RO, BSF, VSEP, and NF to treat CWBD contaminants to obtain a more reasonable conclusion. Regarding energy consumption, both EC~0.18–3.05 kWh/m^3 and VSEP~2.1 kWh/m^3 technologies have high energy demand, and this limits their implementation for large-scale applications unless renewable energy sources are used. When it comes to the recovery and reusing of treated wastewater or even discharging it to the marine environment, standards and regulations must be obeyed. Overall, the treatment of CWBD water should be further studied and considered since research related to this topic is limited. Treatment and reuse of CWBD can help overcome the water scarcity problem and achieve a more sustainable environment.

Funding: This research is funded by Ministry of Municipality and Environment in Qatar, Project MME contract no. P2020/1.

Institutional Review Board Statement: Not applicable.

Data Availability Statement: Not applicable.

Acknowledgments: This paper was made possible by funding from Ministry of Municipality and Environment in Qatar, Project MME contract no. P2020/1. The statements made herein are solely the responsibility of the author(s).

Conflicts of Interest: The authors declare no conflict of interest.

Abbreviations

CWBD	Cooling water blowdown
RO	Reverse osmosis
ED	Electrodialysis
MD	Membrane distillation
EC	Electrocoagulation
NF	Nanofiltration
COD	Chemical oxygen demand
BOD	Biological oxygen demand
TDS	Total dissolved solids
TSS	Total suspended solids
BSF	Ballasted sand flocculation
VSEP	Vibratory shear enhanced membrane process

References

1. Dhadake, Y. *Treatmnet of Cooling Tower Blowdown Water Using Electrodialysis*; California State University: Long Beach, CA, USA, 2018.
2. Afshari, F.; Dehghanpour, H. A Review Study on Cooling Towers; Types, Performance and Application. *ALKU J. Sci.* **2019**, 1–10. Available online: https://dergipark.org.tr/en/pub/alku/issue/41490/489143 (accessed on 28 November 2021).
3. Panjeshahi, M.; Ataei, A.; Gharaie, M.; Parand, R. Optimum design of cooling water systems for energy and water conservation. *Chem. Eng. Res. Des.* **2009**, *87*, 200–209. [CrossRef]
4. Abdel-Shafy, H.I.; Shoeib, M.A.; El-Khateeb, M.A.; Youssef, A.O.; Hafez, O.M. Electrochemical treatment of industrial cooling tower blowdown water using magnesium-rod electrode. *Water Resour. Ind.* **2020**, *23*, 100121. [CrossRef]
5. Hafez, O.M.; Shoeib, M.; El-Khateeb, M.A.; Abdel-Shafy, H.; Youssef, A.O. Removal of scale forming species from cooling tower blowdown water by electrocoagulation using different electrodes. *Chem. Eng. Res. Des.* **2018**, *136*, 347–357. [CrossRef]
6. Saha, P.; Bruning, H.; Wagner, T.V.; Rijnaarts, H.H. Removal of organic compounds from cooling tower blowdown by electrochemical oxidation: Role of electrodes and operational parameters. *Chemosphere* **2020**, *259*, 127491. [CrossRef] [PubMed]
7. Saha, P.; Wagner, T.V.; Ni, J.; Langenhoff, A.A.; Bruning, H.; Rijnaarts, H.H. Cooling tower water treatment using a combination of electrochemical oxidation and constructed wetlands. *Process. Saf. Environ. Prot.* **2020**, *144*, 42–51. [CrossRef]
8. Wagner, V.T.; Parsons, J.R.; Rijnaarts, H.H.M.; de Voogt, P.; Langenhoff, A.A.M.; Wagner, T.V.; Parsons, J.R.; Rijnaarts, H.H.M.; de Voogt, P. Technology a Review on the Removal of Conditioning Chemicals from Cooling Tower Water in Constructed Wetlands. *Crit. Rev. Environ. Sci. Technol.* **2019**, *48*, 1094–1125. [CrossRef]
9. Li, X.; Wu, L.; Lu, S.; Yang, H.; Xie, W.; Zhao, H.; Zhang, Y.; Cao, X.; Tang, G.; Li, H.; et al. Treatment of cooling tower blowdown water by using adsorption-electrocatalytic oxidation: Technical performance, toxicity assessment and economic evaluation. *Sep. Purif. Technol.* **2020**, *252*, 117484. [CrossRef]
10. Stratton, C.L.; Lee, G.F. Cooling Towers and Water Quality. *J. Water Pollut. Control Fed.* **1975**, *47*, 1901–1912.
11. Ahmed, J.; Jamal, Y.; Shujaatullah, M. Recovery of cooling tower blowdown water through reverse osmosis (RO): Review of water parameters affecting membrane fouling and pretreatment schemes. *Desalin. Water Treat.* **2020**, *189*, 9–17. [CrossRef]
12. Altman, S.J.; Jensen, R.P.; Cappelle, M.A.; Sanchez, A.L.; Everett, R.L.; Anderson, H.L.; McGrath, L.K. Membrane treatment of side-stream cooling tower water for reduction of water usage. *Desalination* **2012**, *285*, 177–183. [CrossRef]
13. Cheng, Q.; Wang, C.; Doudrick, K.; Chan, C.K. Hexavalent chromium removal using metal oxide photocatalysts. *Appl. Catal. B Environ.* **2015**, *176–177*, 740–748. [CrossRef]
14. Dickerson, B.R.; Vinyard, G.L. Effects of High Levels of Total Dissolved Solids in Walker Lake, Nevada, on Survival and Growth of Lahontan Cutthroat Trout. *Trans. Am. Fish. Soc.* **1999**, *128*, 507–515. [CrossRef]
15. Maulbetsch, J.S.; DiFilippo, M.N. *Performance, Cost, and Environmental Effects of Saltwater Cooling Towers. Environmental Effects of Mining*; California Energy Commission: Sacramento, CA, USA, 2010; pp. 1–356.
16. Weber-Scannell, P.K.; Duffy, L.K. Effects of Total Dissolved Solids on Aquatic Organisms: A Review of Literature and Recommendation for Salmonid Species. *Am. J. Environ. Sci.* **2007**, *3*, 1–6. [CrossRef]
17. USGS. *Phosphorus and Water*; USGS: Reston, VA, USA, 2016.
18. Papadopoulos, E.; Arsenos, G.; Ptochos, S.; Katsoulos, P.; Oikonomou, G.; Karatzia, M.A.; Karatzias, H. Pollutants in Wastewater Effluents Impacts. *J. Hell. Vet. Med. Soc.* **2014**, *65*, 115–120.
19. Clarke, R. *Groundwater: A Threatened Resource*; UNEP Environmental Library, UN Environment Programme: Nairobi, Kenya, 2002; p. 36.
20. Olariu, R. Treatment of Cooling Tower Blowdown Water. The Effect of Biodispersant on the Ultrafiltration Membrane. Master's Thesis, Delft University of Technology, Delft, The Netherlands, 2015.
21. Fendorf, S.; Michael, H.A.; van Geen, A. Spatial and Temporal Variations of Groundwater Arsenic in South and Southeast Asia. *Science* **2010**, *328*, 1123–1127. [CrossRef] [PubMed]

22. Mccarty, K.M.; Hanh, H.T.; Kim, K.-W. Arsenic geochemistry and human health in South East Asia. *Rev. Environ. Health* **2011**, *26*, 71–78. [CrossRef] [PubMed]
23. Ashraf, S.; Abbas, F.; Ibrahim, M.; Rashid, U.; Khalid, S.; Ahmad, R.; Hakeem, K.R.; Majeed, T. Application of GIS for the identification and demarcation of selective heavy metal concentrations in the urban groundwater. *J. Geogr. Sci.* **2015**, *25*, 225–235. [CrossRef]
24. Isa, N.M.; Aris, A.Z.; Lim, W.Y.; Sulaiman, W.N.A.W.; Praveena, S.M. Evaluation of heavy metal contamination in groundwater samples from Kapas Island, Terengganu, Malaysia. *Arab. J. Geosci.* **2014**, *7*, 1087–1100. [CrossRef]
25. Galadima, A.; Garba, Z.N. Heavy Metals Pollution in Nigeria: Causes and Consequences. *Elixir Pollut.* **2012**, *45*, 7917–7922.
26. Löwenberg, J.; Baum, J.A.; Zimmermann, Y.-S.; Groot, C.; Broek, W.V.D.; Wintgens, T. Comparison of pre-treatment technologies towards improving reverse osmosis desalination of cooling tower blow down. *Desalination* **2015**, *357*, 140–149. [CrossRef]
27. Shibli, S.M.A.; Saji, V.S. Corrosion Inhibitors in Cooling Towers. *Chem. Ind. Dig.* **2002**, *15*, 74–80.
28. ChemWorld. We Sell Direct to Everyone. 2016. Available online: https://www.chemworld.com/ (accessed on 28 November 2021).
29. Pereira, L.; Duarte, E.; Fragoso, R. Water Use: Recycling and Desalination for Agriculture. *Encycl. Agric. Food Syst.* **2014**, *5*, 407–424. [CrossRef]
30. Bhardwaj, R.; Sharma, R.; Handa, N.; Kaur, H.; Kaur, R.; Sirhindi, G.; Thukral, A. Prospects of Field Crops for Phytoremediation of Contaminants. In *Emerging Technologies and Management of Crop Stress Tolerance*; Academic Press: Cambridge, MA, USA, 2014; Volume 2, pp. 449–470.
31. Sultana, M.S.; Jolly, Y.; Yeasmin, S.; Islam, A.; Satter, S.; Tareq, S.M. Transfer of Heavy Metals and Radionuclides from Soil to Vegetables and Plants in Bangladesh. In *Soil Remediation and Plants: Prospects and Challenges*; Elsevier: Amsterdam, The Netherlands, 2015; pp. 331–366.
32. Fales, A.L. *The Effects of Industrial Wastes on Sewage Treatment*; New England Interstate Water Pollution Control Commission: Boston, MA, USA, 1970; pp. 970–985.
33. *Membrane Technology*; Lenntech: Delft, The Netherlands, 2020.
34. Gurreri, L.; Tamburini, A.; Cipollina, A.; Micale, G. Electrodialysis Applications in Wastewater Treatment for Environmental Protection and Resources Recovery: A Systematic Review on Progress and Perspectives. *Membranes* **2020**, *10*, 146. [CrossRef] [PubMed]
35. Bazinet, L.; Doyen, A.; Roblet, C. *Electrodialytic Phenomena, Associated Electromembrane Technologies and Applications in the Food, Beverage and Nutraceutical Industries*; Woodhead Publishing Limited: Sawston, UK, 2013; pp. 202–218.
36. Wang, Y.; Jiang, C.; Bazinet, L.; Xu, T. Electrodialysis-Based Separation Technologies in the Food Industry. In *Separation of Functional Molecules in Food by Membrane Technology*; Academic Press: Cambridge, MA, USA, 2019; pp. 349–381.
37. Koeman-Stein, N.; Creusen, R.; Zijlstra, M.; Groot, C.; Broek, W.V.D. Membrane distillation of industrial cooling tower blowdown water. *Water Resour. Ind.* **2016**, *14*, 11–17. [CrossRef]
38. Ma, J.; Irfan, H.M.; Wang, Y.; Feng, X.; Xu, D. Recovering Wastewater in a Cooling Water System with Thermal Membrane Distillation. *Ind. Eng. Chem. Res.* **2018**, *57*, 10491–10499. [CrossRef]
39. Kebria, M.R.S.; Rahimpour, A. *Membrane Distillation: Basics, Advances, and Applications. Advances in Membrane Technologies*; Abdelrasoul, A., Ed.; IntechOpen: London, UK, 2020. [CrossRef]
40. Abraham, T.; Luthra, A. Socio-economic & technical assessment of photovoltaic powered membrane desalination processes for India. *Desalination* **2011**, *268*, 238–248. [CrossRef]
41. Karanasiou, A.; Kostoglou, M.; Karabelas, A. An Experimental and Theoretical Study on Separations by Vacuum Membrane Distillation Employing Hollow-Fiber Modules. *Water* **2018**, *10*, 947. [CrossRef]
42. Kiss, A.A.; Readi, O.M.K. An industrial perspective on membrane distillation processes. *J. Chem. Technol. Biotechnol.* **2018**, *93*, 2047–2055. [CrossRef]
43. Pangarkar, B.; Deshmukh, S.; Sapkal, V. Review of membrabe distillation process for water purification. *Desalin. Water Treat.* **2014**, *57*, 2959–2981. [CrossRef]
44. Bappy, P.; Jabed, M.; Bahar, R.; Ariff, T.F. Low Energy and Low Cost Freshwater Production by Membrane Distillation. In Proceedings of the International Conference on Industrial Engineering and Operations Management, Kuala Lumpur, Malaysia, 8–10 March 2016; pp. 1799–1804.
45. Shirazi, M.M.A.; Kargari, A. A Review on Applications of Membrane Distillation (MD) Process for Wastewater Treatment. *J. Membr. Sci. Res.* **2015**, *1*, 101–112.
46. BioTech Water Researchers. *Advantages and Disadvantages of Reverse Osmosis*; BioTech Water Researchers: San Antonio, TX, USA, 2019.
47. Farahani, M.H.D.A.; Borghei, S.M.; Vatanpour, V. Recovery of Cooling Tower Blowdown Water for Reuse: The Investigation of Different Types of Pretreatment Prior Nanofiltration and Reverse Osmosis. *J. Water Process Eng.* **2016**, *10*, 188–199. [CrossRef]
48. Rabiee, H.; Khalilpour, K.R.; Betts, J.M.; Tapper, N. Energy-Water Nexus: Renewable-Integrated Hybridized Desalination Systems. In *Polygeneration with Polystorage for Chemical and Energy Hubs*; Academic Press: Cambridge, MA, USA, 2019; pp. 409–458.
49. Elaine, H.; Doerge, T.A.; Baker, P.B. Water Facts: Number 6 Reverse Osmosis Units. no. 6.1994. Available online: https://files.knowyourh2o.com/Waterlibrary/privatewell/reverseosmosis.pdf (accessed on 28 November 2021).

50. Macedo, A.T.Z.N.; Pulido, J.M.O.; Fragoso, R. The Use and Performance of Nanofiltration Membranes for Agro-Industrial Effluents Purification. In *Nanofiltration*; IntechOpen: London, UK, 2018; p. 65. [CrossRef]
51. Abdel-Fatah, M.A. Nanofiltration systems and applications in wastewater treatment: Review article. *Ain Shams Eng. J.* **2018**, *9*, 3077–3092. [CrossRef]
52. Shon, H.K.; Phuntsho, S.; Chaudhary, D.S.; Vigneswaran, S.; Cho, J. Nanofiltration for water and wastewater treatment—A mini review. *Drink. Water Eng. Sci.* **2013**, *6*, 47–53. [CrossRef]
53. Shahmansouri, A.; Bellona, C. Nanofiltration technology in water treatment and reuse: Applications and costs. *Water Sci. Technol.* **2015**, *71*, 309–319. [CrossRef]
54. Leong, J.; Tan, J.; Heitz, A.; Ladewig, B.P. Use of vibratory shear enhanced processing to treat magnetic ion exchange concentrate: A techno-economic analysis. *Desalination* **2016**, *383*, 46–52. [CrossRef]
55. Zouboulis, A.I.; Peleka, E.N.; Ntolia, A. Treatment of Tannery Wastewater with Vibratory Shear-Enhanced Processing Membrane Filtration. *Separations* **2019**, *6*, 20. [CrossRef]
56. Balasubramanian, P. A Brief Review on Best Available Technologies for Reject Water (Brine) Management in Industries. *Int. J. Environ. Sci.* **2013**, *3*, 2010–2018.
57. Zouboulis, A.; Petala, M. Performance of VSEP vibratory membrane filtration system during the treatment of landfill leachates. *Desalination* **2008**, *222*, 165–175. [CrossRef]
58. Babu, D.S.; Singh, T.S.A.; Nidheesh, P.V.; Kumar, M.S. Industrial wastewater treatment by electrocoagulation process. *Sep. Sci. Technol.* **2020**, *55*, 3195–3227. [CrossRef]
59. Hernández, I.L.; Díaz, C.B.; Cerecero, M.V.; Sánchez, P.T.A.; Juárez, M.C.; Lugo, V.L. Soft drink wastewater treatment by electrocoagulation—electrooxidation processes. *Environ. Technol.* **2016**, *38*, 433–442. [CrossRef] [PubMed]
60. Mollah, M.Y.; Morkovsky, P.; Gomes, J.A.; Kesmez, M.; Parga, J.; Cocke, D.L. Fundamentals, present and future perspectives of electrocoagulation. *J. Hazard. Mater.* **2004**, *114*, 199–210. [CrossRef]
61. Akyol, A. Treatment of paint manufacturing wastewater by electrocoagulation. *Desalination* **2012**, *285*, 91–99. [CrossRef]
62. El-Ashtoukhy, E.S.Z.; El-Taweel, Y.A.; Abdelwahab, O.; Nassef, E.M. Treatment of Petrochemical Wastewater Containing Phenolic Compounds by Electrocoagulation Using a Fixed Bed Electrochemical Reactor. *Int. J. Electrochem. Sci.* **2013**, *8*, 1534–1550.
63. Zhao, S.; Huang, G.; Cheng, G.; Wang, Y.; Fu, H. Hardness, COD and turbidity removals from produced water by electrocoagulation pretreatment prior to Reverse Osmosis membranes. *Desalination* **2014**, *344*, 454–462. [CrossRef]
64. Mansour, S.; Hasieb, I. Removal of Nickel from Drinking Water by Electrocoagulation Technique Using Alternating Current. *Curr. Res. Chem.* **2012**, *4*, 41–50. [CrossRef]
65. Hashim, K.S.; Shaw, A.; Al Khaddar, R.; Pedrola, M.O.; Phipps, D. Iron removal, energy consumption and operating cost of electrocoagulation of drinking water using a new flow column reactor. *J. Environ. Manag.* **2017**, *189*, 98–108. [CrossRef] [PubMed]
66. Golder, A.; Samanta, A.; Ray, S. Removal of trivalent chromium by electrocoagulation. *Sep. Purif. Technol.* **2007**, *53*, 33–41. [CrossRef]
67. Silva, J.F.; Graça, N.S.; Ribeiro, A.; Rodrigues, A.E. Electrocoagulation process for the removal of co-existent fluoride, arsenic and iron from contaminated drinking water. *Sep. Purif. Technol.* **2018**, *197*, 237–243. [CrossRef]
68. Hashim, K.S.; Alkhaddar, R.; Jasim, N.; Shaw, A.; Phipps, D.; Kot, P.; Pedrola, M.O.; Alattabi, A.W.; Abdulredha, M.; Alawsh, R. Electrocoagulation as a green technology for phosphate removal from river water. *Sep. Purif. Technol.* **2019**, *210*, 135–144. [CrossRef]
69. An, C.; Huang, G.; Yao, Y.; Zhao, S. Emerging usage of electrocoagulation technology for oil removal from wastewater: A review. *Sci. Total Environ.* **2017**, *579*, 537–556. [CrossRef]
70. Islam, S.M.D.-U. Electrocoagulation (EC) technology for wastewater treatment and pollutants removal. *Sustain. Water Resour. Manag.* **2019**, *5*, 359–380. [CrossRef]
71. Moussa, D.T.; El-Naas, M.H.; Nasser, M.; Al-Marri, M.J. A comprehensive review of electrocoagulation for water treatment: Potentials and challenges. *J. Environ. Manag.* **2017**, *186*, 24–41. [CrossRef]
72. Nariyan, E.; Sillanpää, M.; Wolkersdorfer, C. Electrocoagulation treatment of mine water from the deepest working European metal mine—Performance, isotherm and kinetic studies. *Sep. Purif. Technol.* **2017**, *177*, 363–373. [CrossRef]
73. Anwer, E.A.; Majeed, B.A.A. Different Electrodes Connections in Electrocoagulation of Synthetic Blow down Water of Cooling Tower. *Iraqi J. Chem. Pet. Eng.* **2020**, *21*, 1–7. [CrossRef]
74. Kumar, S.; Ghosh, N.C.; Kazmi, A.A. Ballasted sand flocculation for water, wastewater and CSO treatment. *Environ. Technol. Rev.* **2016**, *5*, 57–67. [CrossRef]
75. Kumar, S.; Kazmi, A.A.; Ghosh, N.C.; Singh, R. Optimization of Ballasted Sand Flocculation Process for Urban Runoff Treatment. *J. Indian Water Resour. Soc.* **2018**, *37*. Available online: http://117.252.14.250:8080/jspui/handle/123456789/4895 (accessed on 28 November 2021).
76. Kazmi, A.A.; Kumar, S.; Ghosh, N.C. Ballasted Sand Flocculation Technology for Storm Water Run Off Treatmnet. In Proceedings of the 21st International Water Technology Conference, Port Said, Egypt, 28–30 June 2018; pp. 28–30.
77. Veolia Water Technology. Actiflo®Sand-Ballasted Water Clarification and Softening Processes. Available online: https://www.oilandgasonline.com/doc/actiflo-ballasted-water-clarification-system-0001 (accessed on 28 November 2021).
78. Sauvignet, P. Sand-Ballasted Flocculation Technology and Mobile Facilities. *Quarry Manag.* **2003**, *30*, 17–18.

79. Treguer, R.; Blair, B.; Klaper, R.; Royer, S.; Magruder, C. Evaluation of Actiflo® Carb Process for the Combined Removal of Trace Organic Compounds and Phosphorous during Wastewater Tertiary Treatment. *Proc. Water Environ. Fed.* **2012**, *2012*, 7176–7196. [CrossRef]
80. Akhter, M.; Habib, G.; Qamar, S.U. Application of Electrodialysis in Waste Water Treatment and Impact of Fouling on Process Performance. *J. Membr. Sci. Technol.* **2018**, *8*, 1000182. [CrossRef]
81. Mahmoudi, H. *Encyclopedia of Membranes*; Springer: Berlin/Heidelberg, Germany, 2015; pp. 1–2.
82. Qasim, M.; Badrelzaman, M.; Darwish, N.N.; Darwish, N.A.; Hilal, N. Reverse osmosis desalination: A state-of-the-art review. *Desalination* **2019**, *459*, 59–104. [CrossRef]
83. Trishitman, D.; Cassano, A.; Basile, A.; Rastogi, N.K. Reverse osmosis for industrial wastewater treatment. In *Current Trends and Future Developments on (Bio-) Membranes*; Elsevier: Amsterdam, The Netherlands, 2020; pp. 207–228.
84. Singh, R. Desalination and On-site Energy for Groundwater Treatment in Developing Countries Using Fuel Cells. In *Emerging Membrane Technology for Sustainable Water Treatment*; Elsevier: Amsterdam, The Netherlands, 2016; pp. 135–162.
85. Culkin, B. Eccentric Drive Mechanism. U.S. Patent US5014564A, 14 May 1991. Available online: https://patents.google.com/patent/US5014564A/en (accessed on 28 November 2021).
86. EMIS. Energie-En Milieu-Informatiesysteem Voor Het Vlaamse Gewest. 2020. Available online: https://emis.vito.be/nl (accessed on 28 November 2021).
87. Al-Karaghouli, A.; Kazmerski, L.L. Energy consumption and water production cost of conventional and renewable-energy-powered desalination processes. *Renew. Sustain. Energy Rev.* **2013**, *24*, 343–356. [CrossRef]
88. Chaoui, I.; Abderafi, S.; Vaudreuil, S.; Bounahmidi, T. Water desalination by forward osmosis: Draw solutes and recovery methods—Review. *Environ. Technol. Rev.* **2019**, *8*, 25–46. [CrossRef]
89. Kim, I.S.; Hwang, M.; Choi, C. Membrane-Based Desalination Technology for Energy Efficiency and Cost Reduction. In *Desalination Sustainability (A Technical, Socioeconomic, and Environmental Approach)*; Elsevier: Amsterdam, The Netherlands, 2017; pp. 31–74.
90. Mulyanti, R.; Susanto, H. Wastewater Treatment by Nanofiltration Membranes. *IOP Conf. Ser. Earth Environ. Sci.* **2018**, *142*, 012017. [CrossRef]
91. Strathmann, H. *Encyclopedia of Membranes*; Springer: Berlin/Heidelberg, Germany, 2015. [CrossRef]
92. Binot, P.; Dahl, C.P.; Zuback, J.E. Methods of Treating Water Via Ballasted Flocculation and. Decantation. Patent CA2471034C, 24 April 2012.
93. Sharma, G.; Choi, J.; Shon, H.K.; Phuntsho, S. Solar-powered electrocoagulation system for water and wastewater treatment. *Desalin. Water Treat.* **2011**, *32*, 381–388. [CrossRef]
94. Hafez, O.M.; Shoeib, M.A.; El-khateeb, M.A.; Abdel-shafy, H.I.; Youssef, A.O. *Pt Graphical Abstract Sc B*; Surface Coating Department, Central Metallurgical Research and Development Institute: Cairo, Egypt, 2018.
95. Bushart, S.; Tran, P.; Asay, R. *Low-Level Liquid Waste Processing Pilot Studies Using a Vibratory Shear Enhanced Process (Vsep) for Fultration*; WM Symposia, Inc.: Tucson, AZ, USA, 2002; pp. 1–8.
96. Wateronline. End of Pipe Effluent Treatment for Pulp and Paper Mills 2021. Available online: https://www.wateronline.com/doc/end-of-pipe-effluent-treatment-for-pulp-and-p-0001 (accessed on 28 November 2021).
97. Jin, X.; Li, E.; Lu, S.; Qiu, Z.; Sui, Q. Coking Wastewater Treatment for Industrial Reuse Purpose: Combining Biological Processes with Ultrafiltration, Nanofiltration and Reverse Osmosis. *J. Environ. Sci.* **2013**, *25*, 1565–1574. [CrossRef]
98. VSEP. Applications—New Logic Research. Available online: https://www.vsep.com/technology/applications/ (accessed on 19 June 2021).
99. New Logic Research. Membrane Filtration of Cooling Tower Blowdown Application. Available online: https://www.vsep.com/pdf/Membrane-Treatment-of-Cooling-Tower-Blowdown.pdf (accessed on 28 November 2021).
100. Fayad, N. The Application of Electrocoagulation Process for Wastewater Treatment and for the Separation and Purification of Biological Media. Master's Thesis, Université Clermont Auvergne, Clermont-Ferrand, France, 2018; pp. 5–226.
101. Westerling, K. ED vs. RO the Benefits of Electrodialysis for Desalination. 2015. Available online: https://www.wateronline.com/doc/ed-vs-ro-the-benefits-of-electrodialysis-for-desalination-0001 (accessed on 28 November 2021).
102. Mendez, E. Sustainable Desalination: Membrane Distillation Delivers Greener Clean Water—Filtration + Separation. 2012. Available online: https://www.filtsep.com/desalination/features/sustainable-desalination-membrane-distillation/ (accessed on 28 November 2021).
103. Cui, H.; Huang, X.; Yu, Z.; Chen, P.; Cao, X. Application progress of enhanced coagulation in water treatment. *RSC Adv.* **2020**, *10*, 20231–20244. [CrossRef]
104. USEPA. *Wastewater Technology Fact Sheet Dechlorination*; Environmental Protection Agency: Washington, DC, USA, 2000; pp. 1–7.
105. Samcotech. *Reverse Osmosis Vs Nanofiltration Membrane Process: What Is the Difference?* Samcotech: New York, NY, USA, 2017.
106. Singh, K.; Lataye, D.H.; Wasewar, K.L.; Yoo, C.K. Removal of fluoride from aqueous solution: Status and techniques. *Desalin. Water Treat.* **2013**, *51*, 3233–3247. [CrossRef]
107. Mondal, S.; Wickramasinghe, S.R. Produced water treatment by nanofiltration and reverse osmosis membranes. *J. Membr. Sci.* **2008**, *322*, 162–170. [CrossRef]
108. Madaeni, S.; Samieirad, S. Chemical cleaning of reverse osmosis membrane fouled by wastewater. *Desalination* **2010**, *257*, 80–86. [CrossRef]

109. EMIS. Nanofiltration. 2010. Available online: https://emis.vito.be/en/bat/tools-overview/sheets/nanofiltration (accessed on 28 November 2021).
110. Al-Amshawee, S.; Yunus, M.Y.B.M.; Azoddein, A.A.M.; Hassell, D.G.; Dakhil, I.H.; Hasan, H.A. Electrodialysis desalination for water and wastewater: A review. *Chem. Eng. J.* **2019**, *380*, 122231. [CrossRef]
111. Subramani, A.; De Carolis, J.; Pearce, W.; Jacangelo, J.G. Vibratory shear enhanced process (VSEP) for treating brackish water reverse osmosis concentrate with high silica content. *Desalination* **2012**, *291*, 15–22. [CrossRef]
112. Massot, A.; Mietton-Peuchot, M.; Peuchot, C.; Milisic, V. Nanofiltration and Reverse Osmosis in Winemaking. *Desalination* **2008**, *231*, 283–289. [CrossRef]
113. Zeynali, R.; Akbari, M.; Ghasemzadeh, K.; Jalilnejad, E.; Basile, A. *Chapter 4—Achievements in High Pressure Membrane Processes Nf and Ro for Wastewater and Water Treatment*; Elsevier: Amsterdam, The Netherlands, 2020.
114. Water, Produced, and Bilge Water. Using Vsep to Treat Produced Water. 1993. Available online: https://www.vsep.com/pdf/ProducedWater.pdf (accessed on 28 November 2021).
115. Biniaz, P.; Ardekani, N.T.; Makarem, M.A.; Rahimpour, M.R. Water and Wastewater Treatment Systems by Novel Integrated Membrane Distillation (MD). *ChemEngineering* **2019**, *3*, 8. [CrossRef]
116. Luo, W.; Phan, H.V.; Li, G.; Hai, F.; Price, W.E.; Elimelech, M.; Nghiem, L.D. An Osmotic Membrane Bioreactor–Membrane Distillation System for Simultaneous Wastewater Reuse and Seawater Desalination: Performance and Implications. *Environ. Sci. Technol.* **2017**, *51*, 14311–14320. [CrossRef]
117. Wang, P.; Chung, T.S. Recent Advances in Membrane Distillation Processes: Membrane Development, Configuration Design and Application Exploring. *J. Membr. Sci.* **2015**, *474*, 39–56. [CrossRef]
118. Ghyselbrecht, K.; Van Houtte, E.; Pinoy, L.; Verbauwhede, J.; Van der Bruggen, B.; Meesschaert, B. Treatment of RO concentrate by means of a combination of a willow field and electrodialysis. *Resour. Conserv. Recycl.* **2012**, *65*, 116–123. [CrossRef]
119. Bartels, C.; Wilf, M.; Andes, K.; Iong, J. Design considerations for wastewater treatment by reverse osmosis. *Water Sci. Technol.* **2005**, *51*, 473–482. [CrossRef]
120. Luminor. Reverse Osmosis Water Filtration Overview. 2021. Available online: https://www.luminoruv.com/education/reverse-osmosis (accessed on 28 November 2021).
121. Van Der Bruggen, B.; Vandecasteele, C. Removal of Pollutants from Surface Water and Groundwater by Nanofiltration: Overview of Possible Applications in the Drinking Water Industry. *Environ. Pollut.* **2003**, *122*, 435–445. [CrossRef]
122. Tibebe, D.; Kassa, Y.; Bhaskarwar, A.N. Treatment and characterization of phosphorus from synthetic wastewater using aluminum plate electrodes in the electrocoagulation process. *BMC Chem.* **2019**, *13*, 107. [CrossRef]
123. Uludag-Demirer, S.; Olson, N.; Ives, R.; Nshimyimana, J.P.; Rusinek, C.A.; Rose, J.B.; Liao, W. Techno-Economic Analysis of Electrocoagulation on Water Reclamation and Bacterial/Viral Indicator Reductions of a High-Strength Organic Wastewater—Anaerobic Digestion Effluent. *Sustainability* **2020**, *12*, 2697. [CrossRef]
124. Pirsaheb, M.; Khosravi, T.; Sharafi, K.; Mouradi, M. Comparing operational cost and performance evaluation of electrodialysis and reverse osmosis systems in nitrate removal from drinking water in Golshahr, Mashhad. *Desalin. Water Treat.* **2015**, *57*, 5391–5397. [CrossRef]
125. Kullab, A.; Martin, A. Membrane distillation and applications for water purification in thermal cogeneration plants. *Sep. Purif. Technol.* **2011**, *76*, 231–237. [CrossRef]
126. Young, J.C.; Edwards, F.G. Factors affecting ballasted flocculation reactions. *Water Environ. Res.* **2003**, *75*, 263–272. [CrossRef] [PubMed]
127. Birnhack, L.; Nir, O.; Lahav, O. Establishment of the Underlying Rationale and Description of a Cheap Nanofiltration-Based Method for Supplementing Desalinated Water with Magnesium Ions. *Water* **2014**, *6*, 1172–1186. [CrossRef]
128. Elazhar, F.; Touir, J.; Elazhar, M.; Belhamidi, S.; El Harrak, N.; Zdeg, A.; Hafsi, M.; Amor, Z.; Taky, M.; Elmidaoui, A. Techno-economic comparison of reverse osmosis and nanofiltration in desalination of a Moroccan brackish groundwater. *Desalin. Water Treat.* **2014**, *55*, 2471–2477. [CrossRef]
129. US Department of the Interior Bureau of Reclamation. *Brine-Concentrate Treatment and Disposal Options Report—Part 1*; US Department of the Interior Bureau of Reclamation: Washington, DC, USA, 2009; pp. 1–46.
130. Qureshi, A.S. Challenges and Prospects of Using Treated Wastewater to Manage Water Scarcity Crises in the Gulf Cooperation Council (GCC) Countries. *Water* **2020**, *12*, 1971. [CrossRef]
131. Chin, D.A. *Water-Quality Engineering in Natural Systems: Fate and Transport Processes in the Water Environment*; John Wiley & Sons: Hoboken, NJ, USA, 2013; Volume 50.
132. EPA. *Guidelines for Water Reuse*; EPA: Washington, DC, USA, 2012.
133. IAEA. *Executive by-Law for the Environment Protection Law. Supreme Council for the Environment & Natural Reserves Decree Law No. 30*; Supreme Council for the Environment & Natural Reserves: Doha City, Qatar, 2002.
134. Al-Muzaini, M.; Mendonca, V.; Al-Sariri, T.; Al-Jabri, M. Waste Water Discharge into the Marine Environment in the Sultanate of Oman: Sources and Standards—An Overview of the Current Situation. In Proceedings of the 3rd Marine Waste Water Discharges (MWWD), Catania, Italy, 27 September–2 October 2004.
135. Abu Dhabi Specification. *Environmental Specifications for Land-Based Liquid Discharges to the Marine Environment*; Abu Dhabi Quality & Conformity Council: Abu Dhabi, United Arab Emirates, 2017.

sustainability

MDPI

Article

Wastewater Discharge through a Stream into a Mediterranean Ramsar Wetland: Evaluation and Proposal of a Nature-Based Treatment System

Jesús de-los-Ríos-Mérida [1,*], Francisco Guerrero [2,3], Salvador Arijo [4,5], María Muñoz [1,5,6], Inmaculada Álvarez-Manzaneda [7], Jorge García-Márquez [4], Begoña Bautista [1,5], Manuel Rendón-Martos [8] and Andreas Reul [1,5,*]

1 Departamento de Ecología y Geología, Universidad de Málaga, Campus de Teatinos, s/n, 29071 Málaga, Spain; mariamunoz@uma.es (M.M.); bbautista@uma.es (B.B.)
2 Departamento de Biología Animal, Biología Vegetal y Ecología, Universidad de Jaén, Campus de Las Lagunillas, s/n, 23071 Jaén, Spain; fguerre@ujaen.es
3 Centro de Estudios Avanzados en Ciencias de la Tierra, Energía y Medio Ambiente, Universidad de Jaén, Campus de Las Lagunillas, s/n, 23071 Jaén, Spain
4 Departamento de Microbiología, Universidad de Málaga, Campus de Teatinos, s/n, 29071 Málaga, Spain; sarijo@uma.es (S.A.); j.garcia@uma.es (J.G.-M.)
5 Instituto de Biotecnología y Desarrollo Azul (IBYDA), Loma de San Julián, n°2, Barriada de San Julián, 29004 Málaga, Spain
6 Centro Oceanográfico de Málaga, Instituto Español de Oceanografía, Puerto Pesquero, s/n, 29640 Fuengirola, Spain; maria.munoz@ieo.es
7 Departamento de Ecología, Universidad de Granada, Campus de Fuentenueva, s/n, 18071 Granada, Spain; miams@ugr.es
8 Consejería de Agricultura, Ganadería, Pesca y Desarrollo Sostenible de la Junta de Andalucía, Reserva Natural Laguna Fuente de Piedra, 29520 Fuente de Piedra, Spain; manuel.rendon@juntadeandalucia.es
* Correspondence: jrmerida@uma.es (J.d.-l.-R.-M.); areul@uma.es (A.R.)

Citation: de-los-Ríos-Mérida, J.; Guerrero, F.; Arijo, S.; Muñoz, M.; Álvarez-Manzaneda, I.; García-Márquez, J.; Bautista, B.; Rendón-Martos, M.; Reul, A. Wastewater Discharge through a Stream into a Mediterranean Ramsar Wetland: Evaluation and Proposal of a Nature-Based Treatment System. *Sustainability* **2021**, *13*, 3540. https://doi.org/10.3390/su13063540

Academic Editor: Anuska Mosquera Corral

Received: 23 February 2021
Accepted: 10 March 2021
Published: 23 March 2021

Abstract: Impacts on wetlands are becoming more pressing every day. Among them, habitat loss, overexploitation of aquifers and changes in land use are considered the most important. However, the impacts linked to wastewater discharges are increasing worldwide. In this context, this study analyses the impacts of input of wastewater to a Mediterranean Ramsar temporary wetland (Fuente de Piedra, south of Spain). To this end, systematic sampling was carried out in the Charcón stream which receives water from a wastewater treatment plant (WWTP) and discharges it into the wetland. The results showed a slight decrease in the nutrient concentrations, particularly for nitrogen compounds. Heterotrophic and fecal bacteria concentration, as well as phytoplankton and zooplankton abundance and biomass, all significantly decreased from the treatment plant to the wetland. When comparing the effect of this discharge with other similar occurring to the same wetland, it was evident that the Charcón stream was responsible for a greater impact. At this point, it is relevant to note that the main difference among both treated wastewater discharges lies in the different water retention time once the wastewater was released from the WWTP. In fact, we recommend an increase in the water retention time by building seminatural ponds, together with the use of biofilters, which will notably contribute to improve the processes of assimilation of nutrients and to decrease the impact generated in the wetland by this spill.

Keywords: cultural eutrophication; management; seminatural ponds; temporary wetlands; wastewater; water quality

1. Introduction

Inland freshwater ecosystems are hotspots of biodiversity as they show higher relative species richness than marine or terrestrial ecosystems [1]. They provide important ecosystem services, with a global estimated annual economic value of USD 44,000 per hectare [2]. In spite of their importance, almost 50% of inland water environments have been lost during the twentieth century, which makes them one of the most threatened ecosystems [1]. The global Living Planet Index (LPI) shows that 60% of species have shown population declines between 1970 and 2014, with freshwater species showing an 83% decline [3]. These authors and others indicate that the biggest drivers of biodiversity decline are habitat loss, habitat degradation, and overexploitation [4,5]. In this sense, wetlands biodiversity is mostly affected by changes in land use and eutrophication processes [6–9]. One of the main problems of the discharge of wastewater to aquatic ecosystems is the increase in eutrophication [10], a current problem suffered by wetlands around the world that cannot be resolved despite the existence of legislation for wetland protection [11]. Cultural eutrophication can be caused, among others, by (i) changes in the land use of the catchment area, where farming and the use of fertilizers increase the runoff of nutrient rich water into the wetlands [8,9], and (ii) the direct spill of urban wastewater into the wetland [12,13]. Attending to the second one, within the EU urban wastewater is supposed to be treated according to the urban wastewater Council Directive [14]. This requires that wastewater delivered into sensitive freshwater and estuaries should not be higher than 1 $mg \cdot L^{-1}$ of total phosphorus (TP) and 10 $mg \cdot L^{-1}$ of total nitrogen (TN). Wastewater treatment is a critical issue in Spain and supervision and improvement of wastewater treatment as well as reuse are urgently needed [15]. This is especially true for the effluents of the wastewater treatment plant (WWTP) that spill into sensitive inland waters, such as the Ramsar wetlands. When this occurs, the administration tries to find solutions to improve the quality of the water discharged, as is the case of building artificial wetlands which can decrease concentration of TP, TN and fecal bacteria [12]. Nevertheless, sometimes the wastewater is directly discharged in those sensitive wetlands. This is the case of Fuente de Piedra (south of Spain), a Ramsar wetland that receives the wastewater of two nearby villages, Fuente de Piedra and Humilladero. From the first one, the wastewater runs through a set of seminatural ponds constructed during a Life project (LIFE03NAT/E/000055) which improved considerably the water quality of the wastewater before reaching the wetland [12]. However, the wastewater from Humilladero village reaches the wetland directly through a temporary stream, named Charcón. Thus, the present study is focused on the wastewater selfdepuration processes that occurs during the transit through the Charcón stream connecting the Humilladero WWTP to the Ramsar wetland. These processes have been evaluated through the study of a series of biotic and abiotic variables along the stream, which has allowed us to compare the results with those previously obtained of the discharge from Fuente de Piedra village to the same wetland [12], with the aim to propose nature-based facilities in order to achieve Sustainable Urban Water Management [16].

2. Materials and Methods

2.1. Study Area and Sampling Sites

Fuente de Piedra is a Mediterranean temporary endorheic ecosystem [17] located in Andalusia (southern Spain). It was included in the list of Ramsar sites in 1983, declared as a Nature Reserve in 1984, and achieved Special Areas of Conservation (SAC) status in 2013 [18,19]. This ecosystem receives the discharge of three wastewater treatment plants (WWTP); two from the Fuente de Piedra village [12] and the other one from the Humilladero village through the Charcón stream, this last being the objective of the present study (Figure 1). The Humilladero WWTP is composed of two aerobic basins and treat the wastewater of a population of approximately 3300 people. Data for biochemical oxygen demand, chemical oxygen demand and suspended solids are available by Andalusian Administration (Table 1).

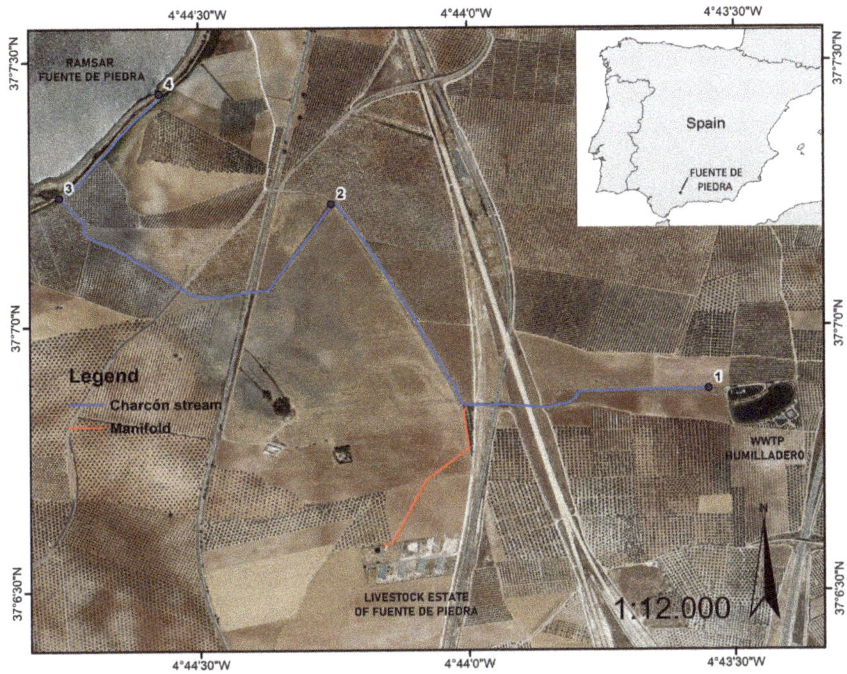

Figure 1. Location of the sampling points (1 to 4) into the Charcón stream, from the Humilladero WWTP (wastewater treatment plant) to Fuente de Piedra wetland. In this figure is also represented a manifold that discharges into the Charcón stream from a livestock estate.

Table 1. Characteristics of input and output wastewater from the wastewater treatment plant of Humilladero. Efficiency (EFF) (data supplied by Junta de Andalucía). BOD5: biochemical oxygen demand for 5 days (mg·L^{-1}); COD: chemical oxygen demand (mg·L^{-1}); SS: suspended solid (mg·L^{-1}). Mean values of the year of study (2018) and values obtained from the day closest to the sample in which data is available.

	BOD5			COD			SS		
Date	**Input**	**Output**	**EFF**	**Input**	**Output**	**EFF**	**Input**	**Output**	**EFF**
Mean 2018	87.4	12.6	71%	252.3	56.2	69%	109.9	59.9	38%
11/06/2018	90	6.4	93%	184	60	67%	92	78	15%

Charcón is a temporary stream that comes from the nearby Sierra de Humilladero and ends at the Fuente de Piedra wetland. Next to this stream is located the WWTP of Humilladero. From this treatment plant to the Fuente de Piedra wetland, the stream flows along 2905 m with an unevenness of 11 m. In this travel, four sampling stations were taken; the first one (point 1) next to the treatment plant and the last (point 4) at the entrance to the Fuente de Piedra wetland (Figure 1). The second sampling station (point 2) is located in an intermediate section of the stream, and the third (point 3) in the entrance to the nature reserve. In order to know the time required for the effluent to travel from the WWTP to the Ramsar wetland, the stream was divided in two homogeneous sections, the first with a length of 1072 m and an average slope of 0.65 cm·m^{-1}, and the second with a length of 1833 m and an average slope of 0.22 cm·m^{-1}. In both sections the water velocity and traveling time of the wastewater was measured by the dilution method using tracers [20]. The samplings were carried out in June 2018, working upstream to avoid contamination, that is, from the Fuente de Piedra wetland (point 4) to the WWTP (point 1).

2.2. Abiotic Variables

At each sampling point, conductivity ($\mu S \cdot cm^{-1}$) and pH were measured with a Hanna Multiparameter sensor HI 9829 (Hanna Instruments, Woonsocket, RI, USA). Three water samples for nutrients concentrations were taken into sterile polyethylene vials and immediately frozen (−20 °C). In the laboratory, dissolved inorganic phosphorus (DIP) was determined by using the molybdenum blue method [21] and TP was measured after the digestion with potassium persulfate of unfiltered and filtered water (Whatman GF/F), respectively, [22]. Ammonium (NH_4^+) was measured by phenate method [23], nitrates (NO_3^-) was analyzed using the ultraviolet spectrophotometric screening method [22] and nitrite (NO_2^-) was determined using the sulfanilamide method [23]. Lastly, TN was analyzed by ultraviolet method of digested unfiltered and filtered water, respectively, [22].

2.3. Biotic Variables

For biotic variables three samples were also taken in the same sampling point for the evaluation of the biotic variables. Total chlorophyll-a concentration (Chl-a) and phytoplankton composition were estimated with a submersible FluoroProbe (bbeMoldaenke GmbH, Schwentinental, Germany), which discriminates between four main phytoplankton groups (i.e., diatoms and dinoflagellates, blue green algae, green algae, and cryptophytes) [24,25]. Abundance of phytoplankton < 20 µm equivalent spherical diameter (ESD), was measured by passing the sample through a BD Acurri C6 flow cytometer, counting at least 10,000 events. Abundance and size of phytoplankton cells (5 and 100 µm ESD) and zooplankton (250 and 1000 µm ESD) were analysed with a FlowCAM (Fluid Imaging Technologies, Inc. Scarborough, ME, USA) using the auto image mode. For phytoplankton, 30 mL of the sample preserved with formol (4% f.c.) were passed through a 100 µm flow cell and 1 mL was analysed with a 100-fold magnification (10× objective). For zooplankton, 50 mL of the samples preserved with formaldehyde (4% f.c.) were passed through a 1000 µm flow cell and analysed with a 20-fold magnification (2× objective).

The enumeration of heterotrophic cultivable microorganisms was carried out by adding 0.1 mL of serial dilution of the samples in Tripticase soy agar (Oxoid Ltd., Wade Road, Basingstoke, UK) plates. The plates were cultured at 22 °C for 48 h (ISO 6222:1999). Coliforms and fecal streptococci concentrations were determined by water filtration through sterile nitrocellulose filters (47 mm diameter, 0.45 µm pore size; Millipore Corp., Bedford, MA, USA). Membranes were incubated in Chromocult coliform agar (Merck, Darmstadt, Germany) at 37 °C, 24 h for the determination of total coliforms and *Escherichia coli* (ISO 9308-1:2000), or in m-*Enterococcus* agar (Merck) at 37 °C, 48 h for the determination of fecal streptococci (ISO 7899-2:2001). After incubation, colonies were counted on each medium, and concentrations of the different groups of microorganisms were determined.

For the identification of the fecal streptococci species, isolates were identified to species level by the amplification and sequentiation of a fragment of 16S rDNA. The Thermo ScientificGeneJET Genomic DNA Purification Kit (Thermo Fisher Scientific, Waltham, MA, USA) was used for the extraction of total genomic DNA from bacteria. Afterward, this fragment was amplified using the universal primers SD-Bact-0008-a-S20 (5′ AGA GTT TGA TCC TGG CTC AG 3′) and SD-Bact-1492-a-A-19 (5′ GGT TAC CTT GTT ACG ACT T 3′) [26]. Polymerase chain reactions were carried out in a 50 µL reaction mixture that included 5 pmol of each primer, 0.2 mM dNTPs mix, 10x DreamTaq Buffer, 2 mM $MgCl_2$, 1.25 U DreamTaq DNA Polymerase (Thermo Fisher Scientific, Waltham, MA, USA) and 1 µL of colony DNA (100 ng/µL). The PCR profile was as follows: 2 min at 95 °C and 35 cycles of 30 s at 95 °C, 40 s at 52 °C and 1.3 min at 72 °C and a final step 5 min at 72 °C. Polymerase chain reaction products were electrophoresed on a 1% agarose gel and visualized via ultraviolet transilumination. PCR products were sequenced by Macrogen Spain (Madrid, Spain). The resulting sequences were compared with those in the GenBank database (www.ncbi.nlm.nih.gov/genbank/ accessed on 16 March 2021) by using the BLAST program (www.ncbi.nlm.nih.gov/blast accessed on 16 March 2021). Finally, water acute toxicity test was carried out by modification of method described by Johnson [27] using the bioluminescent bacteria *Vibrio fischeri* and the Microtox® M500 test (Microbics Corporation, Carlsbad, CA, USA). Microtox test is widely used for the toxicity assessment of environmental samples and is based on the measurement of *V. fischeri* bioluminescence inhibition after sample exposure at various contact times [28]. Cuvettes with 1 mL of each water sample (without dilute, or diluted at 1/50 and 1/100 in 2% saline solution) were maintained at 15 °C. *V. fischeri* growth for 24 h in TSA with 2% of NaCl (TSAs), was suspended in 2% saline solution at 0.6 of optical density (600 nm) and maintained at 5 °C prior use. Then, the cuvettes with the water samples and controls were inoculated with 20 µL of the *V. fischeri* suspension. The samples were incubated at 15 °C for 15 and 45 min. After incubation, bioluminescence of *V. fischeri* in each cuvette was measured by Microtox luminometer. Bioluminescence in each sample was relativized with bioluminescence in control cuvettes, obtaining a percent decrease of bioluminescence. The inhibition of the luminescence was assumed to be correlated with the toxicity of the samples. The percent reduction in bioluminescence of *V. fischeri* produced by the samples were recorded as median effective concentration (EC50) values.

2.4. Statistical Analysis

Residual normality (Shapiro–Wilk test) and homogeneity of variances (Levene test) were checked before performing the statistical analysis of the data. Differences in pH, TN, Chl-a, phytoplankton composition and phytoplankton 1–20 µm abundance between sampling stations were tested by using one-way ANOVA test, followed by Bonferroni's post hoc test. Since all other variables did not satisfy homoscedasticity assumptions (Levene test, $p < 0.05$) or normality distribution (Shapiro–Wilk test, $p < 0.05$), Kruskal–Wallis and post hoc Dunn's test with Bonferroni adjustment were carried out to test the differences between sampling stations. To perform this statistical analysis, the Statistica 7.1 software (StatSoft Inc., Tulsa, OK, USA) was used.

3. Results

3.1. Abiotic Variables

Results obtained shown significant conductivity differences (Table 2) from point 1 to point 4 (Kruskal–Wallis test, $p < 0.001$) with a significant increase among point 1 and points 3 and 4, and among point 2 and point 4 (Dunn–Bonferroni post hoc, $p < 0.05$). The pH in contrast decreased significantly (Table 2) from the wastewater treatment plant spill (point 1) to the Ramsar inflow at point 4 (one-way ANOVA, $p < 0.001$) with a significant decrease in each sampling point (Bonferroni post hoc, $p < 0.001$).

Table 2. Conductivity (Cond.) and pH measured at the four sampling points in the Charcón stream. Kruskal–Wallis test Median (25, 75 percentiles) and one way ANOVA (mean ± standard deviation) Different letters show significant differences between sampling points (mean ± standard deviation).

Abiotic Variables	Point 1	Point 2	Point 3	Point 4
Cond. ($\mu S \cdot cm^{-1}$)	2549 (2545; 2569) ^{a}	2585 (2579; 2585) ^{ab}	2657 (2657; 2656) ^{bc}	2668 (2686; 2688) ^{c}
pH	8.12 ± 0.03 ^{a}	7.72 ± 0.02 ^{b}	7.63 ± 0.03 ^{c}	7.36 ± 0.01 ^{d}

The time required for the effluent to travel the total length of the stream from the WWTP to the Ramsar wetland was aproximately 9.5 h. In this time, TP and TN decreased from the spilling point (point 1) to the Ramsar inflow point (point 4, Table 3) by 12% and 30% respectively. However, these decreases were not significant (Kruskal–Wallis test and one-way ANOVA, respectively, $p > 0.05$). Similarly, nitrogen dissolved compounds decreased in general, while the phosphorus compounds increased (Table 3). These changes in dissolved nutrients were significant for DIP (Kruskal–Wallis test, $p < 0.05$), among point 1 and point 4 (Dunn–Bonferroni post hoc, $p < 0.05$). Regarding to the nitrogen forms, there were not significant differences for NO_3^- and a significant decrease was found for NO_2^- and NH_4^+ (Kruskal–Wallis test, $p < 0.05$). Concerning NO_3^- concentration, it significantly changed among points 3 and 4 (Dunn–Bonferroni post hoc, $p < 0.05$). In addition, NO_2^-

showed significant differences among the sampling points 2 in comparison to points 4 (Dunn–Bonferroni post hoc, $p < 0.005$), while NH_4^+ showed a significant decrease among point 1 and point 4 (Dunn–Bonferroni post hoc, $p < 0.05$).

Table 3. Nutrient concentrations measured in the sampling points in the Charcón stream. Kruskal–Wallis test Median (25, 75 percentiles) and one way ANOVA (mean ± standard deviation). Different letters show significant differences between sampling points.

Nutrient Concentration (mg·L^{-1})	Point 1	Point 2
Dissolved Inorganic Phosphorus	0.70 (0.68; 0.70) ^{a}	0.75 (0.68; 0.76) ^{ab}
Total Phosphorus	1.45 (1.42; 1.66) ^{a}	1.44 (1.41; 1.47) ^{ab}
Nitrates	0.56 (0.56; 0.57) ^{ab}	0.47 (0.45; 0.50) ^{ab}
Nitrites	0.51 (0.48; 0.59) ^{ab}	0.55 (0.54; 0.55) ^{a}
Ammonium	10.92 (9.97; 13.6) ^{a}	9.27 (9.18; 9.92) ^{ab}
Total Nitrogen	15.48 ± 1.56 ^{a}	13.92 ± 2.63 ^{a}
Nutrient Concentration (mg·L^{-1})	**Point 3**	**Point 4**
Dissolved Inorganic Phosphorus	0.86 (0.82; 0.87) ^{ab}	1.04 (0.97; 1.04) ^{b}
Total Phosphorus	1.48 (1.47; 1.54) ^{a}	1.34 (1.30; 1.36) ^{a}
Nitrates	1.14 (1.12; 1.17) ^{a}	0.10 (0.09; 0.13) ^{b}
Nitrites	0.16 (0.16; 0.16) ^{ab}	0.02 (0.02; 0.02) ^{b}
Ammonium	8.55 (8.50; 8.64) ^{ab}	7.68 (6.61; 8.22) ^{b}
Total Nitrogen	10.53 ± 2.86 ^{a}	10.16 ± 2.73 ^{a}

3.2. *Biotic Variables*

Chl-a concentration (Figure 2a) decreased significantly from point 1 to point 4 (one-way ANOVA, $p < 0.05$); although, this decrease was not uniform since a slight increase was observed in point 2. The differences of the successive sampling stations were significant as well as among all the sampling points (Dunn–Bonferroni post hoc, $p < 0.001$). Regarding the phytoplankton groups (Figure 2a), significant changes were also observed (one-way ANOVA, $p < 0.05$). Green algae showed similar Chl-a concentration at point 1 and point 2 and decreased significantly at point 3, reaching the lowest concentration at point 4 (Dunn–Bonferroni post hoc, $p < 0.001$). The Chl-a concentration of blue green algae increased significantly from point 1 to point 2 and decreased significantly to point 3 and point 4 (Bonferroni post hoc, $p < 0.001$). Similarly, diatoms and dinoflagellates Chl-a concentration increased significantly from point 1 to point 2 and decreased significantly from point 3 to point 4 (Bonferroni post hoc, $p < 0.05$). The Chl-a concentration of cryptomonads showed the lowest concentrations of the four groups and was similar at the first three sampling points, decreasing significantly at sampling point 4 (Bonferroni post hoc, $p < 0.05$). Phytoplankton abundance for 1–20 µm ESD showed significant differences (one-way ANOVA, $p < 0.05$), with similar abundances at sampling points 1 and 2, and a significant decrease from point 2 to the half at sampling point 3 and from point 3 to the lowest abundance at sampling point 4 (Bonferroni post hoc, $p < 0.001$). Concerning biovolume of phytoplankton 5–100 µm ESD, significant differences among the sampling points were detected (Kruskal–Wallis test, $p < 0.05$), but it was not possible to determine which sampling stations were significant different (Dunn–Bonferroni post hoc, $p > 0.05$) (Figure 2b). Finally, although a decrease in zooplankton biovolume, can be observed from point 1 to point 2 (Figure 2b) and there were detected significant differences among the sampling points (Kruskal–Wallis test, $p < 0.05$), the post hoc analysis was not able to determine these differences (Dunn–Bonferroni post hoc, $p > 0.05$).

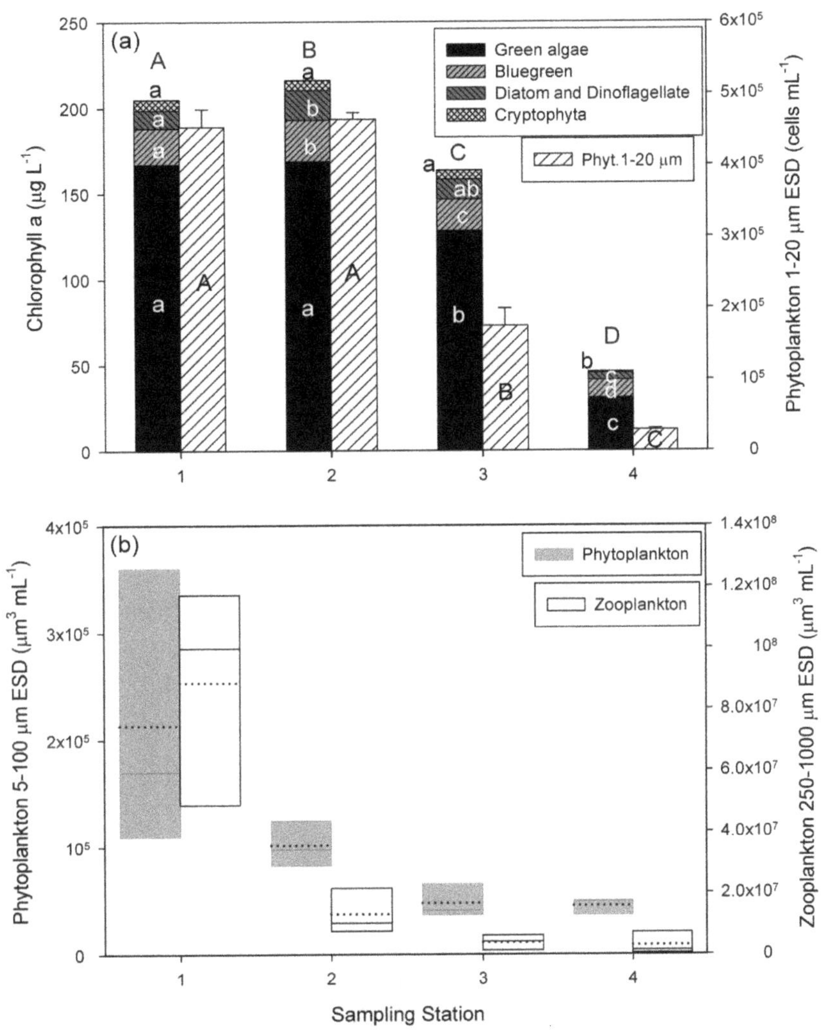

Figure 2. (**a**) Chl-a concentration (stacked mean) of the four phytoplankton groups, and abundance of phytoplankton < 20 µm in the four sampling stations at the Charcón stream. Capital letters indicates significant differences between total Chl-a concentration and lowercase letter indicates significant differences of the respective phytoplanktonic groups among the sampling points. (**b**) Biovolume of phytoplankton (5–100 µm ESD) and zooplankton (250–1000 µm) in the four sampling stations at the Charcón stream. The horizontal continuous line indicates the median and dotted line the mean.

Significant differences were tested in microorganism (Figure 3) by Kruskal–Wallis test ($p < 0.05$). A progressive and significant decrease in the concentration of heterotrophic bacteria was seen between points 1 and 4 (Dunn–Bonferroni post hoc, $p < 0.05$). In the case of total coliforms, no significant differences were observed among any point (Kruskal–Wallis test, $p > 0.05$), while *E. coli* showed a significant decrease in their titer from point 2 to point 4 (Dunn–Bonferroni post hoc, $p < 0.05$). In the case of fecal streptococci, concentration values showed a significant peak at point 2 (Dunn–Bonferroni post hoc, $p < 0.05$). At point 4 fecal streptococci concentrations were between the values reached at point 1 and point 3. Finally, the *E. coli*/fecal streptococci ratio showed a progressive decrease in its values from point 1 to 4 that was significant among point 3 and point 4 (Dunn–Bonferroni post hoc, $p < 0.05$).

A total of 24 strains were isolated from m-*Enterococcus* agar for their identification. The 25% were identified as *Enterococcus hirae* and the 16.7% as *E. faecium*. The rest were identified as *E. mundtii*, *E. haemoperoxidus*, *E. plantarum* and *Bacterium* BEL B14 (each of them at 8.3%). Furthermore, a strain of *E. casseliflavus*, *E. durans* and 4 strains of *Enterococcus* sp. were isolated. The water acute toxicity assessment showed no decrease in *V. fischeri* bioluminescence, regardless of the dilution used or the incubation time; thus, indicating absence of toxicity. Furthermore, a significant activation of bioluminescence compared to control was detected in the samples at dilution 1:50 (data not shown).

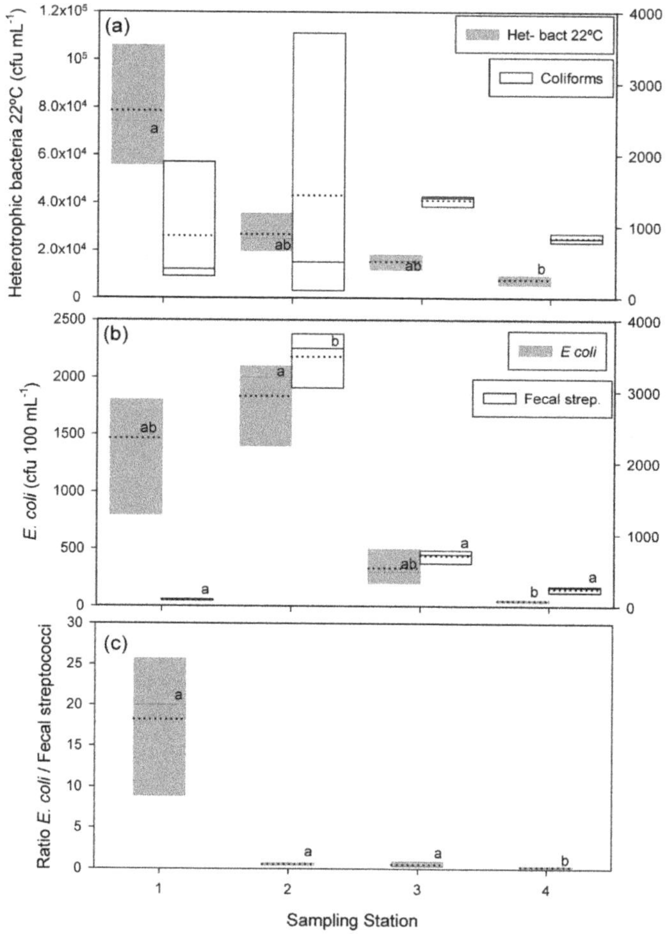

Figure 3. Bacterial concentration measured in all the sampling points in the Charcón stream. (**a**) Heterotrophic bacteria growth at 22 °C and total coliforms (cfu mL^{-1}). (**b**) *E. coli* and fecal streptococci (cfu 100 mL^{-1}). (**c**) Ratio *E. coli*/fecal streptococci. The horizontal continuous line indicates the median and dotted line the mean. Different letters show significant differences between sampling points into each bacterial group.

4. Discussion

Impacts on wetlands due to wastewater discharges are increasing worldwide and becoming a real problem that has been not resolved yet, despite legislation for the protection of wetlands [11]. In this study, we analyze the effect of a direct discharge of wastewater from a village located in the watershed of the Fuente de Piedra Ramsar site through a stream named Charcón. The results show that in general, along the Charcón stream, the measured environmental variables show few significant changes downstream of the discharge. In fact, the main variables related to the cultural eutrophication of aquatic ecosystems (TN and TP) show similar values at the outlet of the wastewater treatment plant and at the in-put to the wetland. Only significant changes in NO_3^- due to their decrease at input to the wetland as well as a significant increase in DIP and significant decreases in nitrite and ammonia concentrations were observed. Attending to the biological variables, a decrease in Chl-a concentration, abundance and biovolume of the phytoplankton and zooplankton community and in the abundance of indicator microorganisms is observed. Comparing these results with those obtained in another of the discharges that pours into the same wetland [12], it is observed that the differences are important. The water coming out the WWTP, travels the 2905 m transect from the spill to the Ramsar wetland in less than 10 h. This is a very short time for self-depuration process through the stream in comparison to those observed with the use of constructed wetlands [12] and it is unsurprising that few variables showed downstream changes in our studies. In this sense, Søballe and Kimmel [29] separates the ecological structure and function of natural rivers, river impoundments, and natural lakes, where natural lakes and river are in opposite extremes. According to these authors, algal abundance per unit of P is lower in rivers than in impoundments and lakes as long as residence time are different. The authors conclude that residence time is a useful system–level index and it has similar ecological implication for rivers, lakes and reservoirs. This explains why, due to longer residence time, effective P and N reduction by biological processes were observed in the adjacent pond system coupled to the wastewater plant of Fuente de Piedra village [12] in comparison to the low nutrient reduction observed in this study in the Charcón stream, with a lower residence time.

Indicator microorganisms are used to assess the effectiveness of water and wastewater treatment processes [30]. These microorganisms are decisive to determine the degree of fecal pollution and the load of organic matter of the water. According to data from the Andalusian administration (Table 2), the treated wastewater discharged into the Charcón stream has a load of suspended solids above the reference levels of the European Directive 91/271/EC [14] (EC 1991). BOD5 and COD values are also high, although within the limits of such legislation. Thus, it is reasonable that indicators microorganisms of fecal contamination can still be isolated in the outflow water (point 1). High heterotrophic and fecal concentration bacteria (coliforms and fecal enterococci) are observed at the spill to the Charcón stream. Strong positive correlations between Chl-a and bacterial production rates suggest that bacteria are mainly controlled by organic substrates released during phytoplankton photosynthesis [31]. Total coliforms decrease no significantly along the Charcón stream, while mesophilic heterotrophs and *E. coli* decrease from points 1 and 2 to point 4. On the other hand, an increase in the bacterial concentration occurs at sampling point 2, being 42 times higher in the case of fecal streptococci. This significant increase in fecal bacteria at sampling point 2 could be due to livestock estate spill released into the Charcón stream between sampling point 1 and sampling point 2 (Figure 4). After this spill, and following our previous interpretation, from point 2 to point 4 the concentration of fecal bacteria decreases progressively. At the Ramsar inflow, the water shows a concentration of fecal microorganisms suitable for bathing, according to Spanish regulations [32]. These results show that the Charcón stream, unlike what happens with the concentration of nutrients, acts as an efficient natural purification system in relation to microorganisms.

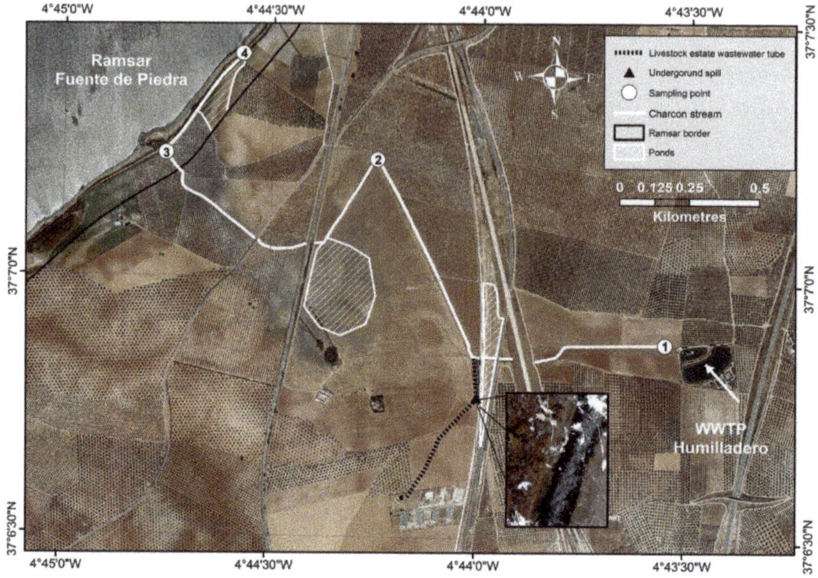

Figure 4. WWTP, Charcón stream, livestock estate spill (inlet) and proposed seminatural ponds system along the Charcón stream.

Wastewater can contain indefinite toxic substances, therefore toxicity tests have to be performed for an integrated evaluation of the environmental impact of contaminants. The toxicity analyses have not shown a decrease in bioluminescence in the different treatments. However, an activation of bioluminescence with respect to controls has been observed, possibly due to the existence of nutrients in the water [33]. It can be concluded that the waters do not have environmental toxic substances in enough concentration as to generate an acute toxic action. Cultural eutrophication is greatly affecting the structure and function of Mediterranean wetlands [34–36]. These ecosystems have been considered as "hotspot" sites for biogeochemical processes [37] due to their intrinsic characteristics. In the case of Fuente de Piedra, an endorheic wetland, it is needed to reduce the entrance of nutrients. This has been efficiently achieved in the other discharge that this wetland receives thanks to seminatural ponds, which increase the residence time of the wastewater [12]. Therefore, in a similar way, it is proposed a system with three seminatural ponds along the Charcón stream, with an area of 21,048 m^2, 63,549 m^2 and 15,250 m^2, respectively, (Figure 4). If the mean depth of the ponds were 1.5 m, 1.0 m and 0.5 m, respectively, the first pond would have 31,572 m^3, the second pond 63,549 m^3 and the third pond 7625 m^3. With a mean flow of 15 $m^3 \cdot h^{-1}$, the residence time in the first, second and third pond would be 88, 177 and 21 days, respectively. These values are higher than those calculated for the seminatural ponds system in the other spill of the adjacent Fuente de Piedra WWTP (65–85 days for the whole pond system) [12]. The site for the first proposed pond is located in an uncultivated area adjacent to a new road (Figure 4). The second proposed pond would be located near to the sampling point 2, where highest bacteria concentration was observed, and downstream of the possible agricultural discharge. In this place is located a temporary wetland disappeared due to the action of a drainage; therefore, in order to recover or restore this wetland, the installation of this pond would be more feasible than that of any other (Figure 4). Finally, the third pond would be located in the Ramsar area, immediately above the wetland. This pond would further enhance water remediation and might facilitate further benefits the Ramsar wetland. In this pond it would be suitable to use helophytic vegetation that plays an excellent role in Mediterranean wastewater purification [38]. The construction of two new ponds and the recovery of a temporary pond will contribute to increasing biodiversity in the study area. In addition, this action will reduce the tourist pressure suffered by the Interpretation Center of the Nature Reserve since visitors, mostly ornithologists, will have more wetlands for bird watching. Furthermore, the presence of a greater number of ponds around the Ramsar wetland will lead to an increase in the environmental heterogeneity, in terms of hydroperiod and salinity, a fundamental aspect since the stability and recovery of some ecosystems depends to some extent on environmental heterogeneity [39]. In addition, this increase in the heterogeneity of the environment will mean an enrichment in the number of species present and also the conservation of numerous species that follow a metapopulation-metacommunity dynamic, as the plankton community [40,41].

Finally, and attending to the restoration of eutrophicated systems, other methods could also be used. Between them, one of the most promising methods is the application of magnetic particles (MPs) for removing P from aquatic ecosystems. MPs have shown a high P removal efficiency in synthetic, natural and wastewaters, see among others, [42–44]. P loaded MPs can be recovered and washed, allowing MPs reuse [42] which reduces the cost of this method. Álvarez-Manzaneda et al. [44] have estimated a final cost of EUR 6601 for the application of MPs (reused four times) in the seminatural pond located in the wastewater spill from Fuente de Piedra village. This cost would be similar to Phoslock®, that has the disadvantage of not being able to recover the P from the system which may be used to make fertilizers, facing the problem of P reserves exhaustion. For all this, the combination of seminatural ponds and application of MPs along the Charcón stream could provide a solution for the release of wastewater by the WWTP in Andalusia.

5. Conclusions

The short residence time of the wastewater in the Charcón stream (<10 h) does not allow an efficient self-depuration process. Therefore, little reduction of TP and TN occurs during the time that the wastewater reaches the Ramsar site. Although no toxic effect has been observed in the water and the concentration of bacteria decreased along the stream, a better purification of the discharge from the wastewater treatment plant is necessary to avoid cultural eutrophication. A combination of bioremediation by increasing the residence time through building seminatural ponds, introducing biofilters with more vegetation such as *Phragmites australis* [45], *Typha latifolia* [46], *Lemna minor* and *Azolla filiculoides* [47], as well as techniques to absorb phosphate by MPs, should be urgently applied to the wastewater spill of the Charcón stream. For this, as a first step, a set of semi-natural ponds system is proposed to be built along the Charcón stream.

Author Contributions: Conceptualization, J.d.-l.-R.-M., F.G., A.R.; methodology, J.d.-l.-R.-M., F.G., S.A., M.M., I.Á.-M., J.G.-M., B.B., M.R.-M., A.R.; software, M.M.; formal analysis, J.d.-l.-R.-M., M.M., I.Á.-M.; resources, F.G., S.A., A.R.; data curation, I.Á.-M., B.B., J.G.-M.; writing—original draft preparation, J.d.-l.-R.-M., F.G., A.R.; writing—review and editing S.A., M.M., I.Á.-M., J.G.-M., B.B., M.R.-M.; visualization, J.d.-l.-R.-M., M.M.; supervision, M.R.-M.; project administration, F.G., A.R., S.A.; funding acquisition, S.A., F.G., A.R. All authors have read and agreed to the published version of the manuscript.

Funding: This study was supported by the Ministry of Science of Spain (Plan Nacional de Investigación, AGL2010-17789) and the European Regional Development Fund. The acquisition of the Flow-CAM by the University of Málaga was co-financed by the 2008–2011 FEDER program for Scientific-Technique Infrastructure (UNMA08-1E005).

Institutional Review Board Statement: Not applicable.

Informed Consent Statement: Not applicable.

Data Availability Statement: Not applicable.

Conflicts of Interest: The authors declare no conflict of interest.

References

1. Millennium Ecosystem Assessment (Program) (Ed.) *Ecosystems and Human Well-Being: Wetlands and Water Synthesis: A Report of the Millennium Ecosystem Assessment*; World Resources Institute: Washington, DC, USA, 2005.
2. Russi, D.; ten Brink, P.; Farmer, A.; Badura, T.; Coates, D.; Förster, J.; Kumar, R.; Davidson, N. *The Economics of Ecosystems and Biodiversity for Water and Wetlands*; Institute for European Environmental Policy & Ramsar Secretariat: London, UK, 2013; p. 85.
3. Grooten, M.; Almond, R. (Eds.) *Living Planet Report. Aiming Higher*; WWF: Gland, Switzerland, 2018.
4. Perrino, E.V.; Musarella, C.M.; Magazzini, P. Management of grazing Italian river buffalo to preserve habitats defined by Directive 92/43/EEC in a protected wetland area on the Mediterranean coast: Palude Frattarolo, Apulia, Italy. *Euro-Mediterr. J. Environ. Integr.* **2021**, *6*, 32. [CrossRef]
5. Brook, B.; Sodhi, N.; Bradshaw, C. Synergies among extinction drivers under global change. *Trends Ecol. Evol.* **2008**, *23*, 453–460. [CrossRef] [PubMed]
6. Brinson, M.M.; Malvárez, A.I. Temperate freshwater wetlands: Types, status, and threats. *Environ. Conserv.* **2002**, *29*, 115–133. [CrossRef]
7. Dudgeon, D.; Arthington, A.H.; Gessner, M.O.; Kawabata, Z.I.; Knowler, D.J.; Lévêque, C.; Naiman, R.J.; Prieur-Richard, A.H.; Soto, D.; Stiassny, M.L.J.; et al. Freshwater biodiversity: Importance, threats, status and conservation challenges. *Biol. Rev.* **2006**, *81*, 163. [CrossRef] [PubMed]
8. García-Muñoz, E.; Gilbert, J.D.; Parra, G.; Guerrero, F. Wetlands classification for amphibian conservation in Mediterranean landscapes. *Biodivers. Conserv.* **2010**, *19*, 901–911. [CrossRef]
9. Gilbert, J.D.; de Vicente, I.; Ortega, F.; García-Muñoz, E.; Jiménez-Melero, R.; Parra, G.; Guerrero, F. Linking watershed land uses and crustacean assemblages in Mediterranean wetlands. *Hydrobiologia* **2017**, *799*, 181–191. [CrossRef]
10. Sánchez-Carrillo, S.; Angeler, D.; Álvarez Cobelas, M.; Sánchez-Andrés, R. Freshwater Wetland Eutrophication. In *Eutrophication: Causes, Consequences and Control*; Ansari, A.A., Singh Gill, S., Lanza, G.R., Rast, W., Eds.; Springer: Dordrecht, The Netherlands, 2010; pp. 195–210. [CrossRef]
11. Dymond, J., Ed. *Ecosystem Services in New Zealand: Conditions and Trends*; Manaaki Whenua Press: Lincoln, New Zealand, 2013.
12. de-los-Ríos-Mérida, J.; Reul, A.; Muñoz, M.; Arijo, S.; Tapia-Paniagua, S.; Rendón-Martos, M.; Guerrero, F. How Efficient are Semi-Natural Ponds in Assimilating Wastewater Effluents? Application to Fuente de Piedra Ramsar, Mediterranean Salt Lake (South of Spain). *Water* **2017**, *9*, 600. [CrossRef]
13. Cunningham, C.; Gharipour, M. Pipe Dreams: Urban Wastewater Treatment for Biodiversity Protection. *Urban Sci.* **2018**, *2*, 10. [CrossRef]
14. Council Directive 91/271/EEC of 21 May 1991 Concerning Urban Waste-Water Treatment. Available online: http://data.europa.eu/eli/dir/1991/271/oj/eng (accessed on 12 February 2021).
15. Jodar-Abellan, A.; López-Ortiz, M.I.; Melgarejo-Moreno, J. Wastewater Treatment and Water Reuse in Spain. Current Situation and Perspectives. *Water* **2019**, *11*, 1551. [CrossRef]
16. Adem Esmail, B.; Suleiman, L. Analyzing Evidence of Sustainable Urban Water Management Systems: A Review through the Lenses of Sociotechnical Transitions. *Sustainability* **2020**, *12*, 4481. [CrossRef]
17. Rodríguez-Rodríguez, M. Hydrogeology of ponds, pools, and playa-lakes of southern Spain. *Wetlands* **2007**, *27*, 819–830. [CrossRef]
18. Official Gazette of the Junta de Andalucía. LAW 1/1984, of January 9, of the Declaration of the Laguna de Fuente de Piedra as an Integral Reserve (Boletín Oficial de la Junta de Andalucía. LEY 1/1984, de 9 de enero, de la Declaración de la Laguna de Fuente de Piedra como Reserva Integral). Available online: https://www.juntadeandalucia.es/boja/1984/4/boletin.4.pdf (accessed on 12 February 2021).
19. Official Gazette of the Junta de Andalucía. Decree 70/2013, of July 2, Which Declares the Laguna de Fuente de Piedra Special Conservation Zone (ES0000033) and Approves the Natural Resources Management Plan of the Laguna de Fuente de Piedra Natural Reserve (Boletín Oficial de la Junta de Andalucía. Decreto 70/2013, de 2 de julio, por el que se declara la Zona Especial de Conservación Laguna de Fuente de Piedra (ES0000033) y se aprueba el Plan de Ordenación de los Recursos Naturales de la Reserva Natural Laguna de Fuente de Piedra). Available online: https://www.juntadeandalucia.es/boja/2013/144/BOJA13-144-00114-12146-01_00030952.pdf (accessed on 12 February 2021).
20. Chapra, S.C. *Surface Water-Quality Modeling*, reissued ed.; Waveland Press: Long Grove, IL, USA, 2008.
21. Murphy, J.; Riley, J. A modified single solution method for the determination of phosphate in natural waters. *Anal. Chim. Acta* **1962**, *27*, 31–36. [CrossRef]
22. Rice, E.W.; American Public Health Association., Eds. *Standard Methods for the Examination of Water and Wastewater*, 22th ed.; American Public Health Association: Washington, DC, USA, 2012.
23. Rodier, J. *Análisis de las Aguas: Aguas Naturales, Aguas Residuales, Agua de mar; quíMica, Fisicoquímica, Bacteriología, Biología*; Omega: Barcelona, Spain, 1989.
24. Beutler, M.; Wiltshire, K.H.; Meyer, B.; Moldaenke, C.; Lüring, C.; Meyerhöfer, M.; Hansen, U.P.; Dau, H. A fluorometric method for the differentiation of algal populations *in vivo* and *in situ*. *Photosynth. Res.* **2002**, *72*, 39–53. [CrossRef]
25. Reul, A.; Martín-Clemente, E.; Melero-Jiménez, I.J.; Bañares-España, E.; Flores-Moya, A.; García-Sánchez, M.J. What Triggers the Annual Cycle of Cyanobacterium *Oscillatoria* sp. in an Extreme Environmental Sulfide-Rich Spa? *Water* **2020**, *12*, 883. [CrossRef]

26. Kim, D.H.; Austin, B. Innate immune responses in rainbow trout (Oncorhynchus mykiss, Walbaum) induced by probiotics. *Fish Shellfish Immunol.* **2006**, *21*, 513–524. [CrossRef]
27. Johnson, B.T. Microtox® Acute Toxicity Test. In *Small-scale Freshwater Toxicity Investigations*; Blaise, C., Férard, J.F., Eds.; Springer: Berlin/Heidelberg, Germnay, 2005; pp. 69–105. [CrossRef]
28. Petala, M.; Tsiridis, V.; Kyriazis, S.; Samaras, P.; Kungolos, A.; Sakellaropoulos, G.P. Evaluation of toxic response of heavy metals and organic pollutants using microtox acute toxicity test. In Proceedings of the 9th International Conference on Environmental Science and Technology, Rhodes Isl, Greece, 1–3 September 2005.
29. Soballe, D.M.; Kimmel, B.L. A Large-Scale Comparison of Factors Influencing Phytoplankton Abundance in Rivers, Lakes, and Impoundments. *Ecology* **1987**, *68*, 1943–1954. [CrossRef]
30. Gerba, C.P. Indicator Microorganisms. In *Environmental Microbiology*, 2nd ed.; Academic Press: San Diego, CA, USA, 2009; pp. 485–499. [CrossRef]
31. Chrost, R.J.; Adamczewski, T.; Kalinowska, K.; Skowronska, A. Inorganic phosphorus and nitrogen modify composition and diversity of microbial communities in water of mesotrophic lake. *Pol. J. Microbiol.* **2009**, *58*, 77–90.
32. Royal Decree 1341/2007, of October 11, on the Management of the Quality of Bathing Water (Real Decreto 1341/2007, de 11 de octubre, sobre la gestión de la calidad de las aguas de baño). Available online: https://www.boe.es/buscar/act.php?id=BOE-A-2007-18581 (accessed on 12 February 2021).
33. Soler-Figueroa, B.M.; Otero, E. The Influence of Rain Regimes and Nutrient Loading on the Abundance of Two Dinoflagellate Species in a Tropical Bioluminescent Bay, Bahía Fosforescente, La Parguera, Puerto Rico. *Estuaries Coasts* **2015**, *38*, 84–92. [CrossRef]
34. de Vicente, I.; Serrano, L.; Amores, V.; Clavero, V.; Cruz-Pizarro, L. Sediment phosphate fractionation and interstitial water phosphate concentration in two coastal lagoons (Albuferas de Adra, SE Spain). *Hydrobiologia* **2003**, *492*, 95–105. [CrossRef]
35. Conde-Álvarez, R.M.; Bañares-España, E.; Nieto-Caldera, J.M.; Flores-Moya, A.; Figueroa, F.L. Submerged macrophyte biomass distribution in the shallow saline lake Fuente de Piedra (Spain) as function of environmental variables. *Anales Jardín Botánico Madrid* **2012**, *69*, 119–127. [CrossRef]
36. Poquet, J.M.; Mezquita, F.; Rueda, J.; Miracle, M.R. Loss of Ostracoda biodiversity in Western Mediterranean wetlands. *Aquat. Conserv. Mar. Freshw. Ecosyst.* **2008**, *18*, 280–296. [CrossRef]
37. Richardson, C.; Vaithiyanathan, P. Biogeochemical Dynamics II: Cycling and Storage of Phosphorus in Wetlands. In *The Wetlands Handbook*; Maltby, E., Barker, T., Eds.; Wiley-Blackwell: Oxford, UK, 2009; pp. 228–249. [CrossRef]
38. Andreo-Martínez, P.; García-Martínez, N.; Almela, L. Domestic Wastewater Depuration Using a Horizontal Subsurface Flow Constructed Wetland and Theoretical Surface Optimization: A Case Study under Dry Mediterranean Climate. *Water* **2016**, *8*, 434. [CrossRef]
39. Palmer, M.A.; Menninger, H.L.; Bernhardt, E. River restoration, habitat heterogeneity and biodiversity: A failure of theory or practice? *Freshw. Biol.* **2010**, *55*, 205–222. [CrossRef]
40. Leibold, M.A.; Holyoak, M.; Mouquet, N.; Amarasekare, P.; Chase, J.M.; Hoopes, M.F.; Holt, R.D.; Shurin, J.B.; Law, R.; Tilman, D.; et al. The metacommunity concept: A framework for multi-scale community ecology: The metacommunity concept. *Ecol. Lett.* **2004**, *7*, 601–613. [CrossRef]
41. Leibold, M.A.; Norberg, J. Biodiversity in metacommunities: Plankton as complex adaptive systems? *Limnol. Oceanogr.* **2004**, *49*, 1278–1289. [CrossRef]
42. de Vicente, I.; Merino-Martos, A.; Guerrero, F.; Amores, V.; de Vicente, J. Chemical interferences when using high gradient magnetic separation for phosphate removal: Consequences for lake restoration. *J. Hazard. Mater.* **2011**, *192*, 995–1001. [CrossRef]
43. Funes, A.; Martínez, F.J.; Álvarez Manzaneda, I.; Conde-Porcuna, J.M.; de Vicente, J.; Guerrero, F.; de Vicente, I. Determining major factors controlling phosphorus removal by promising adsorbents used for lake restoration: A linear mixed model approach. *Water Res.* **2018**, *141*, 377–386. [CrossRef]
44. Álvarez Manzaneda, I.; Guerrero, F.; Cruz-Pizarro, L.; Rendón, M.; de Vicente, I. Magnetic particles as new adsorbents for the reduction of phosphate inputs from a wastewater treatment plant to a Mediterranean Ramsar wetland (Southern Spain). *Chemosphere* **2021**, *270*, 128640. [CrossRef]
45. Milke, J.; Gałczyńska, M.; Wróbel, J. The Importance of Biological and Ecological Properties of Phragmites Australis (Cav.) Trin. Ex Steud., in Phytoremendiation of Aquatic Ecosystems—The Review. *Water* **2020**, *12*, 1770. [CrossRef]
46. Ali, Z.; Mohammad, A.; Riaz, Y.; Quraishi, U.M.; Malik, R.N. Treatment efficiency of a hybrid constructed wetland system for municipal wastewater and its suitability for crop irrigation. *Int. J. Phytoremediat.* **2018**, *20*, 1152–1161. [CrossRef]
47. Amare, E.; Kebede, F.; Mulat, W. Wastewater treatment by Lemna minor and Azolla filiculoides in tropical semi-arid regions of Ethiopia. *Ecol. Eng.* **2018**, *120*, 464–473. [CrossRef]

MDPI
St. Alban-Anlage 66
4052 Basel
Switzerland
Tel. +41 61 683 77 34
Fax +41 61 302 89 18
www.mdpi.com

Sustainability Editorial Office
E-mail: sustainability@mdpi.com
www.mdpi.com/journal/sustainability

www.ingramcontent.com/pod-product-compliance
Lightning Source LLC
Chambersburg PA
CBHW061602210926
51850CB00009B/2191
* 9 7 8 3 0 3 6 5 4 9 8 8 0 *